Nuclear Science and Technology

This interdisciplinary book series publishes monographs and edited volumes in all areas of nuclear science and technology, covering developments in theoretical research, experiments, technology and applications. The series attempts to bring together experts and professionals from all over the world, to provide readers with a library on a wide range of topics in physics, engineering, medicine and environmental sciences. Topics include but are not limited to:

Nuclear physics and interdisciplinary research;

Nuclear energy science and engineering;

Nuclear fusion and high temperature plasmas;

Synchrotron radiation applications, beamline technology;

Accelerator, ray technology and applications;

Nuclear chemistry, radiochemistry, radiopharmaceuticals, nuclear medicine; and Nuclear electronics and instrumentation.

The books are intended for researchers, professionals, and students in the field, as well as those interested in staying current with the latest advances.

Minjun Peng • Zhaofei Tian •
Genglei Xia • Hang Wang

Thermal Analysis of Nuclear Power Plants

Springer

Minjun Peng
Harbin Engineering University
Harbin, Heilongjiang, China

Zhaofei Tian
Harbin Engineering University
Harbin, Heilongjiang, China

Genglei Xia
Harbin Engineering University
Harbin, Heilongjiang, China

Hang Wang
Harbin Engineering University
Harbin, Heilongjiang, China

ISSN 2948-1856 ISSN 2948-1864 (electronic)
Nuclear Science and Technology
ISBN 978-981-96-8982-8 ISBN 978-981-96-8983-5 (eBook)
https://doi.org/10.1007/978-981-96-8983-5

Jointly published with Harbin Engineering University Press

The original submitted manuscript has been translated into English. The translation was done using artificial intelligence. A subsequent revision was performed by the author(s) to further refine the work and to ensure that the translation is appropriate concerning content and scientific correctness. It may, however, read stylistically different from a conventional translation.

Translation from the Chinese Simplified language edition: "Thermal Analysis of Nuclear Power Plants" by Minjun Peng et al., © Harbin Engineering University Press 2023. Published by Harbin Engineering University Press. All Rights Reserved.

This Springer imprint is published by the registered company Springer Nature Singapore Pte Ltd.
The registered company address is: 152 Beach Road, #21-01/04 Gateway East, Singapore 189721, Singapore

Preface

In the 1950s, the world's first nuclear power plant was built in Obninsk, Soviet Union, opening a new era of peaceful use of nuclear energy by humanity. After nearly 70 years of development, as of the end of 2022, there are a total of 422 operating nuclear power units in 33 countries and regions worldwide, and 57 under-construction nuclear power units in 18 countries. The proportion of global nuclear power generation in the power structure is about 9.6%. Nuclear power as a safe, clean, economical, and sustainable energy source has been widely accepted by the international society. Although the Three Mile Island nuclear accident in the United States (1979), the Chernobyl nuclear accident in the Soviet Union (1986), and the Fukushima Daiichi nuclear accident in Japan (2011) have occurred in the history of nuclear power development, causing setbacks in the process of nuclear power development at different times. Based on the rich operating experience and good safety records of nuclear power plants worldwide, as well as the objective requirements of sustainable development in human society, nuclear power will still be developed as a major form of energy in the future.

With the rapid development of the national economy, China's demand for energy is increasing. Due to the depletion of conventional fossil fuels, mainly coal and oil, and the serious environmental pollution caused by their use, vigorously developing nuclear power is one of the important ways to achieve sustainable development, solve energy shortages, and ensure national energy security. The country has clearly put forward the policy of "active, safe, and orderly development of nuclear power," promoting the high-quality development of nuclear power, and helping to achieve the national strategic goals of "carbon peak and carbon neutrality." China began building nuclear power plants in the mid-1980s. As of December 31, 2022, the number of commercial nuclear power units in Chinese Mainland has reached 55. The outline of the 14th Five Year Plan of the country proposes that by 2025, the installed capacity of nuclear power in operation will reach 70 million kilowatts, and by 2030, the installed capacity of nuclear power in operation will reach 120 million kilowatts, accounting for about 8% of the total power generation in the country. China's nuclear power has broad development prospects.

Essentially, a nuclear power plant is an energy system that converts nuclear energy into electrical energy. Reducing the consumption of nuclear fuel and lowering the cost of power generation while obtaining a certain amount of electricity has always been a goal that humans strive to pursue. Through thermal analysis of the thermal system of nuclear power plants, the rationality and effectiveness of energy utilization in energy transmission and conversion processes can be pointed out, and methods and approaches to improve the energy utilization level of the thermal system of nuclear power plants can be proposed.

This book is revised on the basis of *Thermal Analysis of Nuclear Power Plant* (Second Edition) published in 2012. In addition to correcting some errors in some illustrations and formulas in the second edition of the textbook, it also properly supplements and improves theoretical methods and application cases, so as to reflect the latest research trends and scientific and technological achievements in the field of this discipline in a timely manner, highlight the progressiveness and cutting-edge nature of the course teaching content, and broaden students' international vision. At the same time, in accordance with the requirements of ideological and political education in the curriculum, we will enrich the opportunities for nuclear energy development under the dual carbon background, as well as the impact of thermal economy on the application and development of nuclear power plants. This will provide strong support for the implementation of the fundamental task of moral education in graduate education.

Professor Peng Minjun from Harbin Engineering University was responsible for revising the entire book, with the participation of Professor Tian Zhaofei, Professor Xia Genglei, and Associate Professor Wang Hang. The entire book was edited and proofread by Professor Peng Minjun. In the process of writing, the authors referred to or quoted the works of some scholars both domestically and internationally. The authors would like to express their sincere gratitude.

Studying this book requires a foundation in engineering thermodynamics and heat transfer, and a certain level of understanding of nuclear power plants and equipment. This book can serve as a textbook for undergraduate and graduate students majoring in nuclear engineering and thermal energy engineering in higher education institutions, as well as a reference for technical personnel engaged in nuclear power plant work.

Due to the limited knowledge level of the authors, there are inevitably many shortcomings and errors in the book. We sincerely hope that readers can provide valuable feedback.

Harbin, China Minjun Peng
Harbin, China Zhaofei Tian
Harbin, China Genglei Xia
Harbin, China Hang Wang
June 30, 2024

Introduction

This book briefly introduces the theoretical basis of thermal analysis, thermodynamic cycles adopted by different types of nuclear power plants, the main factors affecting the thermal efficiency of cycles, and several commonly used thermal balance analysis methods, focusing on the concept of exergy, calculation methods for exergy, and the application of exergy analysis method in the thermal analysis of pressurized water reactor nuclear power plant.

Through the thermal balance analysis and exergy analysis of pressurized water reactor nuclear power plant thermal system, the rationality and effectiveness of energy utilization in the process of energy transmission and transformation of nuclear power plant is pointed out, and the methods and ways to improve the energy utilization level of thermal system of nuclear power plant is put forward. Further, the basic theory of exergic economic analysis and the method of online thermal economy analysis of nuclear power units also are briefly introduced.

This book can be used as teaching material for graduate students of nuclear science and technology, senior undergraduates of nuclear engineering and nuclear technology, and thermal energy engineering, and can also be used as reference for people engaged in nuclear power technology industry.

Contents

Chapter 1
Introduction

Energy is an important material basis for human society to survive and develop. Every major progress in human civilization is accompanied by the improvement and replacement of energy. Human development and utilization of energy resources have gone through three different stages: the era of firewood, which relies on traditional renewable energy, the coal era, and the oil era. With the development of society and technological progress, we will inevitably enter a new era of new energy and renewable energy utilization.

At present, coal, oil, and natural gas, the three major fossil energies, are still the world's most important primary energy sources. According to the data provided by the *bp Statistical Review of World Energy 2022*, the total global primary energy consumption in 2021 reach 595.15 EJ. Among them, the proportion of oil, natural gas, coal, nuclear energy, hydropower and renewable energy is 30.95%, 24.42%, 26.90%, 4.25%, 6.76% and 6.71% respectively [1].

Electricity plays an extremely important role in modern production and daily life. The application of electricity and the development of the power industry signify a country's level of modern production and the standard of living of its people. Different forms of primary energy in nature (such as coal, oil, natural gas, hydropower, nuclear energy, wind energy, solar energy, geothermal energy, tidal energy, etc.) can be converted into electricity through different types of power plants. Among them, nuclear power, due to its advantages of low resource consumption, small environmental impact, and strong supply capacity, is increasingly understood and accepted by the public, becoming one of the world's three major power supply pillars along with thermal power and hydropower.

M. Peng et al., *Thermal Analysis of Nuclear Power Plants*, Nuclear Science and Technology, https://doi.org/10.1007/978-981-96-8983-5_1

1.1 Characteristics and Development Prospects of Nuclear Power

Nuclear power generation is the use of energy released by nuclear fission or fusion for power generation. Nuclear fission power generation has reached an industrial scale, while controlled thermal nuclear reactions that release nuclear fusion energy on a large scale for peaceful use have not yet achieved industrial application.

1.1.1 The Development History of Nuclear Power in the World

On December 2, 1942, the first nuclear reactor designed and manufactured under the guidance of American physicist Enrico Fermi successfully operated in Chicago, marking the first controlled self-sustaining chain reaction of uranium nuclei and signifying the entry of mankind into the atomic age.

In July 1954, the Soviet Union used its graphite water-cooled reactor technology for producing plutonium for nuclear weapons to build the world's first experimental nuclear power plant (NPP) with an electric power of 5 MW in Obninsk, opening the prelude to the use of nuclear energy for power generation. In December 1957, the United States, based on its pressurized water reactor technology for nuclear submarines, built the world's first prototype pressurized water reactor NPP with an electric power of 90 MW—the Shippingport NPP. In 1956, the UK used its graphite gas-cooled reactor technology for producing military plutonium to build two nuclear power units with a single machine electric power of 46 MW. These experimental and prototype nuclear power units verified the feasibility of nuclear power generation technology, and they are internationally referred to as the first generation of nuclear power units.

In the 1970s, the energy crisis caused by the oil price hike promoted the development of nuclear power. Countries around the world, based on experimental and prototype nuclear power units, have successively developed and built commercial nuclear power units with an electric power of more than 300 MW. The reactor types mainly include pressurized water reactors (PWR) from the United States, Europe, and Japan, boiling water reactors (BWR), light water reactors (VVER/WWER) from Russia, and heavy water reactors (CANDU) from Canada. The majority of the more than 400 nuclear power units currently in commercial operation worldwide belong to this period, known as the second generation of nuclear power units.

On March 28, 1979, the Three Mile Island incident occurred in the United States. In the accident at the Three Mile Island NPP (TMI), a large amount of radioactive material leaked into the environment, damaging about 60% of the fuel rods in the reactor core. Although there were no casualties, it had a profound impact on the development of nuclear power worldwide. After the Three Mile Island accident, the U.S. Nuclear Regulatory Commission (USNRC) strengthened the safety management of NPPs, not only strictly controlling the issuance of new permits, but also proposing many modifications to the equipment and procedures of existing NPPs. On April 26, 1986, an explosion occurred at Reactor No. 4 of the Chernobyl NPP in Ukraine, causing the reactor to be destroyed, a large amount of radioactive material to leak, and serious casualties, large-scale environmental pollution, and population migration, which greatly increased people's worries.

In the 1990s, in order to address the negative impacts of the serious accidents at Three Mile Island and Chernobyl NPPs, the world's nuclear power industry concentrated on research and tackling key problems. The United States and Europe have successively introduced the "Advanced Light Water Reactor User Requirements" document and the "European User Requirements for Light Water Reactor Nuclear Power Plants". Internationally, nuclear power units that meet one of these two documents are usually referred to as third-generation nuclear power units. The third-generation nuclear power units were designed from 1996 to 2010, with safety as the primary reference factor. The main goal is to further improve the safety of the second-generation nuclear power units. The most representative ones are the Advanced Boiling Water Reactor (ABWR), the Passive Advanced PWR (AP1000), and the European Advanced PWR (EPR).

Entering the twenty-first century, the development of global nuclear power has entered a recovery period, and various countries have formulated corresponding nuclear power development plans. On March 11, 2011, due to the tsunami caused by an earthquake, the Fukushima Daiichi and Daini NPPs in Japan were severely affected. Among them, Units 1, 2, and 3 of the Fukushima Daiichi NPP experienced hydrogen explosions in the fuel plant buildings in the following days, and radioactive materials leaked into the environment. The severity of the accident was level 7. The Fukushima NPP accident led Japan and some developed countries to adjust their nuclear power development policies in succession, which had a significant impact on the recovering world nuclear power industry.

According to data provided by the International Atomic Energy Agency (IAEA) PRIS database [2], the growth of the number and installed capacity of nuclear power operating units worldwide over the past 60 years is shown in Table 1.1.

At present, research on the fourth-generation nuclear power system has gradually unfolded, and its reactors and fuel cycles will undergo significant innovations and developments. In 2002, the most promising future reactor concepts were selected internationally, determining the six most promising reactor types from the perspectives of sustainable energy development, economic competitiveness, safety and reliability, and proliferation resistance and external intrusion capabilities, including three fast neutron reactors and three thermal neutron reactors with advanced fuel cycles:

Table 1.1 Growth of the number of nuclear power units and installed capacity in the world from 1960 to 2020 [2]

Year	1960	1970	1980	1990	2000	2010	2020
Operating unit	15	84	245	416	435	441	442
Installed capacity/GW	1.087	17.656	133.037	318.253	349.984	375.277	392.6

1. Sodium-cooled Fast Reactor (SFR)
2. Lead-cooled Fast Reactor (LFR)
3. Gas-cooled Fast Reactor (GFR)
4. Super Critical Water-cooled Reactor (SCWR)
5. Very-High-Temperature gas-cooled Reactor (VHTR)
6. Molten Salt Reactor (MSR).

The main features of the design of the above six fourth-generation nuclear energy systems are improved economics, enhanced safety, minimized waste, and prevention of nuclear proliferation in the fuel cycle.

The world's energy industry is currently facing the dual pressures of sustainable development strategies and ecological environmental protection. Nuclear power, due to its safety, economy, cleanliness, and obvious advantages in fuel supply, will experience a new development climax in the twenty-first century. For a considerable period of time, nuclear power will become the pillar of the power industry.

The utilization of nuclear energy is the inevitable path to solve energy problems. The proportion of nuclear energy in the energy that humans can use will gradually increase, thereby improving the energy structure and hopefully completely solving human energy needs in the future. However, the development and utilization of nuclear energy is a gradual long-term process. According to its scientific and technological difficulties and the prospects for industrialization, it can be roughly divided into three stages: the first stage is the thermal neutron reactor, the second stage is the fast neutron breeder reactor, and the third stage is the controlled fusion reactor. These three stages need to be connected with each other, gradually put into practical use, and achieve industrialization [3].

1.1.2 Characteristics of Nuclear Power Generation

In the world's energy supply, nuclear power already occupies an important position. Compared with conventional energy sources, nuclear power has significant advantages in the following aspects.

1.1.2.1 Nuclear Fuel Has a High Energy Density

Atomic energy can be released through nuclear fission or fusion, and some radioactive nuclides will also release a certain amount of energy during the decay process. Table 1.2 lists the energy values released by several nuclear reactions and combustion chemical reactions.

The energy released by the complete fission of 1 kg of uranium-235 is equivalent to the energy released by the complete combustion of 2400 tons of standard coal or 1570 tons of oil. Under the same mass, the energy released by the deuterium-tritium nuclear fusion reaction is about 4 times that of the uranium nuclear fission reaction.

Table 1.2 Comparison of energy release from nuclear reaction and combustion chemical reaction [4]

Reaction type	Reaction formula	Released energy
Uranium fission	${}^{235}_{92}U + {}^{1}_{0}n \rightarrow {}^{95}_{42}Mo + {}^{139}_{57}La + 2{}^{1}_{0}n + 7{}^{0}_{-1}e$	200 MeV
Hydrogen fusion reaction in the sun	$4{}^{1}_{1}H \rightarrow {}^{4}_{2}He + 2{}^{0}_{1}e$	24.7 MeV
Deuterium tritium fusion reaction in hydrogen bomb	${}^{2}_{1}H + {}^{3}_{1}H \rightarrow {}^{4}_{2}He + {}^{0}_{1}e$	17.6 MeV
Decay of radium nuclei	${}^{226}_{88}Ra \rightarrow {}^{222}_{86}Rn + {}^{4}_{2}He$	4.8 MeV
Decay of co-60	${}^{60}_{27}Co \rightarrow {}^{60}_{28}Ni + {}^{0}_{-1}e$	2.8 MeV
Combustion chemical reaction of carbon atoms	$C + O_2 \rightarrow CO_2$	4.1 eV

A PWR NPP with a capacity of 1000 MWe, running at full power for 300 days, only consumes 25–30 tons of low-enriched uranium (equivalent to 150–180 tons of natural uranium); while a coal-fired power plant of the same capacity, running at full power for 300 days, needs to consume about 3,100,000 tons of coal, requiring the transportation of tens of thousands of tons of fuel and thousands of tons of ash every day.

Due to the high energy density and low fuel consumption of NPPs, their land area is much smaller than that of thermal power plants of the same capacity, and they can significantly reduce the pressure on the transportation system and save a large amount of fuel transportation costs.

1.1.2.2 Nuclear Power Is a Clean Energy Source

NPPs use fissionable heavy nuclei (such as U-235, U-238) as fuel, while thermal power plants use coal, oil, or natural gas as fuel, and the waste they emit to the environment is different. Table 1.3 compares the annual waste emissions of a 1000 MWe NPP with different types of thermal power plants.

From the data listed in the Table 1.3, it can be seen that thermal power plants produce a large amount of smoke dust, CO_2, CO, SO_2, NO_x and some trace elements during operation, causing serious pollution to the environment.

The amount of radioactive waste gas and waste liquid discharged from the daily operation of NPPs is very small, and it is under strict supervision and control. The radiation dose received by the surrounding residents from this is less than 1% of the natural background. At the same time, NPPs do not emit harmful substances such as SO_2, NO_x and smoke dust into the environment, nor do they emit gases such as CO_2 that form the greenhouse effect. Therefore, nuclear power generation is the least greenhouse gas emissions of all kinds of energy generation, nuclear power is a very clean energy.

From the perspective of environmental protection, the use of nuclear energy instead of fossil energy can effectively reduce the emission of CO_2, SO_2 and other chemical combustion products into the atmosphere, thus curbing the development

Table 1.3 Comparison of annual waste emissions from 1000 MWe power plants [5]

Fuel type	Fuel consumption/10^4 t	$CO_2/10^4$ t	$SO_2/10^4$ t	$NOx/10^4$ t	Smoke dust/10^4 t	Lime-ash/10^4 t	Radioactive dose/Bq	Trace element/t
Coal	240	588	4.4	2.2	0.9	45	6.6109	20.8
Nature gas	77	248	0	0.2	0	0	0	0
Oil	104	290	0.2	0	0	0	0	–
Nuclear fuel	0.0024	0	0	0	0	0		0

Note: 1. For coal-fired units, the total sulfur content in coal is calculated at 1.15%, and the coal consumption is 310 g/(kW.h). For gas-fired units (combined cycle), the gas consumption is 140 g/(kW·h). For oil-fired units (combined cycle), the oil consumption is 190 g/(kW·h)

2. The annual operating time of all types of power plants is calculated at 5500 hours

3. The trace elements include arsenic, cadmium, lead, manganese, mercury, nickel, and vanadium

of the greenhouse effect and the formation of acid rain. It is estimated that since 1973, due to the operation of NPPs in the United States, CO_2 emissions have been reduced by about 20×10^8 t, which has played an important role in protecting the environment.

1.1.2.3 Nuclear Fuel Resources Are Abundant

Fossil fuels such as coal, oil and natural gas are non-renewable resources and valuable raw materials for modern chemical and textile industries. According to the references [1], as of the end of 2020, the world's proven oil reserves were 244.4 billion tons, natural gas proved reserves were 188.1 trillion cubic meters, and coal proved reserves were 1074.108 billion tons. At 2020 production levels, the world's Reserves to production ratios (R/P) for oil, gas and coal are 53.5, 48.8 and 139, respectively. Continuing to use fossil fuels as the main energy source will not only cause waste of natural resources and serious environmental pollution, but also aggravate the global energy crisis.

In order to solve the problem of environmental pollution in the process of energy utilization and the energy crisis faced by mankind, and realize the sustainable development of society, human society must strive to improve the efficiency of energy utilization and develop and use other alternative energy. At present, although natural resources such as solar energy, wind energy, water power, geothermal energy and biological energy have been developed and utilized to a certain extent, they are affected and restricted by natural conditions and other objective conditions, and it is impossible to fundamentally solve the energy problem of human society.

After years of research and practice, the use of nuclear fission energy is technically mature and has been applied in industry on a large scale. The use of nuclear power generation provides a new means to supplement and replace conventional energy. Nuclear energy not only has a high energy density, but also is rich in resources. According to the OECD Nuclear Energy Agency (NEA) and the International Atomic Energy Agency (IAEA), global uranium resources in 2019 were 6,147,800 tons, and global uranium mine production reached 48,332 tons in 2021.

The current commercial nuclear power units mainly use thermal reactors, the utilization rate of uranium resources is only about 1%, the fast breeder reactor to be commercialized can increase the utilization rate of uranium resources to 60–70%, which can greatly extend the use time of proven uranium resources.

Once the controlled nuclear fusion technology is commercialized, it will open a new chapter in human energy utilization. The deuterium used in nuclear fusion can be extracted from seawater, and 30 mg of deuterium can be extracted from 1 L of seawater, which can produce about 300 L of energy equivalent to gasoline in nuclear fusion reaction. According to estimates, there is a total of 4.5×10^{13} t of deuterium in the seawater on the earth, if all extracted for nuclear fusion, can provide endless clean energy for human beings.

Table 1.4 Components of standardized power generation costs [6]

Items	Construction cost/%	Fuel cycle cost/%	Operation maintenance fee/%	Total/%
Nuclear power plants	43–70	13–30	17–31	100
Coal-fired power plants	23–45	35–65	6–28	100
Gas-fired power plants	13–33	53–84	3–19	100

Note: The figures in the table consider a 5% discount rate

1.1.2.4 Nuclear Power Is Economically Viable

An important economic indicator of a NPP is the cost required to generate 1 kWh of electricity, which is mainly composed of the investment cost of power plant construction, fuel cycle cost, and operation and maintenance cost. Table 1.4 lists the comparison of power generation costs of nuclear power, coal power, and gas power [6].

NPPs have strict safety requirements, and the system is complex with many equipment, so the construction investment cost of nuclear power is higher than that of thermal power (coal power and gas power), but the fuel cost is much lower than thermal power. Because the fuel cost compensates for the investment cost, the cost of nuclear power may be lower than the cost of thermal power. The U.S. Energy Advisory Committee pointed out after studying the average power generation costs over the past decades that the average power generation costs per kWh of nuclear power, coal power, gas power, and oil power are about 4 cents, 5 cents, 6 cents, and 8 cents respectively. In recent years, international comparisons of the costs of nuclear power and coal power have been made. The cost of coal power in France is 1.75 times that of nuclear power, in Germany it is 1.64 times, in Italy it is 1.57 times, in Japan it is 1.51 times, and in South Korea it is 1.7 times.

With the intensification of the global energy crisis and the gradual maturity of nuclear power technology and the standardization of nuclear power equipment, the cost of nuclear power in all nuclear power countries in the world will generally be lower than the cost of thermal power. In terms of long-term economic benefits, nuclear power has certain advantages.

1.1.2.5 Nuclear Power Is Very Safe

In order to prevent the leakage of radioactive materials, threaten human life safety, and damage the ecological environment, NPPs always implement the safety principle of in-depth defense in the design, construction, and operation process, which includes setting up multiple barriers between radioactive materials and the environment where personnel are located, as well as multi-level defense measures for radioactive materials. Due to the extreme importance of nuclear safety, countries with nuclear power around the world have invested a lot of manpower, material resources, and financial resources in nuclear safety research over the past decades. Nuclear safety technology has made great progress, effectively improving the safety of NPPs and greatly reducing the probability of serious accidents where a large amount of radioactive material is released.

In the nearly half-century history of nuclear power development, the Three Mile Island NPP accident in the United States (March 28, 1979), the Chernobyl NPP accident in the former Soviet Union (April 26, 1986), and the Fukushima NPP accident in Japan (March 11, 2011) have occurred. Among them, the Chernobyl NPP accident caused a large amount of radioactive material to be released into the environment, causing huge losses of life and property. However, the graphite boiling water reactor used by the power plant was only built in the former Soviet Union and does not represent mainstream nuclear reactor technology. These accidents have had a significant impact on the progress of nuclear power development, and some countries have even announced that they will completely abandon nuclear power generation. However, judging from the rich operating experience and good safety record accumulated by existing NPPs in the world, nuclear power generation is far superior to other industrial sectors in terms of casualties in the production process and adverse effects on the environment. The safety of nuclear power is guaranteed. Nuclear power is a safe and clean energy source, which is a recognized conclusion in the world energy industry.

1.1.3 The Prospects for the Development of Nuclear Power in Our Country

Our country is the largest developing country in the world, with its energy structure mainly based on fossil fuels. In 2021, the proportion of various types of energy generation in our country was as follows: thermal power 68.02%, hydropower 15.69%, wind power 7.68%, nuclear power 4.77%, solar power 3.83% [2]. In the same year, the global power structure was: coal power 36.49%, natural gas power 22.16%, hydropower 15.28%, nuclear power 9.94%, wind power 6.59%, solar power 3.72%, oil power 3.10%, other renewable energy power 2.73% [7].

1.1.3.1 Fossil Fuel Reserves Are Limited and Unevenly Distributed

China's production and consumption of coal far exceed that of oil, and the consumption of natural gas and nuclear power is far below the world average. This coal-dominated energy structure is a major cause of high energy consumption intensity, low energy utilization efficiency, and environmental pollution.

China's hydropower and coal resources are relatively abundant, ranking first and third in the world respectively; while high-quality fossil energy is relatively scarce, the proven remaining recoverable reserves of oil and natural gas currently only rank 13th and 17th in the world. Due to the large population, the per capita possession of various energy resources is below the world average, with the per capita remaining recoverable reserves of oil and natural gas only 7.7% and 7.1% of the world average, and the relatively abundant coal reserves only 58.6% of the world average. With the continuous rapid development of China's economy, the demand for energy consumption has surged, the contradiction between supply and demand has become increasingly prominent, and it has become the biggest constraint on the sustainable development of China's economy and society, directly threatening national economic security. From this perspective, China is a country with severe energy resource shortages.

The distribution of energy resources in China is extremely uneven, with 93% of coal reserves distributed in North China, Northwest China, and Northeast China, and most hydropower resources concentrated in the western region, which does not match the national industrial layout and population distribution. This has formed a large-scale energy transportation pattern of "coal from the north to the south, coal from the west to the east", "oil and gas from the west to the east, oil and gas from the northeast to the south, offshore oil and gas landing", and "electricity from the west to the east", as well as a large-scale, long-distance energy transportation system, occupying a large amount of transportation resources. In order to meet the growing demand for electricity from the rapid development of the national economy, and to avoid a large amount of coal transportation from the north to the south and long-distance electricity transmission from the west to the east, in the long run, nuclear power, as a clean energy source, will develop rapidly in coastal economically developed areas with little coal and no hydropower resources.

1.1.3.2 China's Nuclear Power Technology Has Steadily Developed and Has Joined the Ranks of the World's Major Nuclear Power Countries

The development of nuclear power in China has gone through the following stages:

1. Initial stage (early 1970s–1995)

   In 1974, China began to independently design a 300 MW PWR nuclear power unit (CP300). On March 20, 1985, the first batch of concrete was poured for the nuclear island base of the Qinshan NPP. On December 15, 1991, it was first connected to the grid for power generation, and on April 1, 1994, it was put into commercial operation, ending the history of no nuclear power on the Chinese mainland, and also opening the prelude to the development of nuclear power in China, making China the seventh country in the world with the ability to independently design and build NPPs. However, due to the lack of nuclear power technology reserves and insufficient capabilities in China, it was necessary to rely on advanced and mature nuclear power technology from abroad. Therefore, in the 1980s and 1990s, China's nuclear power development adopted a combination of independent research and development and technology introduction. The Guangdong Daya Bay NPP used the M310 commercial nuclear power technology imported from France, construction began in 1987, and it was put into commercial operation in 1994.

2. Moderate Development Stage (1996–2004)

   Although nuclear power is a clean energy source, due to unstable technology and the lingering shadow of the 1986 Chernobyl nuclear accident in the Soviet Union, China has taken a cautious attitude towards the development of nuclear power. Coupled with the relatively sufficient power supply in China at that time, the urgent need for nuclear power was not high. Therefore, the state positioned nuclear power as a supplementary energy source, and the development of NPPs was in a moderate development stage.

   Following the Qinshan NPP, China independently designed and built the Qinshan Second NPP (CP600) in accordance with the principle of "self-reliance and international cooperation". While independently developing nuclear power technology, China steadily introduced, digested, and absorbed advanced international nuclear power technology, actively promoting the localization of million-kilowatt nuclear power technology. On December 15, 2005, the Lingao Phase II NPP, which explored localization based on the introduced M310 nuclear power technology, started construction. On September 15, 2010, and August 7, 2011, the No. 1 and No. 2 units of the Lingao Phase II NPP were officially put into commercial operation.

3. Active Development Stage (2005–2010)

   With the rapid development of China's national economy, the power supply gradually could not meet the growing demand. At the same time, due to the need for environmental protection, the proportion of clean energy began to gradually increase, and the status of nuclear power rose. In 2006, the United States Westinghouse Company's AP1000 technology and France's EPR technology were introduced, and construction began on 4 AP1000 units as localization projects, and 2 EPR units were also started. In October 2007, the Chinese government issued the "Medium and Long-term Development Plan for Nuclear Power (2005–2006)", clearly stating "actively promote the construction of nuclear power", establishing the strategic position of nuclear power in China's economic and energy sustainable development. From then on, China's nuclear power entered a new stage of large-scale development.

4. Safe and Efficient Development Stage (2011–present)

   In 2012, the state issued the "Medium and Long-term Development Plan for Nuclear Power (2011–2020)", adjusting the development goal to: by 2020, the operating nuclear power installed capacity will reach 58 million kilowatts, and the units under construction will reach about 30 million kilowatts.

   Through comprehensive strengthening of independent innovation in nuclear power, implementing major national nuclear power science and technology projects, China's nuclear power technology level has significantly improved, forming domestically produced brands with independent intellectual property rights such as the third-generation PWR "Hualong No.1" and "National Nuclear No.1", high-temperature gas-cooled reactors and fast reactors with fourth-generation features, as well as advanced technologies such as small modular reactors. Currently, China is capable of independently developing third-generation nuclear power technology, forming a relatively complete system of nuclear science and technology research, nuclear power engineering design and verification, equipment manufacturing, nuclear fuel element manufacturing, uranium resource supply, etc., and is one of the few countries in the world with a relatively complete nuclear power industry system.

As of September 30, 2021, China has a total of 52 operating nuclear power units (excluding Taiwan), with a rated installed capacity of 53,485.95 MWe. From January to September 2021, the total electricity generation nationwide was 607.212 billion kilowatt-hours, and the cumulative electricity generation of operating nuclear power units was 302.709 billion kilowatt-hours, accounting for 4.99% of the total electricity generation nationwide. Compared with coal-fired power generation, nuclear power generation is equivalent to reducing the burning of standard coal by 8.65329 million tons, reducing the emission of carbon dioxide by 22.67163 million tons, sulfur dioxide by 735,500 tons, and nitrogen oxides by 640,300 tons.

1.1.3.3 Opportunities for Nuclear Power Development under the "Dual Carbon" Goal

With the rapid increase in global energy consumption, the emissions of environmental pollutants such as CO_2, NOx, and dust particles are increasing year by year, and the pollution of fossil energy to the environment and its impact on global climate are becoming increasingly serious. The international community, in response to global climate change, has promoted countries to take measures to limit or reduce greenhouse gas emissions by signing various conventions. Many countries have added content to address climate change when adjusting energy strategies and formulating energy policies, with a focus on limiting fossil energy consumption and encouraging energy conservation and the use of clean energy. Countries regard nuclear energy, hydropower, wind energy, solar energy, biomass energy and other low-carbon and zero-carbon energy sources as the focus of future development, which is also the main direction of China's future energy development [8].

In September 2020, President Xi Jinping announced at the 75th United Nations General Assembly that China strives to peak its carbon dioxide emissions by 2030 and aims to achieve carbon neutrality by 2060. In December 2020, President Xi Jinping further announced at the Climate Ambition Summit that by 2030, non-fossil energy will account for about 25% of primary energy consumption, and the total installed capacity of wind and solar power will reach more than 1.2 billion kilowatts.

Peaking carbon refers to China's commitment that by 2030, CO_2 emissions will no longer increase and will gradually decrease after reaching the peak. Carbon neutrality refers to enterprises, groups or individuals calculating the total amount of greenhouse gas emissions directly or indirectly produced within a certain period of time, and then offsetting their own CO_2 emissions through afforestation, energy conservation and emission reduction, achieving "zero emissions" of CO_2. Against the backdrop of carbon peaking and carbon neutrality, the process of clean and low-carbon transformation of China's energy and power system will be further accelerated, and nuclear energy, as a nearly zero-emission clean energy, will have a broader development space and is expected to maintain a rapid development trend.

On October 26, 2021, the State Council issued the "Action Plan for Carbon Peaking by 2030", proposing "actively, safely and orderly developing nuclear power", "actively and prudently carrying out nuclear energy heating demonstrations", and "actively promoting the demonstration projects of advanced reactor types such as high-temperature gas-cooled reactors, fast reactors, modular small reactors, and offshore floating reactors, and carrying out demonstrations of comprehensive utilization of nuclear energy". The 2021 "Government Work Report" of China clearly stated, "actively and orderly develop nuclear power on the premise of ensuring safety". The "Fourteenth Five-Year Plan" is a key period and window period for carbon peaking. The country considers from the perspectives of energy supply security, economy and sustainable development, and repositions nuclear power as a main energy source for peaking, creating a new policy opportunity period for the development of nuclear power. China's independently developed third-generation nuclear power units are expected to achieve scaled and batch development at an approval pace of 6–8 units per year. It is estimated that by 2025, China's operating nuclear power capacity will reach about 70 million kilowatts, and by 2030, the operating nuclear power capacity will reach 120 million kilowatts, and nuclear power generation will account for about 8% of the country's total power generation [9].

1.2 Types of NPPs and Their Thermal Systems

There are various forms of nuclear fission reactors, and the reactors used in NPPs can be divided into uranium-plutonium cycle systems and thorium-uranium cycle systems according to the nuclear fuel cycle system. Figure 1.1 lists the types of power reactors for the two fuel cycle systems.

The application of various reactor types in the nuclear power units currently operating and under construction in the world is shown in Table 1.5.

1.2.1 PWR NPP

PWRs have the advantages of compact structure, small volume, high power density, and deep average fuel consumption. Compared with other types of reactors, they have shorter construction periods and lower costs. In addition, PWRs use multiple barriers in their structural design to prevent the leakage of radioactive materials, and the coolant has a negative temperature coefficient, which gives the reactor

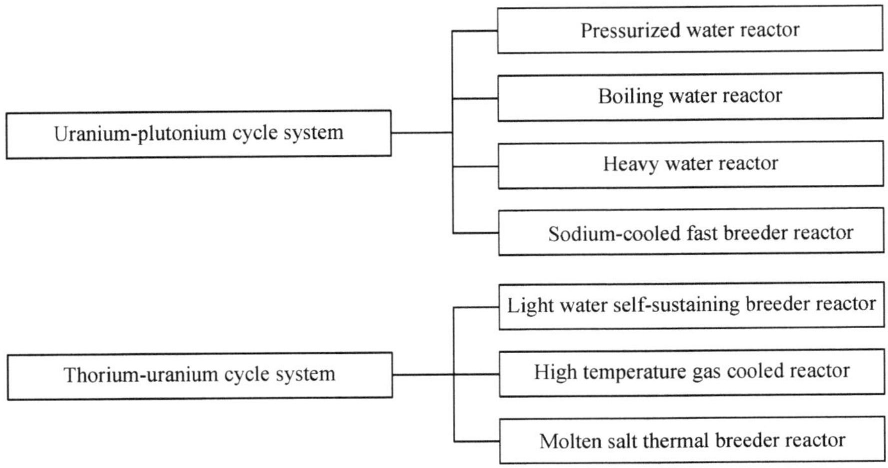

Fig. 1.1 Types of power reactor for NPPs

Table 1.5 Types and share of NPPs in the world (as at 31 December 2021) [10]

Reactor type	Operating units			Units under construction		
	Quantity	Installed capacity/MW	Ratio/%	Quantity	Installed capacity/MW	Ratio/%
Pressurized water reactor (PWR)	303	288,167	69.33	48	52,141	85.71
Boiling water reactor (BWR)	61	61,849	13.96	2	2653	3.57
Pressurized Heavy water Reactor (PHWR)	47	24,314	10.75	3	1890	5.36
Graphite moderated Light water Reactor (LWGR)	11	7433	2.52	0	0	0
Gas cooled Reactor (GCR)	11	6145	2.52	0	0	0
High temperature Gas cooled Reactor (HTGR)	1	200	0.23	0	0	0
Fast breeder Reactor (FBR)	3	1400	0.69	3	1412	5.36
Total	437	389,508	100	56	58,096	100

self-stabilizing and self-adjusting characteristics, thus ensuring good safety. After many years of research and development, PWRs have become a mature type of reactor, and more than half of the NPPs currently operating in the world are of the PWR type [11].

A PWR NPP consists of a reactor building (i.e., safety shell), a primary loop auxiliary system building, a nuclear fuel building, a turbine building, a power distribution building, a main control room, a circulating water pump room, and a

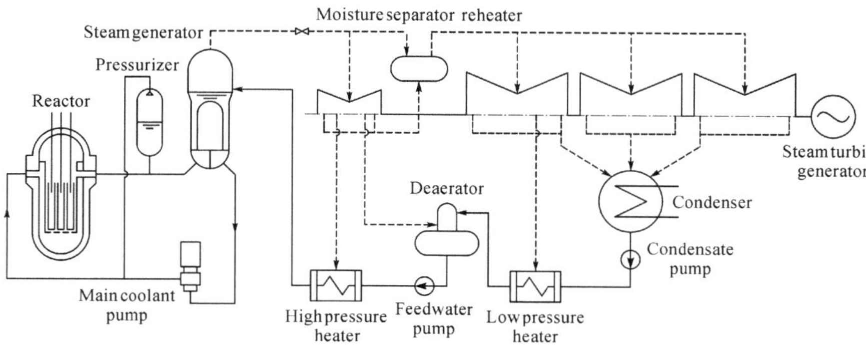

Fig. 1.2 Schematic diagram of thermal system for PWR NPPs

three-waste treatment building. The thermal system of a PWR NPP mainly includes a primary loop nuclear island and a secondary loop conventional island. The principle flow of its thermal system is shown in Fig. 1.2.

The nuclear island mainly includes the reactor, the primary loop coolant system, and the primary loop auxiliary system.

PWRs generally use rod-shaped fuel elements, with the fuel elements encased in zirconium alloy and containing low-enriched U-235 ceramic fuel. High-purity water treated as a coolant and moderator is used. To prevent volume boiling when the coolant flows through the core, the operating pressure of the reactor is usually maintained at a level of 12–17 MPa.

The primary loop system of modern commercial PWR NPPs generally has two to four symmetrical parallel circuits connected to the reactor pressure vessel nozzle. Each circuit consists of a reactor coolant pump, a steam generator, and corresponding pipes. The reactor, primary loop coolant system, and their corresponding auxiliary systems are collectively referred to as the Nuclear Steam Supply System (NSSS). The subcooled coolant circulates in the primary loop system under the drive of the main coolant pump, taking out the heat generated by the fission reaction of the core nuclear fuel, and heating the secondary side feedwater when flowing through the steam generator. The primary loop auxiliary system is used to ensure the normal and safe operation of the reactor and the primary loop coolant system, and provides necessary safety protection measures under accident conditions to prevent the spread and contamination of radioactive materials.

The conventional island part is similar to the thermal system of a conventional thermal power plant, using a Rankine cycle with steam reheating and feedwater reheating. The main equipment includes turbines, generators, condensers, circulating water pumps, condensate pumps, feedwater pumps, feedwater heaters, deaerators, and steam-water separators and reheaters. The secondary side feedwater of the steam generator absorbs the heat released by the primary side coolant to produce saturated steam. The steam pipeline sends the steam to the turbine. After expanding and doing work in the high-pressure cylinder, the low-parameter wet steam discharged enters the steam-water separator and reheater for steam-water separation and reheating. After being heated to a slightly superheated state, it is sent to the low-pressure cylinder to do work. The turbine drives the generator to generate electricity. The exhaust steam discharged from the turbine is condensed into water in the condenser, heated to a certain temperature by the feedwater heater, sent into the steam generator, and starts the next steam-water cycle.

Compared with conventional thermal power plants, PWR NPP turbine units have the following characteristics:

1. The new steam parameters are related to the operating load of the NPP and will vary within a certain range.
2. The new steam is mostly saturated steam with lower parameters.
3. The ideal enthalpy drop of the steam turbine is small, and the volumetric flow rate is large.

   The ideal enthalpy drop of the saturated steam turbine in a NPP is about half that of the superheated steam turbine in a thermal power plant. Therefore, under the same power, the steam volumetric flow rate of the nuclear turbine is 60–90% higher than that of the thermal power plant turbine.
4. Most of the steam turbine blades work in the wet steam area, and the blade surface is easily eroded.

   Due to the erosion of liquid droplets in the steam, the blade surface presents an uneven sponge-like appearance, which not only reduces the stage efficiency, but also changes the vibration and strength characteristics of the blade, and in severe cases, leads to blade fracture.

1.2.2 Boiling Water Reactor NPP

The boiling water reactor, like the PWR, belongs to the light water reactor, using low-enriched uranium as fuel and pressurized light water as a neutron moderator and reactor coolant. The main feature of the boiling water reactor is that it allows the coolant to boil in the core. The steam-water mixture produced in the reactor is dehumidified through the steam-water separator and steam dryer at the top of the reactor pressure vessel. The dried saturated steam can be directly sent to the turbine unit for expansion work. The boiling water reactor directly generates steam for the turbine unit, eliminating the need for a steam generator and regulator, greatly improving the reliability of the NPP. Due to the use of a single cycle loop, the mass flow rate of the power cycle is greatly reduced, reducing the cycle power consumption, which is beneficial to improve the economy of the NPP [11]. Figure 1.3 shows the principle flow chart of the thermal system of the boiling water reactor NPP.

The working pressure of the boiling water reactor is about 5–7 MPa, which is only half of the working pressure of the PWR. Therefore, the design pressure of the equipment and pipelines of the boiling water reactor NPP is relatively low, which can greatly reduce the investment cost of the power plant equipment. However, the power density of the boiling water reactor is lower than that of the PWR, and the

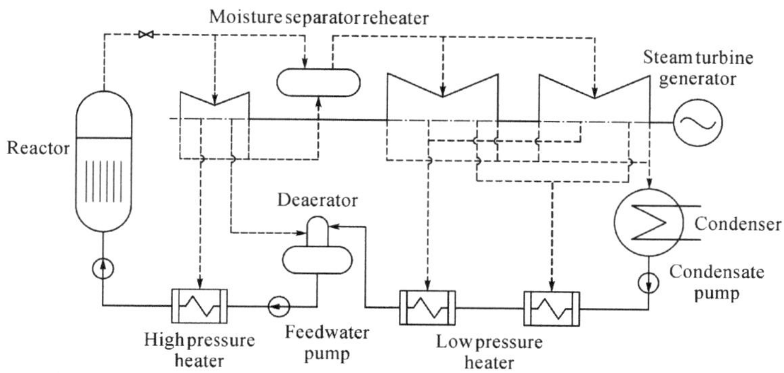

Fig. 1.3 Schematic diagram of thermal system for BWR NPPs

size of the pressure vessel of the boiling water reactor is much larger than that of the PWR under the same power level.

The steam generated by the boiling water reactor will contain a certain amount of radioactivity, which will contaminate the turbine, condenser, and corresponding pipes and valves with radioactivity during the steam-water cycle. Therefore, it is necessary to shield the equipment of the steam system and the water supply system, and include the turbine plant in the radioactive control area, which increases the difficulty of inspection and maintenance.

1.2.3 Heavy Water Reactor NPP

Heavy water reactors use heavy water as a moderator and are widely used as power reactors, nuclear fuel production reactors, and research and test reactors. Because the thermal neutron absorption cross-section of heavy water is very small, heavy water reactors can use uranium of various enrichment levels from natural uranium as nuclear fuel, and there is a possibility of achieving nuclear fuel breeding through the Th-U cycle.

Power-generating heavy water reactors are divided into pressure vessel type and pressure tube type according to the structure of the reactor. The structure of the pressure vessel type heavy water reactor is similar to that of the PWR, and it is limited to using heavy water as a coolant. The pressure tube type heavy water reactor can use heavy water, boiling light water, CO_2 gas, etc. as coolants. The mature commercial heavy water reactor type is the Canadian CANDU type pressure tube heavy water reactor, which uses natural uranium as fuel and pressurized heavy water as a coolant [11]. Figure 1.4 shows the system flow chart of the CANDU reactor NPP.

This type of reactor consumes less natural uranium, can achieve continuous refueling without shutting down, thereby improving the availability and load factor of NPPs, and is beneficial in reducing electricity generation costs. Moreover, the abundance of U-235 in spent fuel is lower than that in diffusion plant tailings, so no

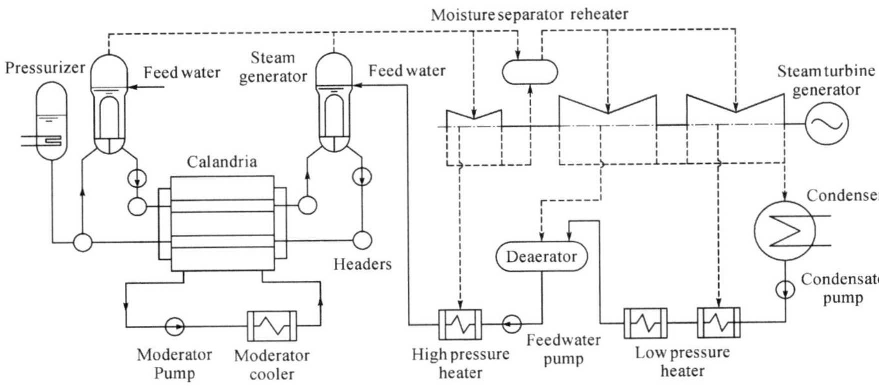

Fig. 1.4 Schematic diagram of a CANDU NPP

post-processing is necessary, which greatly simplifies the fuel cycle while still achieving a high uranium resource utilization rate. However, CANDU reactors require a large amount of heavy water, which increases investment costs, and there are many issues with heavy water sealing and leakage, increasing the difficulty and cost of operation and maintenance.

1.2.4 High-Temperature Gas-Cooled Reactor NPP

High-temperature gas-cooled reactors use chemically inert helium gas with good thermodynamic properties as a coolant. The fuel elements are fully ceramic-coated particles dispersed in a graphite sphere matrix, with high-temperature resistant graphite used as a moderator and core structural material. The core outlet helium gas temperature can reach 950 °C or even higher [11].

High-temperature gas-cooled NPPs mainly use the following two types of thermodynamic cycles:

1. Indirect cycle

   This cycle combines a high-temperature gas-cooled reactor with a steam Rankine cycle. Figure 1.5 shows the principle flow diagram of the thermal system of a high-temperature gas-cooled NPP using a steam thermodynamic cycle. The helium gas coolant flowing through the core carries out the heat released by nuclear fuel fission, heats the secondary side feedwater through a direct steam generator, and generates superheated steam with a temperature of up to 530 °C, which is used to drive the steam turbine generator. The power generation efficiency can reach about 40%.

   The indirect cycle can directly use technically mature steam turbine generator units, and at the same time, it can minimize the possibility of radioactive material

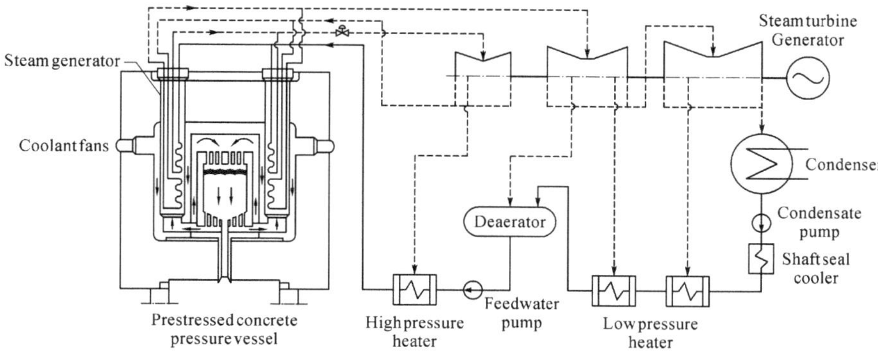

Fig. 1.5 Schematic diagram of the thermal system of an HTGR NPP

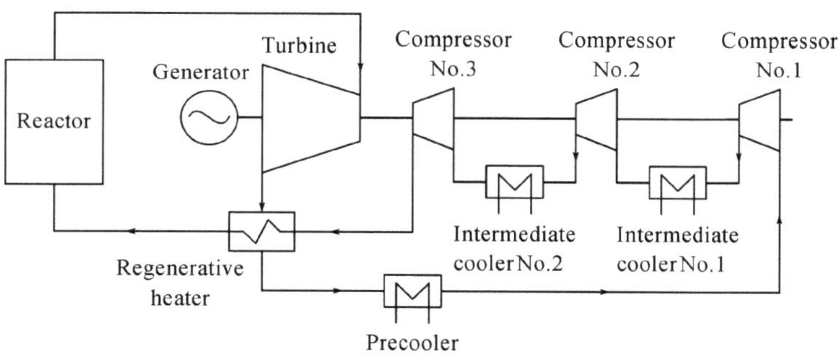

Fig. 1.6 Schematic diagram of the thermal system of HTGR NPP with direct circulation of helium turbine

carried by the helium coolant contaminating the steam turbine generator unit, providing higher safety and reliability.

2. Direct cycle

   This cycle combines a high-temperature gas-cooled reactor with a helium Brayton cycle. The high-temperature helium gas at the outlet of the high-temperature gas-cooled reactor directly enters the helium turbine to do work. The helium gas discharged after work is compressed and heated by the compressor and reheater before entering the core to absorb heat, starting the next cycle. Figure 1.6 shows the schematic diagram of the principle flow of the thermal system of a high-temperature gas-cooled NPP using a helium turbine direct cycle.

The power generation efficiency of a high-temperature gas-cooled NPP using a direct cycle can reach more than 48%. Compared with NPPs using an indirect cycle, it can simplify the energy conversion system, reduce the number of equipment, make the overall volume of the NPP compact, simplify the operation mode, and make the load easy to adjust.

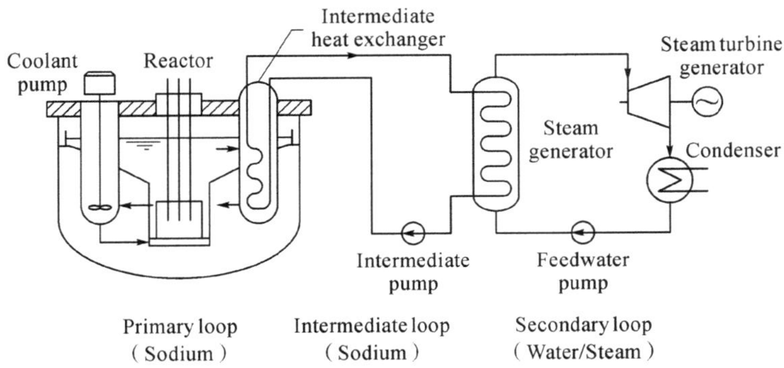

Fig. 1.7 Schematic diagram of the thermal system of the pool sodium-cooled fast reactor NPP

1.2.5 Fast Neutron Reactor NPP

Fast neutron reactors are driven by chain fission reactions caused by fast neutrons with an average energy of 0.08–0.1 MeV, using a mixture of PuO_2 and UO_2 (Pu accounts for 15–30%) powder to make sintered ceramic fuel cores. The core does not need to be filled with a moderator and uses liquid metal or gas as a coolant. Fast reactors truly consume U-238, which accounts for more than 99.2% of natural uranium and is not easily fissionable in thermal neutron reactors. The newly generated fissionable nuclear fuel (such as plutonium) is more than the consumed fissionable nuclear fuel plutonium, which can increase the uranium resource utilization rate to 60–70% [11].

The currently operating fast neutron reactors mainly use liquid sodium as a coolant. Due to the active chemical properties of sodium, in order to prevent the sodium-water reaction that may occur in the steam generator from affecting the reactor core, a sodium-sodium-water/steam three-loop heat transfer system is generally used. Figure 1.7 shows the principle flow chart of the thermal system of a pool-type sodium-cooled fast reactor NPP.

The boiling point of sodium is as high as 881 °C, and a high operating temperature can be obtained in the first loop without pressurization. Fast reactors use austenitic stainless steel as the cladding material and core structural material, the core outlet temperature can approach 550 °C, the temperature and pressure parameters of the three-loop steam can approach the level of thermal power, and the thermal efficiency can exceed 40%.

1.3 Thermal Economic Indicators of NPPs

In order to evaluate the thermodynamic perfection of the thermal system and various thermal equipment of the NPP, the following series of thermal economic indicators are proposed using the PWR NPP as an example, referring to the general method of evaluating thermal economy in thermal power plants [12].

1.3.1 *Gross Efficiency of NPPs*

Also known as the total efficiency of the NPP, represented by the symbol η_{el}, it is an economic indicator to evaluate the perfection of the process of converting reactor thermal energy into electrical energy, defined as the ratio of the generator output electric power N_{el} to the reactor thermal power Q_R:

$$\eta_{el} = \frac{N_{el}}{Q_R} = \eta_1 \eta_{sg} \eta_t \eta_{oi} \eta_{mp} \eta_m \eta_{ge} \tag{1.1}$$

For large-scale PWR NPPs, their gross efficiency is generally 33–35%. The meanings of the efficiencies in the formula are as follows.

1.3.1.1 First Loop Energy Utilization Coefficient η_1

Used to measure the perfection of the energy transfer of the first loop system, defined as the ratio of the thermal power Q_s of the nuclear steam supply system to the thermal power Q_R of the reactor:

$$\eta_1 = \frac{Q_s}{Q_R} \tag{1.2}$$

Its value is generally 0.990–0.995.

1.3.1.2 Steam Generator Efficiency η_{sg}

Used to measure the perfection of energy transfer between the primary and secondary sides of the steam generator, defined as the ratio of the secondary side thermal power Q_{sg} of the steam generator to the thermal power Q_s of the nuclear steam supply system:

$$\eta_{sg} = \frac{Q_{sg}}{Q_s} \tag{1.3}$$

Its value is generally 0.98–0.99.

1.3.1.3 Ideal Cycle Thermal Efficiency η_t

Defined as the ratio of the theoretical power N_a output by the turbine to the secondary side thermal power Q_{sg} of the steam generator:

$$\eta_t = \frac{N_a}{Q_{sg}} \tag{1.4}$$

Its value is generally 0.40–0.54.

1.3.1.4 Turbine Relative Internal Efficiency η_{oi}

When steam expands and does work in the turbine, it will produce throttle losses in the steam engine structure, nozzle losses, moving blade losses, residual speed losses, wet steam losses, steam leakage losses, blower friction losses, etc., these losses are collectively referred to as turbine internal losses, which means that the turbine can only convert part of the available enthalpy drop of steam into work. The smaller the internal loss of the turbine, the higher the perfection of the internal structure of the turbine, which is measured by the relative internal efficiency η_{oi} of the turbine, defined as the ratio of the internal power N_i output by the turbine to the theoretical power N_a:

$$\eta_{oi} = \frac{N_i}{N_a} \tag{1.5}$$

Its value is generally 0.80–0.88.

1.3.1.5 Pipeline Efficiency η_{mp}

The steam required for the operation of the turbine is transported from the steam generator to the turbine through pipes. The exhaust steam from the high-pressure cylinder is then sent to the low-pressure cylinder through pipes and a steam-water separator re-heater. The exhausted steam from the low-pressure cylinder is finally discharged into the condenser through the exhaust port. As the steam flows through these parts, there is a certain degree of throttling and heat loss, which results in a reduction in the actual work done by the steam in the turbine. This energy loss is measured by the pipe efficiency η_{mp}, defined as the ratio of the actual internal power $N_i^{'}$ of the turbine to the output internal power N_i:

$$\eta_{mp} = \frac{N_i^{'}}{N_i} \tag{1.6}$$

Its value is generally 0.98–0.99.

1.3.1.6 Turbine Mechanical Efficiency η_m

This reflects the mechanical friction power consumption between the turbine shaft and the supporting bearings, thrust bearings, and thrust plates, as well as the power consumption of the main oil pump and speed control system. It is defined as the ratio of the output effective power N_e of the turbine to the actual internal power $N_i^{'}$ of the turbine:

$$\eta_m = \frac{N_e}{N_i^{'}} \tag{1.7}$$

The mechanical efficiency of the turbine is generally 0.965–0.99.

1.3.1.7 Generator Efficiency η_{ge}

This reflects the power consumption caused by the friction of the generator shaft and supporting bearings, the friction of the cooling medium inside the generator, copper loss (coil heating), and iron loss (eddy current heating of the magnetized iron core). It is defined as the ratio of the output power N_{el} of the generator to the effective power N_e of the turbine:

$$\eta_{ge} = \frac{N_{el}}{N_e} \tag{1.8}$$

The efficiency of modern large-scale double-water-cooled generators is 96–99%, air-cooled generators is 97–98%, and hydrogen-cooled generators is 98–99%.

1.3.2 Net Efficiency of NPPs

The ratio of the net output power N_{net} of the NPP to the thermal power Q_R of the reactor is called the net efficiency of the NPP, represented by the symbol η_{net}. The net output power N_{net} of the power plant is the output electric power N_{el} of the generator minus the power plant's electric power N_{el}'. The power plant's electricity includes the main coolant pump, feed water pump, and the power consumption of various other machines and equipment, generally accounting for 4–8% of the generator's output power.

$$\eta_{net} = \frac{N_{net}}{Q_R} = \frac{N_{el} - N_{el}'}{Q_R} \tag{1.9}$$

1.3.3 Steam Consumption Rate

The amount of steam consumed by the turbine generator set for every 1 kWh of electricity generated is called the steam consumption rate, represented by the symbol d_0, defined as

$$d_0 = \frac{3600}{H_a \eta_{oi} \eta_m \eta_{ge}} \quad \text{kJ}/(\text{kW}\cdot\text{h}) \tag{1.10}$$

Where H_a is the ideal adiabatic enthalpy drop of the turbine, kJ/kg.

The steam consumption rate d_0 is an important indicator reflecting the ability of steam to do work. For modern large-scale thermal power plants, the steam consumption rate is about 3.0 kg/(kW·h), and the steam consumption rate of a typical PWR NPP is about 6.0 kg/(kW·h). The steam flow determines the shape, size, and weight of the secondary circuit pipes and the main thermal equipment flow parts, and also affects the manufacturing cost of the equipment.

1.3.4 Heat Consumption Rate

The amount of heat consumed by the turbine generator set for every 1 kWh of electricity output is called the heat consumption rate, represented by the symbol q_0, defined as

$$q_0 = d_0 \left(h_{\text{fh}} - h_{\text{fw}} \right) \text{kJ} / \left(\text{kW} \cdot \text{h} \right) \tag{1.11}$$

Where h_{fh} and h_{fw} are the enthalpy of the fresh steam and feed water, respectively, kJ/kg.

The heat consumption rate q_0 collectively reflects the gross efficiency of the power plant. The heat consumption rate of modern large-scale thermal power plants is about 8000 kJ/(kW·h), equivalent to a gross efficiency of about 40–43%. The heat consumption rate of PWR NPPs is about 10,000 kJ/(kW·h), equivalent to a gross efficiency of about 32–35%.

The rated output power of the Daya Bay NPP in Guangdong is 983.8 MW, and the rated thermal power of the reactor is 2904.8 MW. Its main economic indicators are:

The gross efficiency of the power plant $\eta_{\text{el}} = 33.87\%$;
Steam consumption rate $d_0 = 5.61$ kg/(kW·h);
Heat consumption rate $q_0 = 10, 100$ kJ/(kW·h).

1.4 Purpose and Tasks of Thermal Analysis

Due to the long construction cycle and large investment for the NPPs, the economy often becomes one of the main reasons restricting the development of NPPs. In a NPP, the conversion of nuclear energy into mechanical energy is accomplished through a thermal cycle. The chain fission reaction of nuclear fuel generates a high-temperature heat source, and the heat energy is transferred to the working fluid water. The water, after being heated, generates steam and delivers it to the turbine to do work, completing the thermal power conversion. The steam after work is discharged into the condenser, releases heat to the cold source and condenses into water, restoring its initial state, and then re-obtains heat energy from the heat source, thus forming a thermal cycle. In this way, the process of thermal power conversion continues. In these energy conversion processes, there are always various thermal losses of varying amounts and different causes, such as reactor losses, pipeline losses, cold source losses, internal losses of the turbine, mechanical losses, generator losses, etc., causing only a part of the nuclear fission energy to be converted into electrical energy.

The perfection of the thermal cycle has a significant impact on the thermal economic indicators of the NPP. Therefore, analyzing and evaluating the perfection of the thermal power conversion process, and studying ways to improve the thermal economic indicators of the NPP, is an important topic in the design and operation of NPPs. Studying the various links and processes of energy conversion, transmission, and distribution during the operation of the NPP from the perspective of thermodynamics, pointing out the locations, quantities, and causes of energy loss in the thermal system, has practical guiding significance for improving the design of the thermal system of the NPP, reducing fuel consumption, and improving the economy of the power plant.

Marine nuclear power units are mostly used for military purposes and, compared to NPPs, have the following two significant characteristics:

1. There are many operating conditions, frequent load changes, and long-term operation at lower cruising conditions.
2. The thermal cycle is simple, and the thermal equipment requires light weight and small size, which is convenient for installation in the narrow cabin of a ship.

Therefore, while ensuring safety, marine nuclear power units emphasize maneuverability first, and the demand for thermal economy is not as urgent as that of NPPs. However, thermal economy is still an important performance indicator of marine nuclear power units, because higher economy helps to reduce the consumption of nuclear fuel during the operation of the nuclear power unit, extend the full power operation time of the reactor, and thus reduce the number of refueling times during service.

Theoretically, the thermal analysis of marine nuclear power units is no different from the thermal analysis of NPPs. However, marine nuclear power units operate for a long time under cruising conditions, and the economic requirements for cruising conditions are higher than those for rated conditions. Therefore, when conducting a thermal analysis of marine nuclear power units, not only the rated conditions should be considered, but also the cruising conditions.

Thermodynamic analysis of energy systems is based on the first and second laws of thermodynamics. It analyzes and studies the rationality and effectiveness of energy systems in the process of energy transmission and conversion. The aim is to find methods and ways to improve the thermal economy of the power cycle, thereby improving the thermal economy of the entire energy system, reducing energy loss, and increasing energy utilization. The application of thermodynamic analysis methods mainly includes the following situations:

1. Analyze the performance of in-service energy systems.

For example, for an in-service NPP, regular thermal economic evaluations need to be made to provide managers with information reflecting the current operating status of the power plant.

2. Evaluate newly designed energy systems.

   When designing a new energy system, it is not only necessary to determine the performance of the system, but more importantly, to make the energy system have the best overall performance. Through thermodynamic analysis, the main factors and equipment and processes affecting the overall thermal economy of the energy system can be determined, providing a theoretical basis for optimization design.

3. Conduct energy-saving analysis.

   In general system improvements, more attention is paid to external losses caused by accompanying material flow and energy flow losses, but it is more important to correctly analyze the size of internal energy losses in the system, determine their impact on the thermal economic performance of the system, and thus point out the direction for energy-saving work.

Due to the sharp increase in energy demand and the global energy crisis, energy conservation has been highly valued as a global strategic issue. In order to achieve sustainable development, the theory and technology of thermodynamic energy saving have been deeply studied and discussed, with the aim of improving the effective use of energy. Currently, an important direction in the field of thermodynamics is not only to simply analyze the laws of thermodynamics, but also to deeply study methods such as exergy analysis, thermal exergy economics, and comprehensive optimization of process energy systems.

Questions and Exercises

1. What are the advantages of nuclear power generation?
2. Briefly describe the main characteristics of the thermal system of a PWR NPP.
3. What are the main thermal economic indicators of a NPP?
4. What is the main task of thermal analysis?

Chapter 2
Theoretical Basis of Thermal Analysis

2.1 Basic Concepts of Thermodynamics [13, 14]

2.1.1 Thermodynamic System

For the convenience of problem analysis, thermodynamics often separates the object of analysis from surrounding objects and studies the transfer of energy and mass between it and surrounding objects through the interface. This artificially separated object for thermodynamic analysis is called a thermodynamic system, and the surrounding objects are collectively referred to as the environment.

The interface between the thermodynamic system and the environment is called the boundary, which can be real or imaginary. For example, when the working fluid (steam) in the steam turbine is taken as the thermodynamic system, there is a real boundary between the working fluid and the cylinder, but there is no real boundary between the working fluid before and after the inlet or outlet. Here, an imaginary boundary can be artificially set to separate the working fluid in the system from the environment, as shown in Fig. 2.1. In addition, the boundary between the system and the environment can be fixed or movable or deformable.

According to the energy and material exchange situation between the thermodynamic system and the environment, the thermodynamic system can be divided into different types such as closed systems, open systems, isolated systems, adiabatic systems, heat sources, and power sources.

1. Closed system: Only energy (heat or work) is exchanged with the environment, and there is no material exchange, so the mass within the closed system remains constant. It is sometimes also referred to as a control mass. Figure 2.2a illustrates an example of a closed system.
2. Open system: Both energy and material are exchanged with the environment, so the mass and energy in the open system can change. However, this change usually occurs within a defined spatial range, so the open system is also known as a

M. Peng et al., *Thermal Analysis of Nuclear Power Plants*, Nuclear Science and Technology, https://doi.org/10.1007/978-981-96-8983-5_2

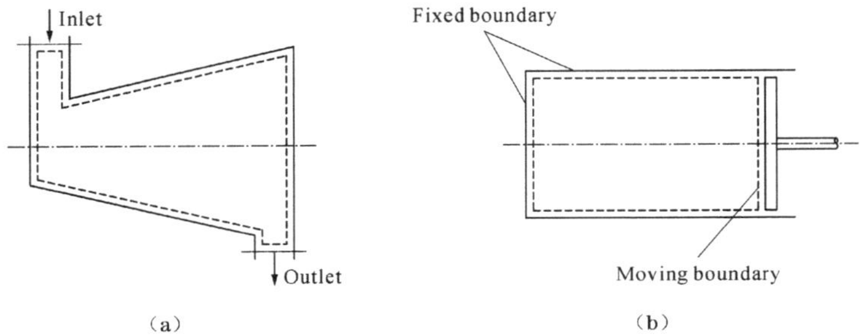

Fig. 2.1 Example of thermal system division: (**a**) Fixed boundary thermal system; (**b**) Moving boundary thermal system

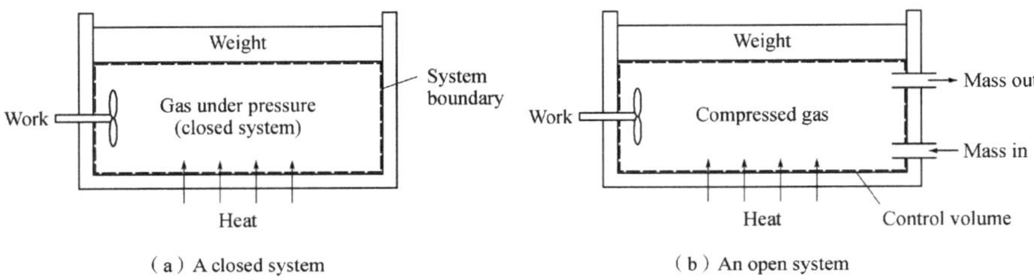

Fig. 2.2 Examples of closed and open systems: (**a**) Closed system; (**b**) Open system [14]

control volume, or a control body. The boundary between the control mass or control volume and the environment is called the control surface. Figure 2.2b illustrates an example of an open system.

3. Isolated system: There is no exchange of energy or material with the environment, and all interactions occur within the system.
4. Adiabatic system: There is no heat exchange with the environment.
5. Heat source: Only heat is exchanged with the environment, and the exchange of finite heat does not cause a change in the system's temperature. Depending on the temperature and function of the heat source, it can be divided into high-temperature heat sources and low-temperature heat sources, also known as heat sources and cold sources.
6. Work source: Only work is exchanged with the environment.

The division of thermodynamic systems should be determined according to specific requirements. For example, the entire nuclear power device can be divided into a thermodynamic system, calculating the consumption of nuclear fuel over a period of time, the output of electrical energy and mechanical energy, and the heat taken away by the cooling water, etc. At this time, the mass of the working substance in the entire nuclear power device does not change, it is a closed system. If only a certain system or device is analyzed, such as the working process of a steam generator, there is not only the material exchange process of the working substance flowing in and out, but also the energy exchange process of heat absorption and heat transfer. At this time, the steam generator is taken as the defined space to form an open system.

2.1.2 Thermodynamic State

The macroscopic physical condition presented by the thermodynamic system at a certain moment during the thermodynamic change process is called the state of the system.

2.1.2.1 State Parameters

The macroscopic physical quantities used to describe the state of the system are called state parameters. There are six commonly used state parameters in engineering thermodynamics, namely temperature, pressure, volume, thermodynamic energy, enthalpy, and entropy. These thermodynamic quantities reflect the properties of the system. Since they are single-valued functions of the state, they are also called state functions. State parameters can be divided into intensive parameters and extensive parameters.

Intensive Parameters

Refers to parameters that are independent of the amount of substance contained in the system at a given state, such as pressure, temperature, specific volume, etc. Intensive parameters do not have additivity, that is, the intensive parameter of the entire system is not equal to the sum of the same-name intensive parameters of each subsystem. The same-name intensive parameters at each point inside a uniform system are the same, while the same-name intensive parameters at each point inside a non-uniform system may not be the same.

Extensive Parameters

Refers to parameters that are related to the amount of substance contained in the system at a given state, such as volume, energy, mass, etc. Extensive parameters have additivity, that is, the extensive parameter of the entire system is equal to the sum of the same-name extensive parameters of each subsystem. Whether the thermodynamic system is uniform or not, the extensive parameters have a definite value. However, the extensive parameters per unit mass have the properties of intensive parameters, such as specific volume, specific enthalpy, specific entropy, etc.

2.1.2.2 Equilibrium State

If a thermodynamic system can always maintain its macroscopic state unchanged under the condition of no external influence, the system is said to be in a thermodynamic equilibrium state, simply referred to as an equilibrium state.

Only when all potential differences (such as temperature potential difference, pressure potential difference, chemical potential difference, etc.) that cause changes in the thermodynamic system state are zero, can the thermodynamic system reach thermodynamic equilibrium (i.e., thermal equilibrium, force equilibrium, phase equilibrium, and chemical equilibrium). It should be pointed out that the equilibrium state is a dynamic equilibrium. When the system reaches the equilibrium state, it is stationary on a macroscopic level, but on a microscopic level, the molecules within the system are still in constant motion. However, the statistical average of molecular motion does not change with time, thus appearing as a constant macroscopic state.

A system in a state of thermal equilibrium, as long as it is not affected by the environment, its state will not change with time, and the equilibrium will not spontaneously break. A system in a state of non-equilibrium, due to heat transfer and displacement between its parts, its state will change with time. The result of the change will inevitably weaken the heat transfer and displacement until it completely stops. Therefore, a system in a state of non-equilibrium, without the influence of external conditions, will always spontaneously tend towards equilibrium.

Conversely, if the system is affected by the environment, it cannot maintain a state of equilibrium. For example, the exchange of heat due to temperature imbalance between the system and the environment, and the exchange of work due to pressure imbalance, will all disrupt the original equilibrium state of the system. The result of the interaction between the system and the environment will inevitably lead the system and the environment to reach a new equilibrium state together.

2.1.3 Process and Cycle

2.1.3.1 Process

The change from one equilibrium state to another through a series of intermediate states in a thermodynamic system is called a process. According to the change rules of state parameters, it can be divided into isothermal process, isobaric process, isochoric process, isenthalpic process, isentropic process, etc.; according to the interaction between the thermodynamic system and the environment, it can be divided into adiabatic process, non-adiabatic process, etc.; according to the reversibility of the process, it can be divided into reversible process and irreversible process.

The mutual conversion of thermal energy and mechanical energy must be completed through the state change process of the working medium. The working medium continuously undergoes heat absorption, expansion, compression, etc. in the thermal equipment, and continuously converts thermal energy into mechanical energy through external work. The processes carried out in actual equipment are all very complex, because all processes are the result of equilibrium being broken. Only when there is heat and force imbalance between the working medium and the environment can the working medium be prompted to change to a new state, so all actual processes are unbalanced. If the working medium can automatically restore equilibrium after the equilibrium is broken, and the time required for recovery (i.e., relaxation time) is very short, when the process is very slow, and the time experienced is much larger compared to the relaxation time, then the working medium has enough time to restore equilibrium during the process, and it will not be far from the equilibrium state at any time. Such a process is called a quasi-equilibrium process. Compared to the relaxation time, the quasi-equilibrium process is an infinitely slow process, so the quasi-equilibrium process is also called a quasi-static process.

In the quasi-equilibrium process, the working medium maintains heat and force balance with the environment at all times, the temperature of the heat source and the working medium are equal at all times, and the force exerted by the working medium on the environment and the resistance of the environment are also equal at all times, so the process can be reversed at any time without conditions. After completing a certain process, if it is possible to make the working medium return to its original state along the same path in reverse, and make the environment involved in the interaction also return to its original state without leaving any changes, this process is a reversible process, and the process that does not meet the above conditions is an irreversible process.

A reversible process is first and foremost a quasi-equilibrium process, which must satisfy the balance conditions of heat and force, and there should be no dissipation effects during the process. The difference between a quasi-equilibrium process and a reversible process is that the quasi-equilibrium process focuses only on the internal balance of the working substance, and whether there is friction has no relation to the internal balance of the working substance. Energy loss may occur during the quasi-equilibrium process; the reversible process, on the other hand, analyzes the overall effect produced by the interaction of the working substance with the environment. It not only requires the internal balance of the working substance, but also requires that the interaction of the working substance with the environment can be reversed unconditionally, and there is no irreversible energy loss during the process. Therefore, a reversible process must be a quasi-equilibrium process, and a quasi-equilibrium process is only one of the necessary conditions for a reversible process.

The reversible process is an extremely important concept in thermodynamics. Its characteristics are that the driving force of the process is infinitesimally small, the speed is infinitesimally slow, and the system is always infinitely close to equilibrium. The reversible process idealizes the actual process, representing the limit situation in which the actual process may proceed. It ignores various irreversible factors such as friction and uneven temperature, pressure, and concentration within the system, simplifies the complex actual process, and facilitates theoretical analysis and calculation. By calculating according to the reversible process and combining with appropriate efficiency, the approximate result of the actual process can be obtained. The reversible process is the standard for measuring or comparing actual processes, reflecting the highest efficiency that energy utilization can possibly reach. Therefore, how to create conditions to make the actual process approach the reversible process is an important consideration for improving production and enhancing technical and economic effects.

Obviously, not all actual processes can be simplified into reversible processes, such as explosions, throttling, and free expansion of gas into a vacuum, etc. These processes are far from the conditions of the reversible process and cannot be treated as reversible processes.

2.1.3.2 Work and Heat

In the thermodynamic process, the energy exchange between the thermodynamic system and the environment can be achieved in two ways, one is work, and the other is heat transfer.

Work is energy transferred in a macroscopically ordered motion under the drive of force. In mechanics, work is defined as the product of force and displacement in the direction of force. If the system produces a small displacement $\mathrm{d}x$ in the direction of force F, the work done by this force is

$$\mathrm{d}W = F\mathrm{d}x \tag{2.1}$$

If the system moves a finite distance, the work done is

$$W = \int_1^2 F\mathrm{d}x \tag{2.2}$$

Work is the energy in the system transfer, it only makes sense during the transfer process. Once the work crosses the boundary of the system, it is transformed into the energy of the system or the environment. Work transfer only occurs when the system state changes, therefore, the size of the work is not only related to the initial and final states of the process, but also related to the nature of the process. Therefore, work is not a state parameter but a process quantity. When the process carried out by the system is different, even if the initial and final states of the system are the same, the work received or output by the system may also be different.

Heat is another form of energy exchange between the system and the environment besides work. When the thermodynamic system is in thermal contact with the environment at different temperatures, they will exchange energy. The energy transferred between the thermodynamic system and the environment relying on the temperature difference is called heat. Heat is also a process quantity.

2.1.3.3 Thermodynamic Cycle

The conversion between thermal energy and mechanical energy is usually achieved by the working substance cycling through the corresponding equipment. The working substance starts from a certain state, goes through a series of processes, and then returns to the initial state. These processes are collectively referred to as the thermodynamic cycle, or simply the cycle. As shown in Fig. 2.3, processes 1-2, 2-3, 3-4, and 4-1 constitute a thermodynamic cycle.

After the working substance completes the cycle and returns to the original starting point, it is possible to continue to operate repeatedly in the same process, thereby continuously doing work. For example, the secondary loop feedwater of the nuclear power device absorbs heat in the steam generator and becomes high-temperature and high-pressure steam, enters the steam turbine to expand and do work, the exhausted steam discharged from the steam turbine condenses into water in the condenser, and after being compressed and pressurized by the feedwater pump, it re-enters the steam generator. In this way, the secondary loop working substance has gone through several processes such as heat absorption, expansion and work, heat release, compression, etc., and has returned to the initial state, forming a thermodynamic cycle. In this way, the heat energy of the steam is continuously converted into mechanical energy and output to the environment.

If every process in the cycle is reversible, then this cycle is called a reversible cycle; a cycle that contains irreversible processes is called an irreversible cycle. Actual thermodynamic cycles are irreversible.

The working substance must go through some expansion and work processes and some compression and power consumption processes in a thermodynamic cycle. The difference between work and power consumption is the possible use of work, called the net work of the cycle, represented by q_0. Similarly, the working substance must also go through some processes of absorbing heat from the heat source and discharging heat to the cold source in a cycle. The heat absorbed by the working substance from the heat source, represented by q_1, is usually obtained by consuming fuel and is the cost required for the thermodynamic engine to do work; the heat discharged by the working substance to the cold source, represented by q_2, is usually given to the natural environment and no longer plays a beneficial role. The economy of the thermodynamic cycle is measured by the thermal efficiency η_t, which is the ratio of the net work of the cycle to the consumed heat, that is

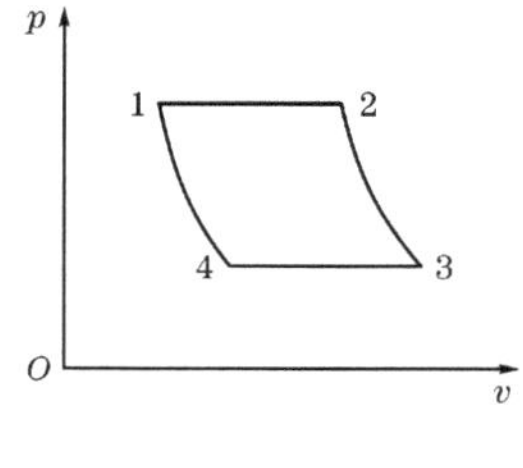

Fig. 2.3 Thermodynamic cycle

$$\eta_t = \frac{q_0}{q_1} = \frac{q_1 - q_2}{q_1} \tag{2.3}$$

The main purpose of studying engineering thermodynamics is to improve the thermal efficiency of the cycle.

Thermodynamic cycles are divided into two major categories: forward cycles and reverse cycles. Any thermodynamic cycle that converts heat energy into mechanical energy is a forward cycle, which cycles in a clockwise direction on the pv diagram, and all heat engines in engineering use forward cycles. Any cycle that consumes energy to extract heat from a low-temperature object and discharges it to a high-temperature object is called a reverse cycle, which cycles in a counterclockwise direction on the $p - v$ diagram, such as refrigeration and heat pumps are all working with reverse cycles.

2.1.4 Forms of Energy and Their Convertibility

All matter has energy, energy is an inherent property of matter, and it is a manifestation of the movement of matter. Energy can neither be created nor destroyed.

2.1.4.1 Forms of Energy and Their Convertibility

The common forms of energy in nature include potential energy, kinetic energy, electric energy, magnetic energy, thermal energy, chemical energy, light energy, atomic energy, etc. In the thermal power system, the main forms of energy are thermal energy, mechanical energy, electric energy, nuclear energy, etc. Various forms of energy can be converted into each other under certain conditions, such as combustion can convert chemical energy into thermal energy; through a heat engine, thermal energy can be converted into mechanical energy, etc. Figure 2.4 shows the mutual conversion relationship between different forms of energy.

Thermal energy is a basic form of energy. In the primary energy sources existing in nature, except for wind power, hydropower, and some ocean energy that can be directly used as mechanical energy, other various energy sources either exist directly in the form of thermal energy, or are first converted into thermal energy through combustion reactions, nuclear reactions, etc., and then utilized.

The main ways to obtain thermal energy include solar energy, geothermal energy, heat release from fuel combustion, heat release from nuclear fission, heat release from nuclear fusion, and electrical energy can also be converted into thermal energy. Mechanical energy can be converted into thermal energy through friction, but this conversion method is considered can be divided into two categorieshould be minimized or eliminated.

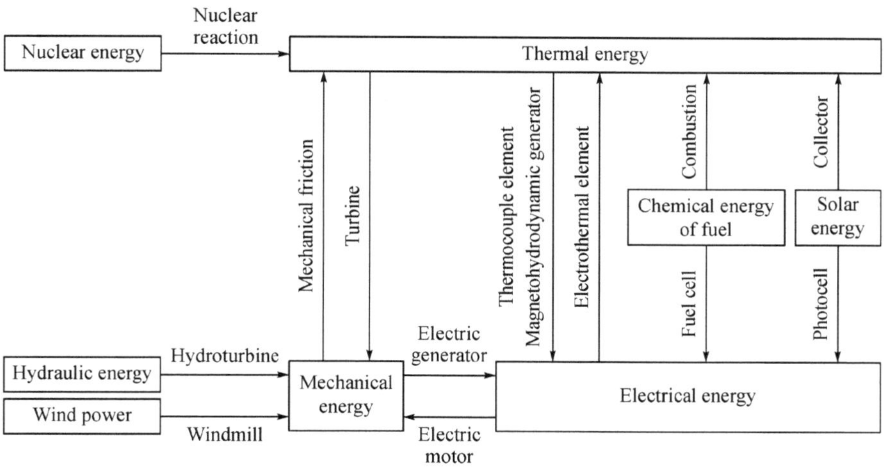

Fig. 2.4 The conversion of different forms of energy into one another

The energy possessed by the macroscopic mechanical operation of an object is called mechanical energy. Apart from a small part that comes from hydropower, wind energy, ocean currents, tides, and waves in some kind of primary energy source, it is basically obtained through some type of heat engine (such as internal combustion engines, gas turbines, steam turbines, etc.) from thermal energy conversion, or obtained through electric motors by electrical energy conversion.

Electrical energy is the ability to work possessed by the flow or accumulation of electric charge. The production of electrical energy is divided into two forms: direct energy conversion and indirect energy conversion. The main ways to directly produce electrical energy include: magnetohydrodynamic power generation that directly converts thermal energy into electrical energy, various batteries or fuel cell power generation that directly converts chemical energy into electrical energy, solar cells that directly convert electromagnetic energy into electrical energy, etc. Indirect energy conversion is the main way of electrical energy production at present, converting mechanical energy into electrical energy through an alternating current generator.

2.1.4.2 Characteristics of Energy Conversion

Different forms of energy have the same and common characteristics reflected by the first law of thermodynamics, but there are certain restrictions when converting from one form to another, which is determined by the second law of thermodynamics. This limitation is extremely important in practice. According to whether the energy conversion is restricted by the second law of thermodynamics, it can be divided into three different qualities of energy.

High-Level Energy—Unlimited Conversion Energy

Such as mechanical energy, electrical energy, nuclear energy, and the natural world's hydropower, wind energy, etc., theoretically can be completely, without any restrictions, converted into other forms of energy, and is not restricted by the second law of thermodynamics during the conversion process. The "quality" and "quantity" of high-level energy are completely unified, and they are especially valuable in technology and economy. These forms of energy have a common feature, that is, entropy equals zero, in other words, they do not use entropy to express, sometimes also called "ordered energy".

Low-Level Energy—Limited Conversion Energy

Such as thermal energy, the thermodynamic energy of the thermodynamic system, etc., cannot be completely converted into other forms of energy, and the conversion ability is restricted by the second law of thermodynamics. The "quality" and "quantity" of low-level energy are often not unified, and the height of its "quality" depends on how much high-level energy can be converted into. This form of energy is always characterized by entropy, usually called "disordered energy" or "entropy energy".

Rigid Energy—Non-convertible Energy

Such as the thermodynamic energy possessed by the atmosphere, the earth, natural water sources and other environmental media, although it can have a considerable amount, it is restricted by the second law of thermodynamics and cannot be converted into other forms of energy under environmental conditions.

The change from high-quality energy to low-quality energy is called energy degradation or energy downgrade. Energy degradation means the loss of workability. The phenomenon of energy degradation is common in thermal processes, such as common heat transfer processes and throttling processes. The former degrades high-temperature heat to low-temperature heat, and the latter degrades high-pressure fluid to low-pressure fluid, both of which have a loss of workability.

The so-called rational use of energy and improving the effective utilization rate of energy is to pay attention to the protection of energy quality, minimize energy degradation, or avoid unnecessary degradation.

In describing energy, it can be broken down into the product of intensity factors and capacity factors. For example, the energy consumed by a moving object is the force multiplied by the distance the object moves in the direction of the force, where the force is the intensity factor and the distance the object moves is the capacity factor. The energy transferred between two objects at different temperatures is the heat transfer temperature difference multiplied by the heat capacity of the heat-absorbing object, where the heat transfer temperature difference is the intensity factor and the heat capacity of the object is the capacity factor.

The intensity factor is a key factor in the movement of matter or energy transfer, and is the driving force of any macroscopic change process, while the capacity factor is the objective effect produced by any process under the action of the intensity factor.

Energy can be transferred from one object to another in two ways: one is by doing work, and the other is by heat transfer. Heat and work are measures of energy transfer, and they are process quantities. Work and heat only exist in the process of energy transfer, and without the process of energy transfer, there is no concept of work and heat.

The process of converting thermal energy into mechanical energy is actually composed of two types of processes: one is the thermodynamic process of energy conversion, in which thermal energy is first transferred and converted into the thermodynamic energy of the working substance, and then the working substance expands to do work and convert the thermodynamic energy into mechanical energy. The thermodynamic state of the working substance changes during the conversion process, and the form of energy also changes. The second is a purely mechanical process, in which the mechanical energy obtained from the conversion of thermal energy is converted into the kinetic energy of the rotating parts of the power machinery.

In various ways of energy transfer, only when the working substance expands to do work can the conversion of thermal energy into mechanical energy be realized, and the mechanical energy produced is equal to the expansion work. Although the process of converting mechanical energy into thermal energy can also be completed by friction, collision, etc., only the conversion process of doing work by compressing the working substance can be reversible. Therefore, the reversible conversion of thermal energy and mechanical energy is always associated with the expansion and compression of the working substance.

2.2 First Law of Thermodynamics [13, 14]

The first and second laws of thermodynamics are the two basic laws of thermodynamics and the theoretical basis of thermodynamics. They are scientific summaries of long-term human practical experience and important laws of nature. Their correctness has been proven by a large number of facts, so they have universal applicability. The mutual conversion relationship between the two forms of energy transfer—heat and work, involves the issue of energy conversion efficiency and improving energy utilization rate. The first and second laws of thermodynamics are the theoretical basis for studying and solving this problem.

2.2.1 Statement of the First Law of Thermodynamics

The first law of thermodynamics is the application of the law of energy conservation and transformation in thermal phenomena. The law of energy conservation and transformation states that all matter in the natural world has energy, energy cannot be created or destroyed, but it can be transformed from one form to another. In the process of energy transformation, a certain amount of one form of energy always corresponds to a certain amount of another form of energy, and the total amount of energy remains constant.

The first law of thermodynamics points out the mutual transformation and total energy conservation of thermal energy and other forms of energy (such as mechanical energy, chemical energy, and electromagnetic energy, etc.). Within the scope of engineering thermodynamics, it mainly involves the mutual transformation and conservation of thermal energy and mechanical energy.

The first law of thermodynamics is usually stated as: heat is a form of energy, and when mechanical energy is converted into heat energy, or heat energy is converted into mechanical energy, the ratio between them is constant.

It can also be stated as: heat can be converted into work, and work can be converted into heat. When a certain amount of heat disappears, a certain amount of work must be produced; when a certain amount of work is consumed, a certain amount of heat corresponding to it must appear.

For any thermodynamic system, the first law of thermodynamics can be expressed as:

$$E_{\text{in}} - E_{\text{out}} = \Delta E \tag{2.4}$$

Where, E_{in} is the energy entering the system, E_{out} is the energy leaving the system, and ΔE is the increment of stored energy in the system.

Energy enters or leaves the system mainly in three forms, namely work, heat transfer, and the energy inherent in the working substance brought in or out by the working substance. Among them, work and heat transfer cannot be regarded as energy possessed by the system, they are both forms of energy transfer, only related to the process.

The energy inherent in the substance is called stored energy. Stored energy is divided into external stored energy and internal stored energy (thermodynamic energy). External stored energy is energy related to the macroscopic movement of the entire thermodynamic system, divided into kinetic energy and potential energy; thermodynamic energy is stored in the thermodynamic system, related to the molecular structure and microscopic movement of the substance, including physical thermodynamic energy, chemical thermodynamic energy, and nuclear energy.

Therefore, the total energy of the system is

$$E = E_{\text{k}} + E_{\text{p}} + U = \frac{1}{2}mc^2 + mgz + U \tag{2.5}$$

or

$$e = e_k + e_p + u = \frac{1}{2}c^2 + gz + u \tag{2.6}$$

Where, E_k, E_p, e_k, e_p are the kinetic energy, potential energy, specific kinetic energy and specific potential energy of the system respectively; U, u are the thermodynamic energy and specific thermodynamic energy of the system respectively; m is the mass of the system; c is the velocity of the system in the reference frame; z is the height of the center of gravity of the system in the reference frame.

According to formula (2.4), formula (2.5) or formula (2.6), if the working substance goes through a thermodynamic cycle and returns to its original state, the change in the system's stored energy is zero, and the energy entering the system (heat absorption) $\mathrm{d}Q$ should be equal to the energy leaving the system (work done) $\mathrm{d}W$, that is,

$$\oint \mathrm{d}E = \oint (\mathrm{d}Q - \mathrm{d}W) = 0 \tag{2.7}$$

2.2.2 *Energy Equation of a Closed System*

A closed system only exchanges energy with the environment without mass exchange. Since the working substance in the closed system is relatively stationary, the change in the state of the system does not cause changes in kinetic energy and potential energy, so only the change in thermodynamic energy is in the total energy of the system. According to the law of conservation of energy, the energy balance equation of the closed system is

$$\Delta U = Q - W \tag{2.8}$$

That is, the change in the system's thermodynamic energy ΔU is equal to the difference between the heat Q absorbed by the system and the work W done by the system.

For the infinitesimal process, the energy balance equation is

$$\mathrm{d}U = \mathrm{d}Q - \mathrm{d}W \tag{2.9}$$

The energy equation derived from the first law of thermodynamics applied to a closed system is the most basic energy equation, called the analytical form of the first law of thermodynamics. It shows that part of the heat added to the working substance is used to increase the thermodynamic energy of the working substance, and is still stored in the working substance in the form of heat energy, and the remaining part is transferred to the environment in the form of work, converted into mechanical energy.

The work done by the system can be divided into two categories:

1. Volume work (or mechanical work): the work done by the system due to volume changes. For any infinitesimal reversible process, it can be written as

$$\mathrm{d}W = p\mathrm{d}V \tag{2.10}$$

Where p is the system pressure; $\mathrm{d}V$ is the volume change of the system.

So when the closed system undergoes a microelement reversible process, the Eq. (2.9) can be written

$$\mathrm{d}Q = \mathrm{d}U + p\mathrm{d}V \tag{2.11}$$

2. Non-volume work: the general term for other work besides volume work, such as electrical work, surface work, magnetic work, etc.

2.2.3 Energy Equation of an Open System

The open system exchanges both energy and matter with the environment. As shown in Fig. 2.5 of the open system, during the process of material flow, the mass entering the system in $\mathrm{d}\tau$ time is $\mathrm{d}m_1$, and the mass leaving the system is $\mathrm{d}m_2$.

The increase in the mass of the working substance in the system is

$$\mathrm{d}m = \mathrm{d}m_1 - \mathrm{d}m_2$$

At the inlet cross section 1-1, the energy brought into the system with the working substance of mass $\mathrm{d}m_1$ is $e_1\mathrm{d}m_1$, and the propulsion work obtained from the subsequent working substance is $p_1v_1\mathrm{d}m_1$.

At the exit cross section 2-2, the energy leaving the system with the working substance of mass $\mathrm{d}m_2$ is $e_2\mathrm{d}m_2$, and the propulsive work done on the environment is $p_2v_2\mathrm{d}m_2$.

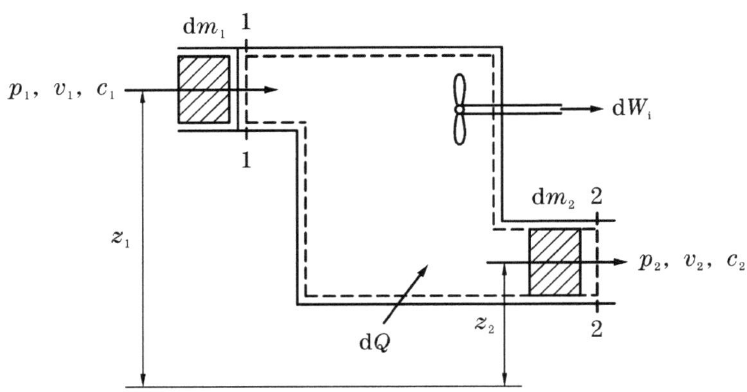

Fig. 2.5 Diagram of an open system

During the time $d\tau$, the open system absorbs heat dQ from the environment through the interface, and the system outputs work dW_i to the environment. The change in the thermodynamic energy of the system is

$$dU=\left[dQ+\left(e_1+p_1v_1\right)dm_1\right]-\left[\left(e_2+p_2v_2\right)dm_2+dW_i\right] \tag{2.12}$$

Considering that the specific enthalpy $h=u+pv$ and specific energy $e=u+\frac{1}{2}c^2+gz$ of the working substance in the system, Eq. (2.12) can be transformed into

$$dQ=dU+\left(h_2+\frac{1}{2}c_2^2+gz_2\right)dm_2-\left(h_1+\frac{1}{2}c_1^2+gz_1\right)dm_1+dW_i \tag{2.13}$$

Equation (2.13) is a general expression of the energy equation for an open system, applicable to any flow process of any working substance, and can also be written in another form:

$$\dot{Q}=\dot{U}+G_2\left(h_2+\frac{1}{2}c_2^2+gz_2\right)-G_1\left(h_1+\frac{1}{2}c_1^2+gz_1\right)+N_i \tag{2.14}$$

Where $\dot{Q}=\frac{dQ}{d\tau}$ is the heat absorption rate of the system; $\dot{U}=\frac{dU}{d\tau}$ is the increase rate of the thermodynamic energy of the system; $G_1=\frac{dm_1}{d\tau}$ is the mass flow into the system; $G_2=\frac{dm_2}{d\tau}$ is the mass flow leaving the system; $N_i=\frac{dW_i}{d\tau}$ is the system output power.

Equation (2.14) is the general form of the energy equation for an unsteady flow open system, applicable to any flow process of any working substance.

For the steady flow process where the thermal parameters and motion parameters of the working substance in the system do not change with time, the mass flow rate at the system inlet and outlet is equal, and the thermodynamic energy of the system does not change with time, that is, G1 = G2 = G = 0, substitute into Eq. (2.14) to get

$$\dot{Q}=G\left[\left(h_2+\frac{1}{2}c_2^2+gz_2\right)-\left(h_1+\frac{1}{2}c_1^2+gz_1\right)\right]+N_i \tag{2.15}$$

or

$$q=\left(h_2-h_1\right)+\frac{1}{2}\left(c_2^2-c_1^2\right)+g\left(z_2-z_1\right)+w_i \tag{2.16}$$

Equations (2.15) to (2.16) are the steady flow energy equations, applicable to any steady flow process of any working substance.

The first law of thermodynamics explains the quantitative relationship of energy during transmission and transformation. When heat is transferred between two objects with different temperatures, the first law states that the heat lost by one object must be equal to the heat gained by the other object, but it does not specify which object the heat will transfer from and to, under what conditions it can be transferred, and to what extent the process will proceed, that is, it does not specify the direction, conditions, and depth of energy transfer. When heat energy and mechanical energy are converted into each other, the first law of thermodynamics only states that there is a certain equivalent relationship between the two in quantity, but it does not specify the direction, conditions, and depth of the conversion.

2.3 The Second Law of Thermodynamics [13, 14]

In 1850, the German physicist Clausius proposed the second law of thermodynamics, pointing out that all actual processes related to thermal phenomena have their spontaneous direction within a finite time and space, that is, the second law of thermodynamics is a law that explains the direction, conditions, and depth of various processes related to thermal phenomena, among which directionality is its fundamental content.

The second law of thermodynamics is stated as: heat cannot spontaneously, without cost, transfer from a low-temperature object to a high-temperature object. The second law of thermodynamics also has other forms of expression, such as: it is impossible to manufacture a heat engine that absorbs heat from a single heat source, converts it all into work, and leaves no other changes.

2.3.1 Carnot Cycle

In 1824, the French engineer S. Carnot proposed the Carnot theorem, pointing out that among all heat engines working between a high-temperature heat source and a low-temperature heat source at a certain temperature, the efficiency of a reversible heat engine is the highest.

The Carnot cycle consists of two isothermal processes and two adiabatic processes, and it is assumed that all are reversible processes. Figure 2.6 shows the representation of the Carnot cycle with an ideal gas as the working substance on the p-v diagram and the T-s diagram.

a-b is the isothermal expansion (heat absorption) process, the working substance absorbs heat q_1 from the high-temperature heat source at the same temperature T_1;

b-c is an adiabatic expansion process, during which the temperature of the working substance decreases from T_1 to T_2, so as to release heat to the low-temperature heat source at the same temperature;

c-d is an isothermal compression (heat release) process, during which the working substance releases heat q_2 to the low-temperature heat source at the same temperature T_2;

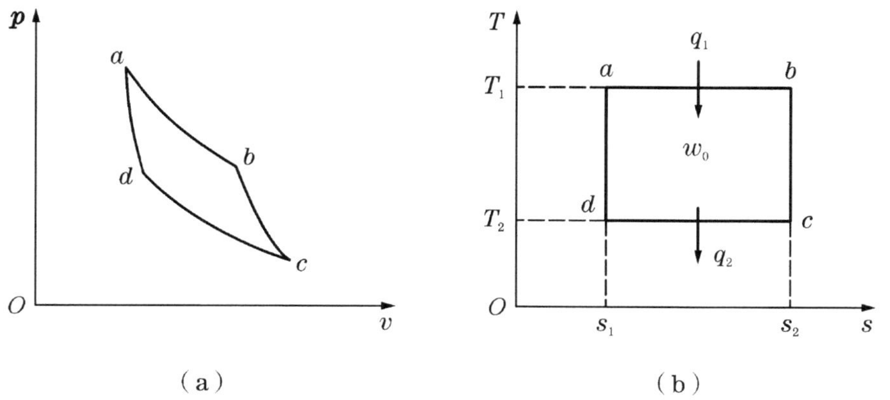

Fig. 2.6 Carnot cycle: (**a**) *p*-*v* diagram; (**b**) *T*-*s* diagram

d-*a* is an adiabatic compression process, during which the temperature of the working substance increases from T_2 to T_1, so as to absorb heat from the high-temperature heat source at the same temperature.

The thermal efficiency of the cycle is defined as

$$\eta_t = \frac{w_0}{q_1} = \frac{q_1 - q_2}{q_1} = 1 - \frac{q_2}{q_1}$$

Therefore, $q_1 = T_1(s_2 - s_1)$, $q_2 = T_2(s_2 - s_1)$, the expression for the efficiency of the Carnot cycle can be obtained

$$\eta_c = \frac{T_1 - T_2}{T_1} = 1 - \frac{T_2}{T_1} \tag{2.17}$$

The Carnot cycle was the first to lay the foundation for the second law of thermodynamics in history. It shows that of the heat obtained from the heat source, only a part can be converted into mechanical work, while the other part of the heat is given to the cold source. The following important conclusions can be drawn from the formula for the thermal efficiency of the Carnot cycle:

1. The thermal efficiency of the Carnot cycle is only determined by the temperatures of the high-temperature heat source and the low-temperature heat source, that is, the temperatures at which the working substance absorbs and releases heat. Increasing T_1 and decreasing T_2 can improve its thermal efficiency.
2. The thermal efficiency of the Carnot cycle can only be less than 1, and it can never be greater than or equal to 1, because it is impossible for T_1 to be infinitely high and T_2 to be zero. That is, it is impossible to convert all heat energy into mechanical energy in a cycle engine.
3. When $T_1 = T_2$, the cycle thermal efficiency is zero. That is to say, in a system with temperature balance, heat energy cannot be converted into mechanical energy.

To use heat energy to generate power, there must be a temperature difference, and there must be a high-temperature heat source with a temperature higher than the environment.

The Carnot cycle determines the limit value that the thermal efficiency of the actual thermal cycle can approach, thereby measuring the perfection of the thermodynamics of the actual thermal cycle. On the other hand, the Carnot cycle points out the direction on how to improve the efficiency of the thermal cycle, that is, to increase the temperature of the working substance as much as possible, and to expand the working substance to a lower temperature as much as possible, and to release heat to the outside at a temperature close to the natural environment.

For any complex cycle, the concept of a generalized (equivalent) Carnot cycle is proposed, that is, to replace the concept of T_1 and T_2 with the average heat absorption temperature T_1 and the average heat release temperature T_2, both of which have the same thermal efficiency.

The Carnot cycle and its thermal efficiency formula have extremely important significance in thermodynamics, and first laid the foundation for the second law of thermodynamics in history. Although no thermal engine that works completely according to the Carnot cycle has been manufactured so far, the Carnot cycle and its thermal efficiency formula show that the fundamental factor determining the cycle thermal efficiency is the temperature of the working substance absorbing and releasing heat, thereby pointing out the direction for improving the thermal efficiency of various thermal power machines.

2.3.2 Entropy and the Principle of Entropy Increase

The German physicist Clausius first proposed the concept of entropy (Entropy) in 1854, which originated from the direction judgment of irreversible processes.

2.3.2.1 Clausius Inequality and the Concept of Entropy

For any irreversible cycle, it can be divided into infinitely many micro-cycles. According to Carnot's theorem, the thermal efficiency of all irreversible heat engines working between the same high-temperature heat source and low-temperature heat source is less than the thermal efficiency of reversible heat engines, i.e., the thermal efficiency of irreversible micro-cycles is

$$\eta_t = 1 - \frac{dQ_2}{dQ_1} < \eta_c = 1 - \frac{T_2}{T_1}$$

That is

$$\frac{dQ_1}{T_1} < \frac{dQ_2}{T_2}$$

If we consider the heat absorption as a positive value and the heat release as a negative value, the above formula can be expressed as

$$\frac{dQ_1}{T_1} < -\frac{dQ_2}{T_2} \text{ or } \frac{dQ_1}{T_1} + \frac{dQ_2}{T_2} < 0$$

Summing up all the micro-cycles, we get

$$\oint \frac{dQ}{T} < 0 \tag{2.18}$$

If we further consider the case of a reversible cycle, the above formula can be expressed in the following more universally meaningful form:

$$\oint \frac{dQ}{T} \leq 0 \tag{2.19}$$

Equation (2.19) is known as the Clausius integral inequality, one of the mathematical expressions of the second law of thermodynamics, and can be used as a criterion for whether a cycle is reversible. The Clausius inequality applies to any cycle and provides a basis for the introduction of the concepts of entropy and entropy balance.

In 1865, Clausius defined the thermodynamic state parameter entropy (entropy), represented by the symbol S, and its definition is

$$dS = \frac{dQ}{T} \tag{2.20}$$

or

$$\Delta S = S_2 - S_1 = \int_1^2 \frac{dQ}{T} \tag{2.21}$$

Where.

dQ is the heat absorbed by the working medium in the reversible process of the micro element;

T is the thermodynamic temperature at which the working medium absorbs heat.

For a reversible adiabatic process, since dQ = 0, we have dS = 0, i.e., in a reversible adiabatic process, the entropy of the working substance remains constant, so a reversible adiabatic process is also called an isentropic process.

The first law of thermodynamics establishes an important thermodynamic parameter—thermodynamic energy, while the second law of thermodynamics establishes another equally important thermodynamic parameter—entropy. Thermodynamic energy is a thermodynamic property associated with the energy of microscopic particles in the system, while entropy is a thermodynamic property associated with the degree of disorder of the motion of molecules within the system.

From a physics perspective, entropy is a description of the degree of disorder of material molecules; the greater the degree of disorder, the greater the entropy. From the perspective of energy and its utilization, entropy is a measure of the degree of irreversible dissipation; the more irreversible energy dissipation, the greater the change in entropy. An increase in entropy means a decrease in the ability to do effective work.

2.3.2.2 Entropy Change in Irreversible Processes

Assume that the system changes from state 1 to state 2 through an irreversible process 1-*A*-2, and then returns to state 1 through a reversible process 2-*B*-1, forming an irreversible cycle 1-*A*-2-*B*-1. According to the Clausius integral inequality, we have

$$\oint \frac{\mathrm{d}Q}{T} = \int_{1-A-2} \frac{\mathrm{d}Q}{T} + \int_{2-B-1} \frac{\mathrm{d}Q}{T} < 0 \tag{2.22}$$

Since entropy is a state parameter, the entropy difference between two states is independent of the process. Therefore, for the reversible process 2-*B*-1, according to Eq. (2.20) that the definition of entropy, the entropy difference (i.e., entropy change) between state 2 and state 1 is

$$\Delta S_{12} = S_2 - S_1 = -\int_{2-B-1} \frac{\mathrm{d}Q}{T} \tag{2.23}$$

Substituting into Eq. (2.22), we get

$$\int_{1-A-2} \frac{\mathrm{d}Q}{T} + \int_{2-B-1} \frac{\mathrm{d}Q}{T} = \int_{1-A-2} \frac{\mathrm{d}Q}{T} - \left(S_2 - S_1\right) < 0 \tag{2.24}$$

That is

$$S_2 - S_1 > -\int_{1-A-2} \frac{\mathrm{d}Q}{T} \tag{2.25}$$

If 1-A-2 is an irreversible adiabatic process, i.e., $\mathrm{d}Q = 0$, then $S_2 - S_1 > 0$. At this time, although the working substance does not absorb heat from the external heat source, the dissipation effect caused by the irreversible factors in the process heats the working substance after a part of the mechanical energy is reconverted into thermal energy within the working substance.

Therefore, in the irreversible process, the entropy change $\mathrm{d}S > \frac{\mathrm{d}Q}{T}$, which can be expressed as

$$\mathrm{d}S = \frac{\mathrm{d}Q}{T} + \mathrm{d}S_{\mathrm{g}} = \mathrm{d}S_{\mathrm{f}} + \mathrm{d}S_{\mathrm{g}} \tag{2.26}$$

Where.

$\mathrm{d}S_{\mathrm{f}} = \frac{\mathrm{d}Q}{T}$ is the entropy flow that represents the entropy change caused by the exchange of heat between the system and the environment;

$\mathrm{d}S_{\mathrm{g}}$ is the entropy production that represents the increase in entropy due to irreversible factors in the process (such as friction, temperature difference heat transfer).

Entropy production is a measure of the degree of irreversibility of the actual thermal process. Whether it is finite potential difference or dissipation effect (resistance, friction, magnetic resistance, etc.) that causes irreversibility, it devalues the system energy in the process, reduces the ability to do work, and converts a part of the effective energy into ineffective energy that can never be converted into work.

2.3.2.3 Entropy Balance Equation of Typical Systems

After a process, the energy, mass, and volume of a thermal system can change, and similarly, entropy can also change. Similar to the energy balance equation, the entropy balance equation is as follows:

$$\Delta S = S_{\mathrm{in}} - S_{\mathrm{out}} + \Delta S_{\mathrm{g}} \tag{2.27}$$

Where ΔS is the entropy change of the system; S_{in} is the entropy entering the system; S_{out} is the entropy leaving the system; ΔS_{g} is the entropy production due to irreversible processes.

Unsteady Flow System

For an open system with unsteady flow, the entropy balance equation can be expressed as:

$$\Delta S = \left(\sum_{j=1}^{n} S_j\right)_{\mathrm{in}} - \left(\sum_{k=1}^{m} S_k\right)_{\mathrm{out}} + \int \frac{\mathrm{d}Q}{T} + \Delta S_{\mathrm{g}} \tag{2.28}$$

The entropy flow entering and leaving the system includes the entropy flow caused by mass flow and the heat entropy flow caused by heat flow. The reason for the generation of entropy in the system is that ordered energy (such as mechanical energy or electrical energy) is dissipated into disordered thermal energy and absorbed by the system, leading to an increase in system entropy. For example, in the irreversible expansion process of a fluid, due to the internal friction of fluid molecules and mechanical friction, some energy that could have been used for mechanical work is dissipated as heat, reducing the actual work done. At the same time, the rise in fluid temperature increases internal entropy. The generation of entropy is not a property of the system, but is related to the irreversibility of the process. The higher the degree of irreversibility of the process, the greater the entropy generated. Only reversible processes do not generate entropy.

Steady Flow System

The system state does not change with time, and there is no entropy accumulated by unsteady flow, that is, $\Delta S = 0$, the entropy balance equation is:

$$\left(\sum_{j=1}^{n} S_j\right)_{\text{in}} - \left(\sum_{k=1}^{m} S_k\right)_{\text{out}} + \int \frac{\mathrm{d}Q}{T} + \Delta S_{\text{g}} = 0 \tag{2.29}$$

For an adiabatic process, $\int \frac{\mathrm{d}Q}{T} = 0$, the Eq. (2.29) becomes:

$$\Delta S_{\text{g}} = \left(\sum_{k=1}^{m} S_k\right)_{\text{out}} - \left(\sum_{j=1}^{n} S_j\right)_{\text{in}} \tag{2.30}$$

For example, when a stream of fluid passes through a throttle valve, because the mass flow rate of the fluid does not change, the above equation becomes:

$$\Delta S_{\text{g}} = S_{\text{out}} - S_{\text{in}}$$

The throttling process is an irreversible process. The greater the pressure drop during throttling, the more entropy is generated, and the greater the degree of irreversibility. The pressure difference is originally the potential for driving force or doing work, but the fluid does not do mechanical work when it flows through the throttle valve. In fact, it has been dissipated as thermal energy, generated by the friction between the fluid and the valve, and absorbed by the fluid itself, increasing the entropy of the fluid.

Closed System

The closed system has no mass exchange with the environment, and the entropy brought in and out is zero, that is, $\left(\sum_{j=1}^{n} S_j\right)_{\text{in}} = 0, \left(\sum_{k=1}^{m} S_k\right)_{\text{out}} = 0$, the entropy balance equation is:

$$\Delta S = \int \frac{\mathrm{d}Q}{T} + \Delta S_{\text{g}} \tag{2.31}$$

If the flow of the working fluid in the closed system is a reversible process, that is, $\Delta S_g = 0$, the Eq. (2.31) becomes

$$\Delta S = \int \frac{dQ}{T} \tag{2.32}$$

Isolated System

For an isolated system that does not exchange energy and mass with the environment, the entropy balance equation is:

$$\Delta S = \Delta S_g \geq 0 \tag{2.33}$$

An isolated system that includes a heat source, a cold source, and a working substance, the entropy change of the system is composed of the entropy changes of the heat source, the cold source, and the working substance, that is:

$$\Delta S = \Delta S_{heatsource} + \Delta S_{coldsource} + \Delta S_{workingsubstance} \geq 0 \tag{2.34}$$

Assuming that the temperatures of the heat source and the cold source are T_1 and T_2 respectively, the heat source transfers heat Q to the cold source without doing work on the outside, then $\Delta S_{heatsource} = -\frac{Q}{T_1}, \Delta S_{coldsource} = \frac{Q}{T_2}$; the working substance neither absorbs heat nor releases heat, so $\Delta S_{workingsubstance} = 0$. Substitute into Eq. (2.34) to get:

$$\Delta S = -\frac{Q}{T_1} + \frac{Q}{T_2} \geq 0 \tag{2.35}$$

Since there is a temperature difference between the heat source and the cold source, the heat transfer process is irreversible, so:

$$\Delta S = \Delta S_g = Q\left(\frac{1}{T_2} - \frac{1}{T_1}\right) > 0 \tag{2.36}$$

2.3.2.4 Principle of Entropy Increase in Isolated Systems

Equation (2.33) is the mathematical expression of the principle of entropy increase in isolated systems. The principle of entropy increase in isolated systems can be stated as: all actual processes (irreversible processes) carried out in an isolated system proceed in the direction of increasing system entropy; only in ideal conditions (reversible processes) does the entropy of the system remain unchanged. Entropy is a non-conservative parameter. Only in ideal reversible processes is entropy conserved. In all actual processes, entropy production is greater than zero; entropy production is a measure of the degree of irreversibility of the process. The greater the degree of irreversibility, the greater the entropy production. The principle of entropy increase applies to isolated systems. An object within the system can interact with other objects in the system, and its entropy can increase, decrease, or remain unchanged, but the entropy of the isolated system will not decrease. It is not difficult to infer that any process that reduces the entropy of an isolated system is impossible to achieve.

The principle of entropy increase describes the direction and degree of spontaneous conversion of thermodynamic energy and other forms of energy, that is, the direction of energy conversion is towards the increase of entropy, and as the energy conversion proceeds, the system tends to equilibrium, and the entropy value reaches its maximum. In this process, although the total energy remains unchanged, the energy available for use or conversion is decreasing.

Given that all actual processes are irreversible, the process always proceeds in the direction of increasing system entropy. Therefore, the increase in system entropy is the criterion for the direction of the process. The direction, conditions, and limits of the thermal process can be judged according to the principle of entropy increase.

Example 2.1 Saturated steam with a flow rate of 4 kg/s and a temperature of 140 °C enters the heat exchanger and heats the air at 20 °C. At the outlet of the heat exchanger, the steam condenses into saturated water at 140 °C. Try to determine the entropy increase of the steam during the heat transfer process.

Solution Take the steam side of the heat exchanger as an open system, and the steam mass flow rate is $G_s = 4\text{kg/s}$.

The inlet of the open system is 140 °C saturated steam, and the corresponding enthalpy and entropy values can be found from the water and steam table as.

$$h_1 = 2{,}733.1\text{kJ} / \text{kg}, s_1 = 6.9284\text{kJ} / (\text{kg} \cdot \text{K})$$

The outlet of the open system is 140 °C saturated water, and its enthalpy and entropy values are

$$h_2 = 589.11\text{kJ} / \text{kg}, s_2 = 1.7390\text{kJ} / (\text{kg} \cdot \text{K})$$

According to the entropy balance equation of the steady flow system.

$$S_{\text{in}} - S_{\text{out}} + \int \frac{\text{d}Q}{T} + \Delta S_{\text{g}} = 0$$

Since the steam releases heat to the outside of the system, the entropy increase expression of the open system is

$$\Delta S_{\text{g}} = S_{\text{out}} - S_{\text{in}} - \int \frac{\text{d}Q}{T} = G_{\text{s}} (s_2 - s_1) + \frac{Q_{\text{s}}}{T_{\text{s}}}$$

The heat released by the steam during the heat transfer process is.

$$Q_s = G_s\left(h_1 - h_2\right) = 4\times\left(2{,}733.1 - 589.11\right) = 8{,}575.96\text{kW}$$

The entropy produced by the steam flowing through the open system is.

$$\Delta S_g = G_s\left(s_2 - s_1\right) + \frac{Q_s}{T_s} = 4\times\left(1.7390 - 6.9284\right) + \frac{8{,}575.96}{140 + 273.15} = 0$$

Example 2.2 A stream of hot air with a flow rate of 10 kg/s is mixed with a stream of cold air with a flow rate of 5 kg/s under adiabatic conditions. The state of the hot air is 0.1 MPa, 500 K, and the state of the cold air is 0.1 MPa, 300 K. Assuming the average specific heat at constant pressure of the air is 1.01 kJ/(kg K), calculate the increase in entropy during the mixing process.

Solution The mixing of the two streams of air is an adiabatic steady flow process, and the air can be regarded as an ideal gas. According to the entropy balance equation of the steady flow system, since $\int \frac{\mathrm{d}Q}{T} = 0$, we have.

$$\Delta S_g = S_{out} - S_{in} = S_m - \left(S_h + S_c\right)$$

Where the subscript "m" represents mixed air, "h" represents hot air, and "c" represents cold air.

The mass flow rate of the two streams of air after mixing is

$$G_m = G_h + G_c = 10 + 5 = 15\text{kg/s}$$

According to the principle of energy conservation, we have

$$G_m h_m = G_h h_h + G_c h_c$$

Substitute the calculation formula of the enthalpy of the ideal gas into the above formula, we get.

$$G_m c_{p,m} T_m = G_h c_{p,h} T_h + G_c c_{p,c} T_c$$

Then the temperature after the cold and hot air mix is.

$$T_m = \frac{G_h c_{p,h} T_h + G_c c_{p,c} T_c}{G_m c_{p,m}}$$

Ignoring the change in the specific heat at constant pressure of the air, that is, taking c_p as a constant, then.

$$T_m = \frac{G_h T_h + G_c T_c}{G_m} = \frac{10\times 500 + 5\times 300}{15} = 433.3\text{K}$$

Therefore, the entropy increase of the two streams of air mixing process is.

$$\begin{aligned}\Delta S_g &= G_m s_m - (G_h s_h + G_c s_c) \\ &= G_h (s_m - s_h) + G_c (s_m - s_c) \\ &= G_h c_p \ln \frac{T_m}{T_h} + G_c c_p \ln \frac{T_m}{T_c}\end{aligned}$$

$$= 10 \times 1.01 \ln \frac{433.3}{500} + 5 \times 1.01 \ln \frac{433.3}{300} = 0.411 \text{kW} / \text{K}$$

This shows that the mixing process of the two streams of air is irreversible.

2.3.3 *Workability and Energy Quality*

The working substance undergoes a thermodynamic cycle, of the heat Q_1 absorbed from the heat source, only a part is converted into work W_s, and the heat discharged into the cold source is Q_2, that is

$$W_s = Q_1 - Q_2 = Q_1 \left(1 - \frac{Q_2}{Q_1}\right) = Q_1 \eta_t \tag{2.37}$$

Only when the thermodynamic cycle is reversible, the thermodynamic cycle efficiency η_t can reach its maximum value, the work done to the outside $W_{s,\max}$ is the maximum, referred to as the workability (or available energy) in Q_1; the part of the heat Q_2 that cannot be converted into work and must be discharged to the cold source in the reversible cycle is referred to as the waste heat (or unavailable energy) in Q_1. At this time, formula (2.37) can be expressed as

$$W_{s,\max} = Q_1 \left(1 - \frac{\overline{T}_2}{\overline{T}_1}\right) \tag{2.38}$$

Where $\overline{T}_1$ and $\overline{T}_2$ are the average heat absorption temperature and average heat release temperature of the thermodynamic cycle respectively.

Mechanical energy and electrical energy, under ideal conditions (no friction loss and resistance loss), can be continuously and completely used to do work; while thermal energy has some waste heat, even under ideal conditions, it cannot be continuously and completely converted into work. Therefore, the "quality" of thermal energy doing work is lower than the "quality" of other forms of energy. Not only that, the quality of thermal energy at different temperatures is also different, the workability of high average heat absorption temperature is large, the workability of low average heat absorption temperature is small; the quality of thermal energy with large workability is high, the quality of thermal energy with small workability is low.

From formulas (2.37) and (2.38), it can be seen that between the same heat source and cold source, when the heat absorbed from the heat source is the same, the work W_s done by the working substance in the irreversible cycle is less than the work $W_{s,max}$ done in the reversible cycle, and the difference is

$$\Delta W = W_{s,max} - W_s = \left(Q_1 - Q_2'\right) - \left(Q_1 - Q_2\right) = Q_2 - Q_2' \tag{2.39}$$

Where Q_2' is the heat discharged into the cold source in the reversible cycle.

From the above formula, it can be seen that the reduced work in the irreversible cycle compared to the reversible cycle, which is the loss of workability caused by the irreversibility of the cycle, has become waste heat discharged into the environment.

Work can be completely converted into heat, but heat cannot be completely converted into work through a heat engine cycle, which has been proven by a large amount of practice. The non-equivalence of work and heat is a basic content of the second law of thermodynamics.

The second law of thermodynamics points out that reducing the irreversibility of a cycle or a process in the same cycle can improve the thermal efficiency of the cycle. Therefore, by analyzing and comparing the different irreversibilities of different cycles or different processes in the same cycle, we can identify the thermal economy of different cycles or different processes, that is, the irreversibility analysis can become the qualitative analysis criterion for whether the cycle or process is energy-saving.

2.4 Finite Time Thermodynamics

Finite time thermodynamics is an extension and expansion of classical thermodynamics, and a new branch of irreversible thermodynamics developed in the 1930s. Finite time thermodynamics is also a cross-disciplinary field of modern thermodynamic theory and one of the theoretical tools for current thermodynamic engineering applications. It mainly studies the laws that energy transfer and irreversible behavior in non-equilibrium systems satisfy in finite time thermodynamic processes.

2.4.1 Concept of Finite Time Thermodynamics [15, 16]

The theoretical basis of engineering thermodynamics research is equilibrium and reversible processes, with reversible processes and cycles being the ultimate goal of engineering thermodynamics research. However, the potential difference between the internal and external of reversible processes and cycles tends to zero, which requires the process to proceed infinitely slowly, the time experienced by the process is infinitely long, leading to the power of the reversible cycle tending to zero.

Taking the heat transfer process as an example, heat is transferred from a high-temperature object to a low-temperature object, and the temperature difference between the two objects is the driving force for heat transfer. Due to the existence of the temperature difference, the heat transfer process is irreversible. Therefore, only when the temperature difference between the two objects tends to zero, the heat transfer process approaches the reversible process. However, transferring a finite amount of heat through an infinitesimal temperature difference requires infinite time or an infinite heat transfer area. If the heat transfer area tends to infinity, it is impossible to achieve in engineering; if the time tends to infinity, the output power will tend to zero.

In thermodynamics, power and efficiency are the two main parameters for measuring the performance of a heat engine. According to classical thermodynamics, the upper limit of the efficiency of a reversible heat engine is Carnot efficiency, but the realization of the Carnot cycle depends on the quasi-static assumption of infinite time, and the corresponding power is zero, which shows that the highest efficiency of the thermal cycle does not correspond to the maximum output power of the thermal cycle. Therefore, how to improve the efficiency of the heat engine while ensuring the power of the heat engine has become an important scientific problem in thermodynamics.

In classical thermodynamics, the first law of thermodynamics describes energy conservation, and the second law describes irreversible processes characterized by entropy increase. These basic laws constitute the core framework of thermodynamics research. Although some basic laws (such as the second law of thermodynamics) are expressed in irreversible processes, classical thermodynamics focuses on the study of equilibrium systems. However, most of the actual thermodynamic processes are irreversible processes, and the system is generally in a non-equilibrium state when undergoing these processes, and reversible cycles cannot be realized in actual engineering.

In order to study the non-equilibrium effects of thermodynamic systems in irreversible processes, the development of thermodynamics in the 1930s expanded from equilibrium to non-equilibrium. During this period, the theory of finite-time thermodynamics was developed, and people began to theoretically study more realistic actual heat engines: heat engines operating in thermodynamic cycles with finite time deviating from the quasi-static assumption. In the finite-time thermodynamic cycle, the process of the heat engine's working substance contacting the heat source and the process of doing work externally are no longer quasi-static reversible processes, and non-zero output power can be achieved.

In the 1950s, French physicist J. Yvon was the first to study the power optimization of steam engines in thermodynamic cycles, and obtained the cycle efficiency when the cycle has the maximum power, that is, the so-called maximum power efficiency (efficiency at maximum power, EMP).

In 1975, Canadian physicists F. Curzon and B. Ahlborn used an internal reversible heat engine model that obeys Newton's law of heat transfer to study the optimization problem of the finite-time Carnot heat engine, and considered the non-equilibrium heat flow when the heat engine contacts the heat source at temperature T_H and the cold source at temperature T_L, and obtained the maximum power efficiency of this heat engine $\eta_{CA} = 1 - \frac{T_L}{T_H}$, the famous Curzon-Ahlborn (CA) efficiency. Curzon and Ahlborn first discussed the concept of finite-time thermodynamics, proposed a problem with practical engineering significance of completing the cycle in a finite time and optimizing the cycle, and opened up a new field for the modern development of macroscopic thermodynamics.

Irreversible thermodynamics generally discusses various irreversibilities in the system, finite-time thermodynamics discusses external irreversibility caused by heat sources and cold sources, and internal irreversibility discusses thermodynamic processes in the cycle (including friction, etc.). The irreversibility of any system cannot be completely eliminated, but it can be minimized to a certain extent. Irreversible thermodynamics has become a powerful tool for studying energy conversion systems. The concept of finite-time thermodynamics that appeared after the innovative research of Curzon and Ahlborn not only answered the above questions, but also solved the research method problems of heat transfer and energy conversion systems.

2.4.2 Finite-Time Carnot Cycle [17, 18]

In 1975, Curzon and Ahlborn applied the concept of finite-time thermodynamics to the Carnot heat engine, considering the heat capacity of the heat source and the cold source as infinite and finite, respectively. The time of the two adiabatic processes of the Carnot cycle is proportional to the time of the two isothermal processes, as shown in Fig. 2.7.

In Fig. 2.7a, the temperatures of the heat source and the cold source are T_H and T_L, respectively, and the working temperature range of the heat engine is $[T_1, T_2]$, and $T_H > T_1$, $T_L < T_2$. The cases of infinite and finite heat capacity of the heat source and the cold source are discussed separately.

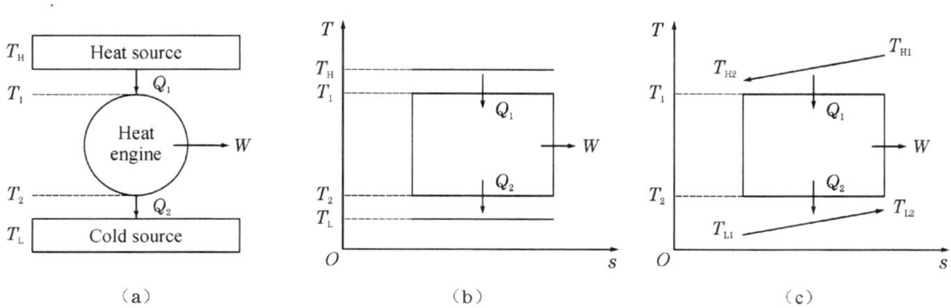

Fig. 2.7 Finite time Carnot cycle: (**a**) Heat engine; (**b**) T-s diagram of infinite heat capacity Carnot cycle; (**c**) T-s diagram of the Carnot cycle with finite heat capacity

2.4.2.1 Infinite Heat Capacity

As shown in Fig. 2.7b, it is assumed that both the heat source and the cold source have infinite heat capacity, and the temperatures T_H and T_L remain constant during the heat transfer process. The heat absorption temperature of the thermodynamic cycle is T_1, and the heat release temperature is T_2. The heat transferred from the heat source to the thermodynamic cycle per unit time is $\dot{Q}_1$, and the heat transferred from the thermodynamic cycle to the cold source is $\dot{Q}_2$. According to the heat transfer relationship, it can be expressed as

$$\dot{Q}_1 = \frac{Q_1}{\tau_1} = K_1 \left(T_H - T_1\right) \tag{2.40}$$

$$\dot{Q}_2 = \frac{Q_2}{\tau_2} = K_2 \left(T_2 - T_L\right) \tag{2.41}$$

Where

Q_1 is the total heat obtained from the heat source during the isothermal heat absorption process of the thermodynamic cycle, kJ;

Q_2 is the total heat discharged to the cold source during the isothermal heat release process of the thermodynamic cycle, kJ;

τ_1, τ_2 are the time used for the isothermal heat absorption and isothermal heat release processes, respectively, s;

K_1, K_2 are the heat transfer coefficients of the isothermal heat absorption and isothermal heat release processes, respectively, kJ/K.

According to the second law of thermodynamics, for reversible processes, we have

$$\frac{Q_1}{T_1} = \frac{Q_2}{T_2} \tag{2.42}$$

Substituting Eqs. (2.40) and (2.41) into Eq. (2.42), we get

$$\frac{\tau_1}{\tau_2} = \frac{K_2 \left(T_2 - T_L\right) T_1}{K_1 \left(T_H - T_1\right) T_2} \tag{2.43}$$

The power output of the thermodynamic cycle is

$$W = \frac{Q_1 - Q_2}{\tau_1 + \tau_2 + \tau_a} = \frac{Q_1 - Q_2}{\gamma \left(\tau_1 + \tau_2\right)} \tag{2.44}$$

Where.

τ_a is the time to complete two adiabatic processes, proportional to the time to complete two isothermal processes, s.

Substituting Eqs. (2.40) and (2.41) into Eq. (2.44), we get

$$W = \frac{K_1 \left(T_H - T_1\right) \tau_1 - K_2 \left(T_2 - T_L\right) \tau_2}{\gamma \left(\tau_1 + \tau_2\right)}$$

Let $\delta_H = T_H - T_1$, $\delta_L = T_2 - T_L$, substitute into the above equation and eliminate τ_1 and τ_2, we get

$$W = \frac{K_1 K_2 \delta_H \delta_L (T_1 - T_2)}{\gamma (K_2 \delta_L T_1 + K_1 \delta_H T_2)} = \frac{K_1 K_2 \delta_H \delta_L (T_H - T_L - \delta_H - \delta_L)}{\gamma \left[K_2 \delta_L T_H + K_1 \delta_H T_L + \delta_H \delta_L (K_1 - K_2) \right]} \quad (2.45)$$

From Eq. (2.45), it can be seen that the cycle power W is a function of two variables δ_H and δ_L. When $\frac{\partial W}{\partial \delta_H} = 0$ and $\frac{\partial W}{\partial \delta_L} = 0$, the maximum value W_{max} of W can be obtained.

According to Eq. (2.45), when $\frac{\partial W}{\partial \delta_H} = 0$, we have

$$K_2 \delta_L T_H (T_H - T_L - \delta_H - \delta_L) = \delta_H \left[K_2 \delta_L T_H + K_1 \delta_H T_L + \delta_H \delta_L (K_1 - K_2) \right] \quad (2.46)$$

When $\frac{\partial W}{\partial \delta_L} = 0$, we have

$$K_1 \delta_H T_L (T_H - T_L - \delta_H - \delta_L) = \delta_L \left[K_2 \delta_L T_H + K_1 \delta_H T_L + \delta_H \delta_L (K_1 - K_2) \right] \quad (2.47)$$

From this we get

$$\delta_L = \sqrt{\frac{K_1 T_L}{K_2 T_H}} \delta_H \quad (2.48)$$

Let $\theta = \frac{\delta_H}{T_H}, \beta = \sqrt{\frac{K_1 T_L}{K_2 T_H}}$, substitute into the Eq. (2.46) or (2.47), we get

$$\left(1 - \frac{K_1}{K_2}\right)\theta^2 - 2(1+\beta)\theta + \left(1 - \frac{T_L}{T_H}\right) = 0 \quad (2.49)$$

The solution for θ is

$$\theta = \frac{(1+\beta) \pm \sqrt{(1+\beta)^2 - \left(1 - \frac{K_1}{K_2}\right)\left(1 - \frac{T_L}{T_H}\right)}}{1 - \frac{K_1}{K_2}} \quad (2.50)$$

After solving, we can get

$$\theta_1 = \frac{1 + \sqrt{\frac{T_L}{T_H}}}{1 - \sqrt{\frac{K_1}{K_2}}} \quad \text{and} \quad \theta_2 = \frac{1 - \sqrt{\frac{T_L}{T_H}}}{1 + \sqrt{\frac{K_1}{K_2}}}$$

According to the definition of θ, $\theta < 1$, so θ_2 is taken as the root of θ, that is

$$\theta = \frac{\delta_{\mathrm{H}}}{T_{\mathrm{H}}} = \frac{1-\sqrt{\frac{T_{\mathrm{L}}}{T_{\mathrm{H}}}}}{1+\sqrt{\frac{K_1}{K_2}}} \tag{2.51}$$

According to Eq. (2.48), we get

$$\frac{\delta_{\mathrm{L}}}{T_{\mathrm{L}}} = \frac{\sqrt{\frac{T_{\mathrm{H}}}{T_{\mathrm{L}}}}-1}{1+\sqrt{\frac{K_2}{K_1}}} \tag{2.52}$$

Substitute Eqs. (2.51) and (2.52) into (2.45), we get

$$W_{\max} = \frac{K_1 K_2}{\gamma}\left(\frac{\sqrt{T_{\mathrm{H}}}-\sqrt{T_{\mathrm{L}}}}{\sqrt{K_1}+\sqrt{K_2}}\right)^2 \tag{2.53}$$

The corresponding thermal efficiency is

$$\eta_{\mathrm{t,C\ A}} = 1-\frac{T_2}{T_1} = 1-\frac{T_{\mathrm{L}}+\delta_{\mathrm{L}}}{T_{\mathrm{H}}+\delta_{\mathrm{H}}} = 1-\sqrt{\frac{T_{\mathrm{L}}}{T_{\mathrm{H}}}} \tag{2.54}$$

As can be seen from Table 2.1, the observed efficiency is closer to the Curzon-Ahlborn efficiency than the Carnot efficiency. Ibrahim et al. (1991) mathematically proved the observations of Curzon and Ahlborn (1975), and plotted the relationship between power output and thermal efficiency of the Carnot cycle, as shown in Fig. 2.8.

From the curve in the figure, it can be seen that the power increases with the increase of thermal efficiency, peaks at the Curzon-Ahlborn point, and then drops sharply. As the thermal efficiency approaches the Carnot efficiency, the power output tends to zero.

Table 2.1 Observed efficiency of actual thermal engine [17]

Power plant type	T_{H}/°C	T_{L}/°C	Carnot efficiency	C-A efficiency	Observed efficiency
Coal-fired steam power plant	565	25	64.44%	40.37%	36%
CANDU HWR nuclear power plant	300	25	47.99%	27.88%	30%
Geothermal steam power plant	200	80	32.50%	17.84%	16%

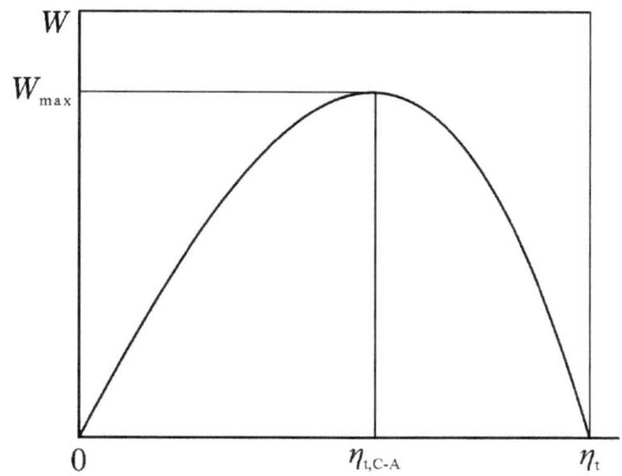

Fig. 2.8 Power-efficiency curve of the Carnot cycle

2.4.2.2 Finite Heat Capacity

As shown in Fig. 2.7c, it is assumed that both the heat source and the cold source have finite heat capacities, and therefore the temperature changes during the heat transfer process. The temperature of the heat source drops from T_{H1} to T_{H2} during the heat release process, and the temperature of the cold source rises from T_{L1} to T_{L2} during the heat absorption process. The heat transfer equations between the heat source and the isothermal heat absorption process of the Carnot cycle, and between the isothermal heat release process of the Carnot cycle and the cold source are

$$\dot{Q}_1 = \frac{Q_1}{\tau_1} = k_H A_H \frac{(T_{H1} - T_1) - (T_{H2} - T_1)}{\ln \frac{T_{H1} - T_1}{T_{H2} - T_1}} \tag{2.55}$$

$$\dot{Q}_2 = \frac{Q_2}{\tau_2} = k_L A_L \frac{(T_2 - T_{L1}) - (T_2 - T_{L2})}{\ln \frac{T_2 - T_{L1}}{T_2 - T_{L2}}} \tag{2.56}$$

Where.

k_H, k_L are the heat transfer coefficients between heat source and working medium, and between working medium and cold source, respectively, kJ/(m^2•K);

A_H, A_L are the heat transfer areas between the heat source and the working fluid, and between the working fluid and the cold source, respectively, m^2.

The heat absorbed from the heat source by the working fluid per unit time and the heat discharged to the cold source by the working fluid are represented as

$$\dot{Q}_1 = \frac{Q_1}{\tau_1} = \dot{m}_H c_{p,H} (T_{H1} - T_{H2}) \tag{2.57}$$

$$\dot{Q}_2 = \frac{Q_2}{\tau_2} = \dot{m}_{\mathrm{L}} c_{p,\mathrm{L}} \left(T_{\mathrm{L}2} - T_{\mathrm{L}1} \right) \tag{2.58}$$

By combining Eqs. (2.55) and (2.57), and Eqs. (2.56) and (2.58), we can get

$$T_{\mathrm{H}2} = T_1 + \left(T_{\mathrm{H}1} - T_1 \right) \exp\left(-\frac{k_{\mathrm{H}} A_{\mathrm{H}}}{\dot{m}_{\mathrm{H}} c_{p,\mathrm{H}}} \right) \tag{2.59}$$

$$T_{\mathrm{L}2} = T_2 - \left(T_2 - T_{\mathrm{L}1} \right) \exp\left(-\frac{k_{\mathrm{L}} A_{\mathrm{L}}}{\dot{m}_{\mathrm{L}} c_{p,\mathrm{L}}} \right) \tag{2.60}$$

Ignoring the time used for the two adiabatic processes in the Carnot cycle that are carried out very quickly, the total cycle time is

$$\tau_{\mathrm{t}} = \tau_1 + \tau_2 = \frac{Q_1}{\dot{m}_{\mathrm{H}} c_{p,\mathrm{H}} \left(T_{\mathrm{H}1} - T_{\mathrm{H}2} \right)} + \frac{Q_2}{\dot{m}_{\mathrm{L}} c_{p,\mathrm{L}} \left(T_{\mathrm{L}2} - T_{\mathrm{L}1} \right)} \tag{2.61}$$

Substitute Eqs. (2.59) and (2.60) into Eq. (2.61), we get

$$\tau_{\mathrm{t}} = \frac{Q_1}{\mathrm{HE} \left(T_{\mathrm{H}1} - T_1 \right)} + \frac{Q_2}{\mathrm{LE} \left(T_2 - T_{\mathrm{L}1} \right)} \tag{2.62}$$

The expressions for HE and LE are as follows

$$\mathrm{HE} = \dot{m}_{\mathrm{H}} c_{p,\mathrm{H}} \left[1 - \exp\left(-\frac{k_{\mathrm{H}} A_{\mathrm{H}}}{\dot{m}_{\mathrm{H}} c_{p,\mathrm{H}}} \right) \right] \tag{2.63}$$

$$\mathrm{LE} = \dot{m}_{\mathrm{L}} c_{p,\mathrm{L}} \left[1 - \exp\left(-\frac{k_{\mathrm{L}} A_{\mathrm{L}}}{\dot{m}_{\mathrm{L}} c_{p,\mathrm{L}}} \right) \right] \tag{2.64}$$

The output power of the thermodynamic cycle is

$$W = \frac{Q_1 - Q_2}{\tau_{\mathrm{t}}} \tag{2.65}$$

Considering that $\frac{Q_1}{T_1} = \frac{Q_2}{T_2}$, we get

$$Q_1 = \frac{W \tau_{\mathrm{t}} T_1}{T_1 - T_2} \tag{2.66}$$

$$Q_2 = \frac{W \tau_{\mathrm{t}} T_2}{T_1 - T_2} \tag{2.67}$$

Substitute Eqs. (2.66) and (2.67) into Eq. (2.62), we can get:

$$\tau_t = \frac{W\tau_t T_1}{\mathrm{HE}(T_{H1} - T_1)(T_1 - T_2)} + \frac{W\tau_t T_2}{\mathrm{LE}(T_2 - T_{L1})(T_1 - T_2)} \tag{2.68}$$

Let $\delta_H = T_{H1} - T_1$, $\delta_L = T_2 - T_{L1}$, then the output power is

$$W = \frac{\mathrm{HE}\delta_H\delta_L\mathrm{LE}(T_{H1} - T_{L1} - \delta_H - \delta_L)}{\mathrm{HE}\delta_H T_{L1} + \mathrm{LE}\delta_L T_{H1} + \delta_H\delta_L(\mathrm{HE} - \mathrm{LE})} \tag{2.69}$$

From Eq. (2.69), for a set of specific operating parameters, T_{H1}, T_{L1}, HE and LE are constants, and the power W is a function of δ_H and δ_L. By optimizing W with respect to δ_H and δ_L, we get

$$T_1 = C_1\sqrt{T_{H1}} \quad \text{and} \quad T_2 = C_1\sqrt{T_{L1}} \tag{2.70}$$

The expression for the coefficient C1 is

$$C_1 = \frac{\sqrt{\mathrm{HE}\cdot T_{H1}} + \sqrt{\mathrm{LE}\cdot T_{L1}}}{\sqrt{\mathrm{HE}} + \sqrt{\mathrm{LE}}} \tag{2.71}$$

Substitute Eqs. (2.70) and (2.71) into Eq. (2.69), we can get the maximum output power of the heat engine:

$$W_{max} = \frac{\mathrm{HE}\cdot\mathrm{LE}\left(\sqrt{T_{H1}} - \sqrt{T_{L1}}\right)^2}{\left(\sqrt{\mathrm{HE}} + \sqrt{\mathrm{LE}}\right)^2} \tag{2.72}$$

The cycle thermal efficiency at the maximum output power is

$$\eta_t = 1 - \sqrt{\frac{T_{L1}}{T_{H1}}} \approx \eta_{C\ A} \tag{2.73}$$

From Eq. (2.73), it can be seen that the cycle thermal efficiency at the maximum output power only depends on the inlet temperatures of the heat source fluid and the cold source fluid.

2.5 Thermodynamic Analysis Method of Energy Systems [19, 20]

The thermodynamic analysis of energy systems is based on the first and second laws of thermodynamics, analyzing and studying the rationality and effectiveness of energy transfer and conversion in energy systems. The aim is to find methods and ways to improve the thermal economy of power cycles, thereby improving the thermal economy of the entire energy system, reducing energy loss, and improving energy utilization.

The rationality of energy utilization refers to whether the way of using energy conforms to scientific principles, that is, whether the utilization of energy matches the energy level, and there is no high energy and low use; the effectiveness of energy utilization refers to the degree to which energy is effectively utilized, that is, how much of the supplied energy is utilized. In order to perform energy analysis on actual energy-using equipment and systems, a set of methods for formulating analysis models, establishing energy balance equations, and establishing energy use evaluation criteria, namely energy analysis methods, are needed.

In the history of energy application science, two types of energy analysis methods have been formed, one is the first law of thermodynamics analysis method, commonly known as energy balance analysis method, thermal balance analysis method, enthalpy analysis method; the other is the second law of thermodynamics analysis method, including entropy analysis method and exergy analysis method.

2.5.1 Energy Balance Analysis Method

The energy balance analysis method is based on the first law of thermodynamics, and analyzes and reveals the conversion, transmission, utilization, and loss of energy in the energy system in terms of quantity based on the principle of energy balance.

The energy balance analysis method was formed more than 100 years ago. It is a traditional energy analysis method and is still widely used in engineering. Because it quantifies its thermal power conversion effect by cycle thermal efficiency, device efficiency, and equipment efficiency, it is also called the efficiency method. When analyzing, the energy item is often expressed in terms of enthalpy, and it is also commonly referred to as the enthalpy analysis method.

The basic content of the energy balance analysis method includes:

1. According to the thermodynamic model of the energy system, perform energy balance calculations for the system;
2. Calculate the thermal efficiency of the energy system and evaluate the pros and cons of the energy system;
3. Calculate the heat loss of the main equipment and processes in the energy system, and determine the distribution of the heat loss rate of the energy system.

For any device or process in the energy system, its energy balance equation is

$$E_{sup} = E_{obt} + E_{loss} \tag{2.74}$$

Where E_{sup}, E_{obt} and E_{loss} are the supply energy, obtained energy and energy loss, respectively.

The evaluation index for the thermal economy of the energy system is thermal efficiency, defined as

$$\eta_t = \frac{E_{obt}}{E_{sup}} = \frac{E_{sup} - E_{loss}}{E_{sup}} \quad (2.75)$$

The energy balance analysis method has the following problems:

1. Energy balance only analyzes the balance in quantity during the conversion and transmission of energy, without considering the difference and change in the quality of energy.

   For example, high-pressure steam becomes low-pressure steam through throttling. Since the enthalpy value does not change before and after the throttling process, from the perspective of energy balance, it can be considered that there is no energy loss. But in fact, this is a typical irreversible process, and there is a loss of the working ability of the working medium (steam), which cannot be reflected by the energy balance method.
2. Energy balance cannot truly reflect the rationality of energy utilization and should not be used as a measure of the thermodynamic perfection of energy utilization.

   For example, the thermal efficiency of industrial boilers can be as high as 80% or more, while the thermal efficiency of thermal power plants can only reach up to 40%. However, we cannot say that the energy utilization rate of industrial boilers is twice that of thermal power plants. Therefore, using thermal efficiency to evaluate the thermodynamic perfection of equipment or devices will result in errors.
3. Energy balance cannot accurately and essentially reflect the problems and weak links in the energy use process.

   For example, according to the law of thermal power equivalence, one degree of electricity is equal to 3600 kJ of heat, but if one degree of electricity is used to drive a heat pump, then it can provide several times or tens of times the heat of 3600 kJ. Such a result is incomprehensible from the perspective of energy balance.
4. It is difficult to judge the level of comprehensive energy utilization.

   For example, in two processes that use both thermal and electrical energy simultaneously, when their efficiencies are equal but the amount of energy used is not, the comprehensive rational energy use level of their respective energy use processes cannot be correctly described by the energy balance method.

   The expression characterizing the relationship of energy quantity established based on the first law of thermodynamics only reveals the conversion, transfer, utilization, and loss of energy in quantity, but does not involve the change in the quality of energy. The energy-saving approach pointed out by the energy balance equation is to reduce the heat discharged into the environment, but it does not reveal the root cause of this part of the loss. Therefore, only by studying the quantity of energy and considering the quality of energy at the same time can these problems be satisfactorily solved.

2.5.2 Entropy Analysis Method

The entropy analysis method is based on the first and second laws of thermodynamics, focusing on the changes in workability during various thermal processes. By calculating the entropy production, ideal work, and lost work of the thermodynamic system or process, the loss of workability is determined, the causes of the loss are analyzed and found, and ways and measures to save energy, reduce consumption, and improve energy utilization are proposed.

The second law of thermodynamics states that the irreversibility of a process inevitably leads to a decrease in workability. After introducing the state parameter entropy, it further determines that the entropy increase of the process system (all substances involved in energy exchange, including all substances within the energy, and substances contained in the environment, are called the process system) is a measure of the irreversibility of the process. In this way, the increase in entropy of the process system is linked to the loss of workability, and the following relationship is established by thermodynamics:

$$\Delta I_{12} = T_0 \Delta S_{\mathrm{g}} \tag{2.76}$$

Where.

ΔI_{12} is the loss of work capacity during an irreversible process by which a thermodynamic system changes from state 1 to state 2;

ΔS_{g} is the entropy production of thermodynamic system in process 1-2;

T_0 is the thermodynamic temperature of the environment.

Entropy and entropy increase are additive, and the entropy increase of a complex thermodynamic system is equal to the sum of the entropy increases of each subsystem:

$$\Delta S_{\mathrm{g}} = \sum_{k=1}^{n} \left(\Delta S_{\mathrm{g}} \right)_k \tag{2.77}$$

The criterion for evaluating the energy use status of a thermodynamic system in entropy analysis is effective efficiency, defined as

$$\eta_{\mathrm{e}} = \frac{W}{W_{\max}} \tag{2.78}$$

Where

$W_{\max}$ is the theoretical work done by the system;

W is the effective output work of the system, it is equal to the theoretical work done by the system minus the loss of work ability.

The entropy analysis method can obtain the distribution of workability loss rates in various subsystems of complex thermodynamic systems and the second law efficiency of thermodynamic systems. It should be noted that because the definition of entropy is relatively abstract, it is difficult to calculate the entropy production of some actual thermal processes, so this method is less used in engineering.

2.5.3 Exergy Analysis Method

For a long time, the thermodynamic analysis of thermodynamic systems generally used the energy balance method based on the first law of thermodynamics. Since energy has both quantity and quality attributes, the energy balance method only focuses on the balance of energy in terms of quantity during the conversion and transfer process, and does not consider the differences and changes in the quality of energy, so it cannot accurately, essentially, and comprehensively reflect the weak links in the energy use process, and it is difficult to correctly evaluate the level of comprehensive energy utilization. In order to correctly point out the direction and ways of energy saving, a new thermodynamic analysis method—exergy analysis has gradually developed and is increasingly valued.

The exergy analysis method is based on the first and second laws of thermodynamics. On the basis of energy balance in the energy system, exergy balance analysis is carried out to determine the exergy loss of each device and process, find out the causes of exergy loss and the weak links in energy utilization, and point out the direction for energy saving, consumption reduction, and improvement of energy utilization rate.

The main content of the exergy analysis method includes:

1. Carry out energy balance for the energy system, determine the material flow, energy flow input and output of the system, and the state parameters of each material flow;
2. Calculate the exergy of material flow and heat flow;
3. Determine the exergy loss of the thermodynamic process from the exergy balance equation;
4. Determine the exergy efficiency of the equipment or process.

The main thermodynamic indicator of exergy analysis is exergy efficiency, defined as

$$\eta_{\mathrm{e}} = \frac{E_{\mathrm{x,obt}}}{E_{\mathrm{x,sup}}} \tag{2.79}$$

Where.

$E_{\mathrm{x,\,sup}}$ is the exergy supply to the system by environment;

$E_{\mathrm{x,\,obt}}$ is the system obtained exergy, and it is equal to the exergy supply to the system by environment minus the system exergy loss.

The reason why exergy analysis method has received widespread attention and rapid development is mainly because:

1. Exergy value is a unified measure of the ability of energy to do work. The larger the exergy value of energy, the more part of the energy can be converted into useful work, and the more we should try to save this type of energy.
2. Exergy efficiency is a unified indicator to measure the perfection of the thermodynamics of equipment or systems. The closer the exergy efficiency is to 1, the more perfect the thermodynamic process of the equipment or system.
3. Using the exergy analysis method can accurately reveal the location and cause of exergy loss in complex thermodynamic systems, pointing out the direction for improving equipment and saving energy.
4. Using exergy efficiency as the objective function of energy-using equipment or systems can obtain corresponding optimization conditions.

The exergy analysis method itself is also constantly developing and improving. One important direction of development is to try to form a feasible exergy analysis and economic analysis combined exergy economics method in system design, making the system design technically feasible and economically reasonable, in line with the principles of energy level matching and reasonable use of exergy. Exergy economics has achieved rapid development in economic optimization and improvement of thermodynamic systems, making exergy analysis practically applied in engineering.

2.5.4 Comparison of the Two Analysis Methods

The energy balance analysis method is guided by the first law of thermodynamics, applies the energy equation, and evaluates the energy utilization rate of the process and device from the quantity relationship of energy conversion. Its main indicator is the first law efficiency, that is, thermal efficiency. The energy balance analysis method can only calculate the energy discharge loss, but it is powerless for revealing the energy loss caused by the irreversible process (work loss, exergy loss). The level of the first law efficiency is not enough to explain the perfection of the process and device in energy utilization, so energy-saving measures cannot be formulated based on the energy balance calculation results alone.

Entropy analysis method and exergy analysis method are both guided by the first and second laws of thermodynamics, based on the entropy balance equation and exergy balance equation, and evaluate the energy utilization rate of the thermodynamic system and thermodynamic process from the quality (grade) of energy conversion and the utilization degree of exergy. Their main indicators are thermodynamic efficiency and second law efficiency (that is, exergy efficiency). These two methods can calculate the size and distribution of the workability loss (exergy loss) of the thermodynamic system and thermodynamic process, find out the irreversible loss and the cause, point out the real weak link in energy utilization, and facilitate the formulation of correct and effective energy-saving measures.

Through the entropy analysis method and exergy analysis method, the workability loss of the thermodynamic system and process can be calculated, but knowing only the absolute quantity of energy loss is not enough to fully reflect the perfection of the device and process in thermodynamics. It is also necessary to know the relative size of its energy loss. Therefore, it is necessary to further determine the efficiency of the process and system.

Thermal efficiency represents the ratio of various forms of energy obtained in the process to the various forms of energy consumed. Although both the numerator and the denominator involve energy, their quality can be different, so this efficiency inevitably leads to the equal observation of high-quality energy and low-quality energy. For this reason, thermal efficiency is not an ideal and unified scientific measure to measure and evaluate the rationality of energy utilization in the process. Because energy has different qualities, in order to obtain the true efficiency of the process, the numerator and denominator must belong to the same category, that is, they must be compared with equivalent energy, which is exergy efficiency. Because the same amount of exergy, regardless of what kind of material flow or energy flow it possesses, theoretically, they are all of the same quality, the same use value, and the same amount of energy, they are all strictly the same category.

From the discussions of energy balance analysis, entropy analysis, and exergy analysis, it can be seen that although the energy balance analysis method has its defects, it still does not lose its importance, because it is the basis of thermodynamic system design, and it is also the basis of entropy analysis and exergy analysis; the results obtained by entropy analysis and exergy analysis are consistent, but exergy analysis is more convenient and clear than entropy analysis, and its application is also more extensive.

Example 2.3 uses energy balance analysis, entropy analysis, and exergy analysis to perform thermal analysis on a certain type of steam power device, and the calculation results are shown in Table 2.2.

Table 2.2 Thermodynamic analysis of a specific steam power plant

Computational item	Energy balance analysis		Entropy analysis		Exergy analysis	
	calculated value/MW		calculated value/MW		calculated value/MW	
Total energy supply from fuel	3704.0	100.00	3226.6	100.00	3226.6	100.00
Plant outputs effective power	839.7	22.67	841.1	26.07	839.8	26.03
Energy loss in Boiler	555.6	15.00	1929.8	59.81	1929.6	59.81
Internal loss of steam turbine	236.0	6.37	220.0	6.82	221.0	6.85
Friction loss of steam turbine	13.2	0.36	13.2	0.41	13.2	0.41
Energy loss in condenser	2006.4	54.70	163.3	5.06	164.0	5.08
Plant effective efficiency	22.67%		26.07%		26.03%	

From the table, it can be seen that the results obtained by the entropy analysis method and the exergy analysis method, which belong to the second law of thermodynamics analysis method, are very close. If the errors brought about by different calculation methods are considered, it can basically be considered that there is no difference in the calculation results of these two methods. The calculation result of the energy balance analysis method is quite different from that of the entropy analysis method and the exergy analysis method, mainly reflected in the meaning of energy loss and its distribution.

From the results of energy balance analysis, the largest energy loss in the steam power device is in the condenser, followed by the boiler, and then the steam turbine. The energy loss here refers to the amount of energy not used by the steam power device; from the calculation results of entropy analysis and exergy analysis, the largest part of energy loss is in the boiler, followed by the steam turbine, and then the condenser, where the energy loss refers to the loss of energy's ability to do work.

Obviously, according to the results of energy balance analysis, only by significantly reducing the heat discharged into the environment in the condenser can the energy utilization rate of the steam power device be effectively improved. But in fact, the temperature at which the condenser releases heat to the environment is only slightly higher than the ambient temperature, so the ability of the discharged heat to do work is very low; for the boiler, because the chemical energy of the fuel is converted into heat energy through combustion, and there is a large temperature difference between the flue gas and the steam heat exchange, although the amount of energy lost to the environment is not much, the ability of energy to do work drops significantly due to the decrease in energy quality, which is the main reason for the low thermal economy of the steam power device.

Example 2.4 We uses energy balance analysis and exergy analysis to perform thermal analysis on the LaSalle County boiling water reactor nuclear power station, and the calculation results are shown in Figs. 2.9 and 2.10 [21].

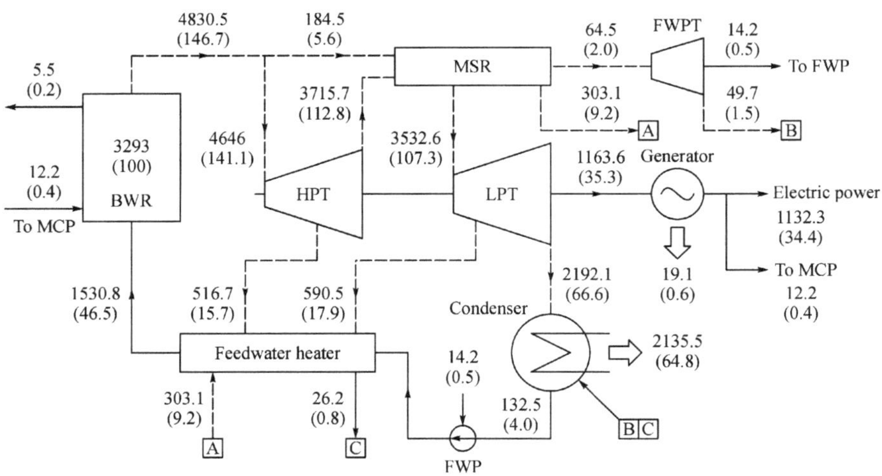

Fig. 2.9 Energy flow diagram of LaSalle County BWR NPP

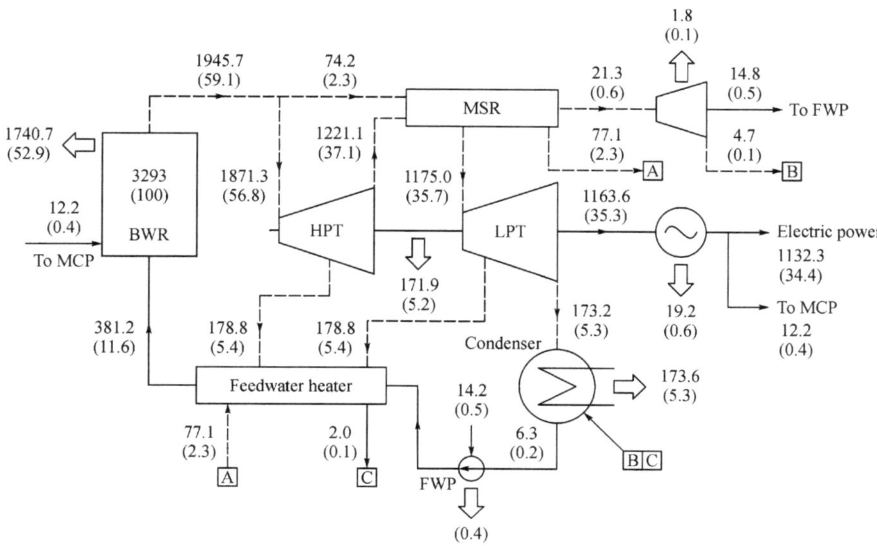

Fig. 2.10 Exergy flow diagram of LaSalle County BWR NPP

Table 2.3 Comparison of thermal analysis results of LaSalle County BWR NPP

	Energy balancing analysis		Exergy analysis	
computational item	Energy loss/MW	Loss ratio/%	Exergy loss/MW	Loss ratio/%
Boiling water reactor	5.5	0.2	1740.7	52.9
Main steam turbine			171.9	5.2
Generator	19.1	0.6	19.2	0.6
Main condenser	2135.5	64.8	173.6	5.3
Pump and other equipment				0.4
Feed water heater				1.1
Total		65.6		65.6
Plant efficiency	34.4%		34.4%	

Figure 2.9 shows the energy flow diagram of thermal analysis, which gives the absolute values of energy flow of each main equipment and its share of the reactor thermal power (the unit of the number in the brackets in the figure is %).

Figure 2.10 shows the exergy flow diagram of thermal analysis, which gives the absolute values of the exergy flow of each main equipment and its share of the reactor thermal power (the unit of the value in the brackets in the figure is %).

Two analysis methods were used to calculate the energy loss of the main equipment in the LaSalle County boiling water reactor nuclear power station, and the results are listed in Table 2.3.

As can be seen from Table 2.3, the main differences in the calculation results of the two analysis methods lie in the boiling water reactor and the main condenser.

The energy balance analysis method does not consider the quality characteristics of energy, only focuses on the balance of energy in quantity, so it cannot make an objective and true evaluation of the rationality and effectiveness of energy system and thermal equipment energy utilization.

By comparing the energy balance analysis method and the exergy analysis method, the following main conclusions can be drawn:

1. The use of exergy efficiency can correctly and comprehensively evaluate the energy utilization rate of thermal systems or processes, and make correct judgments on energy-saving potential.
2. Based on the distribution of exergy loss in various equipment or processes in the thermal system, the weak links in the energy use process can be scientifically diagnosed.
3. According to the cause of exergy loss, the correct way to improve the thermal economy of the thermal system can be pointed out.

Questions and Exercise

1. How many types of thermal systems are divided in the thermal analysis process? What are their characteristics?
2. What is a thermodynamic equilibrium state? What are the necessary conditions to reach a thermodynamic equilibrium state?
3. What are reversible and irreversible processes? What is the relationship between the thermodynamic process and the thermodynamic cycle?
4. How many types of energy can be divided according to their functional strength? What are their characteristics?
5. Briefly describe the basic content of the first and second laws of thermodynamics.
6. Briefly describe the basic meaning of the principle of entropy increase.
7. Briefly describe the concept of internal reversible cycle in finite time thermodynamics.
8. How many types of common thermodynamic analysis methods are there? Try to compare their differences.

Chapter 3
The Exergy Analysis Method of Thermodynamic Process

In order to measure the energy possessed by a system in terms of both quantity and quality, British scientist Tait first used the concept of "Availability" in 1868 to determine the convertible and non-convertible parts of heat. In 1871, British scientist J.C. Maxwell used the term "Available Energy" and proposed in 1873 to represent the available energy of the system with the reversible net work when the closed system reaches a silent state.

In 1873, American scientist Josiah W. Willard Gibbs first derived the formula for the thermodynamic energy exergy of the closed system that is widely used today. In 1889, French scientist Giuy analyzed the available energy with the total reversible shaft work, and derived the relationship between available energy loss and entropy increase. In 1898, Stodola of Switzerland studied the stable material flow of great significance in engineering practice, derived the maximum technical work of stable material flow, and confirmed the relationship between the loss of total output work and entropy increase. This relationship is known as the Giuy-Stodola principle, which lays the foundation for calculating available energy loss. Because the concept given is abstract and difficult to understand, it has not attracted widespread attention.

In 1941, American scientist James Prescot Keenan systematically introduced the concepts of available energy and power loss, summarizing the theories of his predecessors and forming a more complete theoretical system. It wasn't until 1950 that J.H.Keenan and G.Darrieus simplified these concepts, proposed the calculation formula for the steady flow system's exergy, and used it to analyze the thermodynamic second law efficiency of steam power systems.

In 1956, Z.Rant first unified the meaning of exergy. He divided energy into two parts according to Plank's theory: the part that can be converted under certain environmental conditions and the part that cannot be converted. He used the Greek word "ergo" (meaning work or force) with the prefix "ex" (meaning from within, external) to name the convertible part of energy as "Exergy", and named the non-convertible part as "Anergy". In China, these are translated as "Exergy" and

M. Peng et al., *Thermal Analysis of Nuclear Power Plants*, Nuclear Science and Technology, https://doi.org/10.1007/978-981-96-8983-5_3

"Anergy". In the 1960s, Bosnjakovic and H.D.Baehr extensively discussed exergy in their works and conducted exergy analysis on power plants.

The emergence of a global energy crisis has highlighted the importance of effective energy utilization. Scientists have conducted extensive research, gradually improving and developing exergy analysis theory. In Russia, Germany, Japan, and Western countries, this thermodynamic analysis method has been widely applied in energy utilization and in industries such as power, low-temperature engineering, refrigeration, heat pumps, chemical engineering, and metallurgy.

3.1 Basic Concept of Exergy

3.1.1 Exergy and Anergy

3.1.1.1 Definition of Exergy

The concept of exergy is based on the first and second laws of thermodynamics and is a measure of the maximum work capability that energy theoretically possesses. In thermodynamics, there are several ways to define exergy, the most common of which are:

1. For a stable environment, the maximum available work that a unit mass of steady-flow working substance can do when it reversibly changes from its current state to a state in equilibrium with the environmental medium.
2. A certain form of energy or a certain state of matter, after a completely reversible change process (such as heat transfer, mass transfer, chemical reactions, etc.), finally reaches a state completely in balance with the environment. The maximum work that this energy or matter can do in this reversible process is called exergy.
3. Exergy is the maximum work obtained by the outside world when the thermodynamic system or energy flow interacts with the environmental medium to reach a complete balance.

In the definition of exergy, three constraints should be noted:

1. The process of matter changing from a given state to an environmental state is completely reversible.
2. The end state of the work process should be in complete equilibrium with the surrounding environment, including thermal equilibrium, force equilibrium, and chemical equilibrium.
3. In the work process, there are no other heat sources or power sources besides the environment.

According to the Carnot cycle, whether it is the working substance in a closed system or the flowing working substance in an open system, as long as there is a thermodynamic imbalance between the thermodynamic state of the working substance and the environmental medium (for example, the temperature or pressure of the working substance is higher than the environmental medium), the working substance has the ability to do useful work.

By definition, if the environmental state is certain, the size of exergy depends only on the state of the working substance, therefore, exergy is essentially a state parameter. Obviously, the more exergy contained in energy, the greater its ability to convert into useful work, and the higher its utilizable value.

The above-mentioned exergy is also known as thermodynamic exergy. At present, the concept of exergy has been extended to call the maximum useful work in various situations as exergy. For example, the ability of heat to do work is called thermal exergy, and all mechanical energy is available energy, called mechanical exergy. It should be noted that thermal exergy and mechanical exergy are not thermodynamic state parameters.

3.1.1.2 Characteristics of Exergy

Exergy is a type of energy, with the dimension and properties of energy, but its meaning is not exactly the same as energy. Generally speaking, the quantity and quality of energy are not unified, while exergy represents the part of energy where quantity and quality are unified. All forms of energy contain a certain amount of exergy, regardless of the form of energy, the exergy contained in it reflects the unified part of quantity and quality in their own energy. In this way, exergy can be used to evaluate and compare various different forms of energy.

From the first and second laws of thermodynamics, it can be directly derived that the system exergy has a definite value in any given state. In the actual process, the work obtained is less than the reduced exergy of the system, which means that some exergy has not been converted into work and has disappeared in the irreversible process. This is an important difference between exergy and energy. Energy does not disappear, but exergy can disappear. It only obeys the conservation law in the reversible process. In all other cases, it may disappear partially or even completely, dissipating and losing in the irreversible process. Table 3.1 compares the characteristics of energy and exergy.

Table 3.1 The comparison of the characteristics of energy and exergy [22, 23]

Energy E	Exergy E_x
1. Related only to the parameters of the substance or energy flow, not to the parameters of the ambient medium 2. Follows the law of conservation in any process and cannot arise and disappear without cause 3. The energy always has a non-zero value equal to the corresponding value of $E = mc^2$ 4. Interconversion of different forms of energy follows the second law of thermodynamics	1. Related to both environmental media parameters and system parameters 2. Follows conservation laws only in reversible processes, and can vanish partially or completely in a variety of practically irreversible processes 3. The value of exergy can be equal to zero (when in perfect equilibrium with the ambient medium, the value of exergy is zero) 4. The value of exergy the different forms of energy are completely equivalent

3.1.1.3 The Relationship Between Exergy and Anergy

The concept opposite to exergy is anergy. Under a certain environmental baseline, any form of energy that can theoretically be converted into the maximum available work is called exergy, and the part that cannot be converted into useful work is called anergy. Therefore, the total energy of the system E_t is composed of exergy E_x and anergy A_n, that is,

$$E_t = E_x + A_n \tag{3.1}$$

Different forms of energy contain different proportions of exergy and anergy, and some forms of energy may have zero exergy or anergy. For example, mechanical energy and electrical energy are all exergy, and their anergy is zero; thermal energy under natural environmental conditions and heat input or output from the environment are all anergy, and their exergy is zero.

After introducing the concepts of exergy and anergy, the concept of energy has fundamentally changed. The first and second laws of thermodynamics can be described with the concepts of exergy and anergy as follows:

First Law of Thermodynamics—The total amount of exergy and anergy is conserved in the process of energy conversion and transmission.

Second Law of Thermodynamics—All actual processes (i.e., irreversible processes) are processes of decreasing exergy and increasing anergy, i.e., actual processes always proceed in the direction of converting exergy into anergy.

According to the second law of thermodynamics, the process of energy conversion proceeds in the direction of decreasing exergy and increasing anergy. The decrease in exergy not only indicates the loss of energy quantity, but also the devaluation of energy quality, which is the real sense of energy loss. The so-called energy conservation is essentially exergy conservation; the so-called energy crisis is essentially an exergy crisis. Therefore, in the process of energy use, that is, in the process of exergy being converted into anergy, we should fully and effectively play the role of exergy, minimize unnecessary and unreasonable exergy losses, and avoid the precious exergy easily and freely transforming into anergy.

3.1.2 Environment and Environmental State

Humans live in the natural environment on the surface of the earth, and the energy transfer and conversion processes achieved by humans are all carried out under environmental conditions. The thermodynamic processes carried out by the thermodynamic system often interact with the environment (exchange of energy and mass), and when the thermodynamic system changes to reach thermodynamic equilibrium with the environment, it is impossible to change again, that is, the depth of the thermodynamic process is limited by the environment.

In order to express the ability of the system to function in a certain state, a reference state (also known as the base state, dead state, or silent state) must first be determined, and it is defined that the ability of the system to function at the base state is zero. Since the system is always in the surrounding natural environment, all changes are carried out in the environment, so to determine the base state, the environmental state must first be determined.

Exergy is a relative quantity based on the environment as a reference. The "environment" referred to here is a specific part of the outside world outside the system. Generally speaking, everything outside the system is called the outside world, which can have various kinds of outside worlds, and the outside world itself may be changing and unstable. For the convenience of analyzing and studying problems, the environment is regarded as a huge system that is stationary, in a state of thermodynamic equilibrium (thermal equilibrium, force equilibrium, and chemical equilibrium), and its intensity parameters (such as pressure, temperature, etc.) and chemical composition remain unchanged (even when absorbing or releasing energy and matter). The surrounding atmosphere, river water, lake water, and sea water can be regarded as the environment, and sometimes the outer part of the earth's crust can also be regarded as the environment. This is because the earth's surface, oceans, and atmosphere are almost infinitely large compared to the system, and the intensity parameters of the environment will not change due to the transfer of energy and mass with the system. Therefore, it can be considered that the "environment" is an ideal outside world that is infinitely vast, internally uniform and relatively stationary, composed solely of the earth's surface, oceans, and atmosphere. Which substances are chosen as the environment can be determined based on the location and nature of the thermodynamic process.

The state of any thermodynamic system's environment in a state of thermodynamic equilibrium is called the environmental state, and the parameters of the thermodynamic system in the environmental state are represented by the subscript "0", such as p_0, T_0, h_0, S_0, etc.

The equilibrium between the system and the environment can be divided into two types according to its nature: constrained equilibrium and unconstrained equilibrium.

Constrained equilibrium refers to the system and the environment reaching thermal and force equilibrium. The equilibrium between the closed system and the environment belongs to this category. At this time, there is no mass transfer between the system and the environment, there will be no mixing and diffusion, and no chemical reactions will occur, so this type of equilibrium does not involve the issue of chemical equilibrium caused by changes in concentration and composition. The state when the system and the environment reach a constrained equilibrium is called the "constrained dead state" or "physical dead state".

Unconstrained equilibrium refers to a state where the system and the environment have reached thermal, force, and chemical equilibrium, that is, a state of complete equilibrium. The state when the system and the environment reach an unconstrained equilibrium is called "unconstrained dead state" or "chemical dead state". For unconstrained equilibrium, the material composition of the environment and its components are crucial, because not only should the temperature and pressure of the system be equal to the temperature and pressure of the environment, but the material composition of the system should also have the same parts as the material composition of the environment.

The environmental medium has the following two characteristics:

1. Its parameters are independent of the parameters of the system being analyzed. That is to say, the environmental medium should be large enough compared to the system so that any changes caused by the system's actions on it can be ignored.
2. Its components are in a state of complete equilibrium. For the actual environmental medium, there are temperature gradients, pressure gradients, and chemical potential gradients. In solving most thermodynamic problems, these situations can be completely ignored.

In most thermodynamic problems, the parameters of the environmental medium can be considered constant, but in some special cases, they may also be variables.

3.1.3 Basic Properties of Exergy

Exergy has the following properties:

1. Exergy has the property of energy. Like energy, exergy can be a state quantity or a process quantity. Generally speaking, the exergy contained in the energy of the implied form is a state quantity, and the exergy contained in the energy of the migration form is a process quantity.
2. Exergy has equivalence. It is a physical quantity used to measure the ability of various forms of energy to convert, indicating the possible limit values during energy conversion. From a thermodynamic point of view, the exergy of various forms of energy is equivalent.
3. Exergy has comparability. Exergy reflects the value of energy from the perspective of quantity and quality, representing the unified part of quantity and quality in energy.
4. Exergy has relativity. Exergy is a relative quantity taken with the environment as a benchmark. Its value is zero in a quiescent state, and its value is not zero when it deviates from the quiescent state.
5. Exergy has divisibility. Exergy can be decomposed into physical exergy, kinetic exergy, potential exergy, chemical exergy, etc., and physical exergy can be decomposed into mechanical exergy and thermal exergy; chemical exergy can also be decomposed into diffusion exergy and reaction exergy.
6. In a reversible process, exergy has conservation. The total value of exergy remains constant.
7. In an irreversible process, exergy has non-conservation. Inevitably, a part of exergy will transform into anergy.

3.2 Classification and Calculation of Exergy

3.2.1 *Classification of Exergy and its Unbalanced Potential*

Exergy is a measure of the functional ability of system theory. Because there is a certain unbalanced potential between the system and the environment, the system has a certain functional ability. According to the different unbalanced potentials, exergy can be divided into three major categories: heat exergy, material exergy, and power source exergy, each of which can be further divided into several types of exergy, as shown in Table 3.2.

Electric energy, mechanical energy, hydraulic energy, wind energy, etc. belong to high-quality energy, unlike thermal energy, they are not restricted by the second law of thermodynamics and can be completely converted into available work, so the exergy value of these energies is equal to the quantity of energy.

3.2.2 *Mechanical Exergy*

The kinetic and potential energy possessed by a thermodynamic system undergoing macroscopic motion are both forms of mechanical energy, which can be fully converted into work. Therefore, the kinetic and potential energy of the thermodynamic system are all exergy, referred to as "kinetic exergy" and "potential exergy" respectively.

Table 3.2 Classification of exergy and its unbalanced potential [22]

Type	Subclasses		Unbalanced potential
heat exergy (cooling exergy)			Temperature differences
Material Exergy	Mechanical Exergy	Kinetic Exergy	Speed difference between the system and the environment
		Potential Exergy	Height difference between the system and the environment
	Thermodynamic Exergy	Physical Exergy	Temperature and pressure differences between the system and the environment
		Chemical Reaction Exergy	Chemical potential difference between the substance and the environmental reference material
		Diffusion exergy	Concentration difference
Power Source Exergy	Electrical Power Exergy		Potentiometric difference
	Water Conservancy Exergy		Water level difference
	Wind Power Exergy		Wind pressure difference
	Soil Fertility Exergy		Difference in pressure between underground and surface layers
	Wavy Exergy		Difference between the wind pressure acting on the sea surface and the atmospheric pressure

3.2.2.1 Kinetic Exergy

If the kinetic energy of the system under the reference state (i.e., the state in equilibrium with the environment) is taken as zero, all the kinetic energy of the system is exergy, represented as

$$E_{x,\mathrm{k}} = \frac{1}{2} mc^2 \tag{3.2}$$

Where m is the quality of the system;
c is the speed at which the system moves relative to the environment.

3.2.2.2 Potential Exergy

When the potential energy under the system's reference state is taken as zero, all the potential energy of the system is exergy, represented as

$$E_{x,\mathrm{p}} = mgz \tag{3.3}$$

Where g is the gravitational acceleration;.
z is the height of the system relative to the environmental datum.

3.2.3 *Heat Exergy*

For a system at a temperature higher than the environment, during reversible changes under given environmental conditions, the maximum useful work that can be done by the heat transferred through the boundary, i.e., the maximum work capability of the heat relative to the environment, is called thermal exergy.

3.2.3.1 Calculation of Heat Exergy

If the system is taken as the heat source and the environment as the cold source, imagine a series of infinitesimal Carnot heat engines working between this heat source and cold source, ensuring that the system undergoes reversible state changes while releasing heat Q. Suppose the heat provided by the system to the infinitesimal Carnot heat engine is $\mathrm{d}Q$, according to the efficiency formula of the Carnot cycle, the maximum useful work done by this part of the heat is

$$\mathrm{d}W_{\max} = \left(1 - \frac{T_0}{T}\right)\mathrm{d}Q \tag{3.4}$$

Where $\left(1-\frac{T_0}{T}\right)$ is the efficiency of the Carnot cycle working between T and T_0, also known as the "Carnot factor η_c".

The total heat provided by the system to a series of infinitesimal Carnot heat engines during reversible changes is $Q = \int \mathrm{d}Q$, the maximum useful work that can be done is

$$W_{\max} = \int \mathrm{d}W_{\max} = \int \left(1-\frac{T_0}{T}\right)\mathrm{d}Q = Q - T_0 \Delta S$$

The heat exergy is

$$E_{x,Q} = W_{\max} = Q - T_0 \Delta S \tag{3.5}$$

The heat anergy is

$$A_{n,Q} = Q - E_{x,Q} = T_0 \Delta S \tag{3.6}$$

For thermostatic heat sources, there are

$$E_{x,Q} = \left(1-\frac{T_0}{T}\right)Q \tag{3.7}$$

$$A_{n,Q} = \frac{T_0}{T}Q \tag{3.8}$$

Figure 3.1 shows the representation of heat exergy on the *T-s* diagram, where (a) is the variable temperature situation, and (b) is the constant temperature situation.

The area 1-2-3-4-1 in the figure represents the heat exergy of the heat Q absorbed by the process 1-2 relative to the environmental baseline T_0, while the area 3-5-6-4-3 represents the part of the heat Q that cannot be converted into exergy, i.e., heat anergy.

3.2.3.2 Properties of Heat Exergy

1. Heat exergy is the maximum useful work that heat can be converted into, and is an inherent characteristic of the heat itself. When the system releases heat, it also releases the exergy in that heat; conversely, when the system absorbs heat, it also absorbs the exergy in that heat.
2. The size of the heat exergy is not only related to the heat Q, but also to the environmental temperature T0 and the heat source temperature T.
3. Heat anergy is related not only to T_0, but also to S, so the entropy change of the system during the reversible heat transfer process can be used as a measure of heat anergy.
4. Heat exergy, heat anergy, and heat Q are all process quantities.

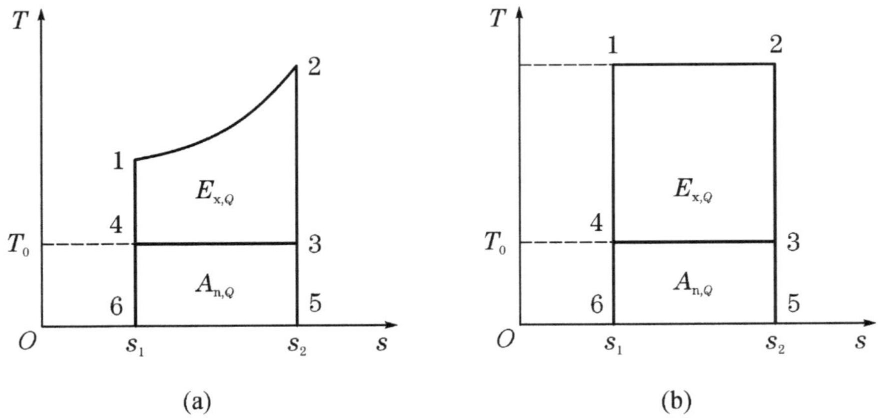

Fig. 3.1 The representation of heat exergy on the *T-s* diagram: (**a**) Varied temperature during heat transfer; (**b**) Constant temperature during heat transfer

5. The heat Q transferred by the system under physical quiescent conditions cannot be converted into useful work, and all the heat is heat anergy.
6. The same amount of heat Q has different exergy under different temperature conditions.

Example 3.1 A steadily operating heating boiler, with an average furnace temperature of 1250 °C, and a thermal load of 20 × 10⁶ kW, assuming the ambient temperature is 25 °C, calculate the exergy flow of the furnace heat source.

Solution Assuming that the furnace temperature remains constant during the stable operation of the boiler, it is a constant temperature heat source, then its exergy is

$$E_x = Q\left(1-\frac{T_0}{T}\right) = 2.0\times10^7 \times\left(1-\frac{25+273.15}{1\ 250+273.15}\right) = 1.608\ 5\times10^7\ \text{kW}$$

Example 3.2 A tank with steam as the insulation heat source, the heating steam temperature is 150 °C, the heat release is 45 kW, the oil temperature in the tank is maintained at 60 °C, if the ambient temperature is 20 °C, without considering heat dissipation loss, calculate the exergy released by the heating steam and the exergy absorbed by the oil.

Solution The temperature of the heating steam remains constant during the heat release process, so the exergy of the released heat is

$$E_{x,\text{steam}} = Q\left(1-\frac{T_0}{T_{\text{steam}}}\right) = 45\times\left(1-\frac{20+273.15}{150+273.15}\right) = 13.82\text{kW}$$

The oil maintains a constant temperature during the heat absorption process, so the exergy of the absorbed heat is

$$E_{x,\text{oil}} = Q\left(1-\frac{T_0}{T_{\text{oil}}}\right) = 45\times\left(1-\frac{20+273.15}{60+273.15}\right) = 5.40\ \text{kW}$$

3.2.4 *Cooling Exergy*

In engineering, the heat exchanged between a system below the ambient temperature and the outside world is called cooling. Heat spontaneously transfers from a high-temperature object to a low-temperature object, or it can be said that cooling spontaneously transfers from a low-temperature object to a high-temperature object, that is, the direction of cooling flow is opposite to that of heat flow.

Using a reverse cycle, by consuming a certain amount of mechanical energy, heat can be extracted from the low-temperature system and discharged to the high-temperature system, so the reverse cycle is also called the refrigeration cycle. In the refrigeration cycle, the heat Q is removed from the low-temperature system, which is equivalent to equivalently releasing the cooling Q' to the low-temperature system. The direction of the cooling flow is opposite to that of the heat flow, that is, $Q' = -Q$.

The maximum workability of cooling relative to the environment is called cooling exergy. Similar to heat exergy, the exchange of cooling exergy also accompanies the exchange of cooling.

3.2.4.1 Calculation of Cooling Exergy

The temperature T of the low-temperature system is lower than the ambient temperature T_0. If there is a reversible heat engine between the environment and the low-temperature system, in the micro reverse Carnot cycle, the heat $\mathrm{d}Q$ is absorbed from the low-temperature system, and the heat $\mathrm{d}Q_0$ is released to the environment. The minimum useful work consumed by the reversible heat engine is $\mathrm{d}W_{\min}$, according to the expression of the reverse Carnot cycle refrigeration coefficient

$$\varepsilon = \frac{\mathrm{d}Q}{\mathrm{d}W_{\min}} = \frac{T}{T_0 - T} \tag{3.9}$$

It can be obtained

$$\mathrm{d}W_{\min} = \left(\frac{T_0}{T} - 1\right)\mathrm{d}Q \tag{3.10}$$

The energy balance equation of the refrigeration cycle is

$$dQ_0 = dW_{min} + dQ = \frac{T_0}{T} dQ \tag{3.11}$$

That is, the heat discharged into the environment in the reverse cycle is equal to the sum of the heat absorbed from the low-temperature system and the mechanical energy consumed by the outside world.

Refer to the derivation method of heat exergy to obtain the expressions of cooling exergy and cooling anergy.

Assuming the environment is a high-temperature heat source, the system is a low-temperature heat source, and there is a reversible heat engine between them. In the micro forward cycle, the heat $-dQ_0$ is absorbed from the environment, and the heat $-dQ$ is discharged to the low-temperature system. Then the maximum available work done to the outside world in the micro cycle is

$$dW_{max} = -\left(1 - \frac{T}{T_0}\right) dQ_0 = \left(\frac{T}{T_0} - 1\right) dQ_0 \tag{3.12}$$

$$dW_{max} = (-dQ_0) - (-dQ_0) = -dQ_0 - dQ' \tag{3.13}$$

Combining the above two formulas, we can get

$$dW_{max} = \left(\frac{T_0}{T} - 1\right) dQ' \tag{3.14}$$

Therefore, in the entire cycle process, the maximum useful work output, that is, the cooling exergy is

$$E_{x,Q'} = W_{max} = \int \left(\frac{T_0}{T} - 1\right) dQ' = T_0 \Delta S - Q' \tag{3.15}$$

Since all the energy discharged into the environment has been converted into anergy, therefore, the cooling anergy is

$$A_{n,Q'} = -\int dQ_0 = \int \left(\frac{T_0}{T}\right) dQ' = T_0 \Delta S \tag{3.16}$$

If the temperature of the low-temperature system remains constant, then

$$E_{x,Q'} = \left(\frac{T_0}{T} - 1\right) Q' \tag{3.17}$$

$$A_{n,Q'} = \frac{T_0}{T} Q' \tag{3.18}$$

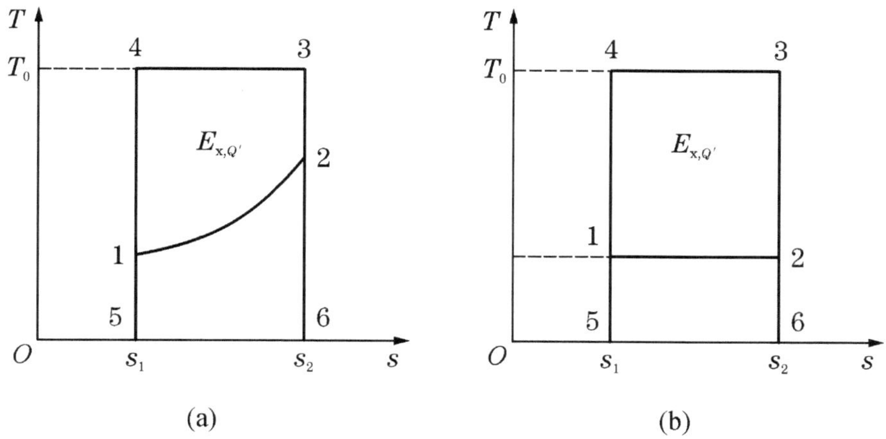

Fig. 3.2 The cooling exergy on the *T*-*s* diagram: (**a**) varied temperature; (**b**) constant temperature

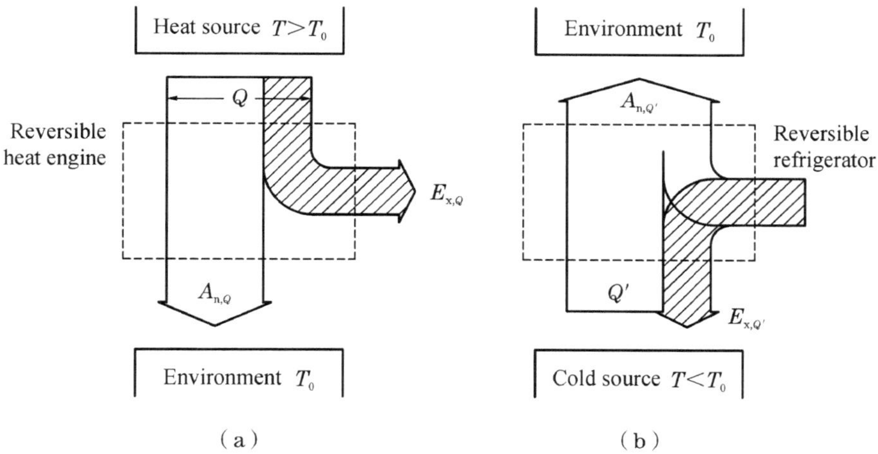

Fig. 3.3 Comparison of the heat exergy and the cooling exergy: (**a**) Heat exergy; (**b**) Cooling exergy

Figure 3.2 represents the cooling exergy on the *T*-*s* diagram, where (a) represents the temperature change of the low-temperature system, and (b) represents the constant temperature of the low-temperature system.

The area 1-2-6-5-1 in the figure represents the cooling capacity, and the area 1-2-3-4-1 represents the cooling exergy.

Figure 3.3 compares the heat exergy with the cooling exergy.

The difference between cooling exergy and heat exergy is that while the system releases heat, it also releases heat exergy. The direction of heat and heat exergy is always the same, and the value of heat exergy is always less than the value of heat. However, when the system absorbs cooling capacity, it releases cooling exergy. The direction of cooling capacity and cooling exergy is always opposite, and when $T < \frac{T_0}{2}$, the absolute value of cooling exergy $E_{x,Q'}$ is greater than the absolute value of cooling capacity Q'.

3.2.4.2 Properties of Cooling Exergy

Cooling exergy and cooling anergy have the following characteristics:

1. Cooling exergy is the maximum useful work that cooling capacity can do under given environmental conditions. It is also an inherent characteristic of cooling capacity, regardless of whether a reversible heat engine is used.
2. Since cooling capacity is also heat and is a process quantity, cooling exergy and cooling anergy are also process quantities.
3. The value of cooling exergy is not only related to the size of the cooling capacity, but also to the temperature of the cold storage and the environment. Under a given environmental temperature T_0, the cooling exergy of a certain cooling capacity increases as the temperature T of the cold storage decreases.

3.2.5 Physical Exergy

The maximum useful work provided by the system when it reversibly changes from its current state to a constrained equilibrium state with the environment is the physical exergy of the system. That is, the exergy possessed by the system due to the difference in temperature and pressure from the environment is called physical exergy.

3.2.5.1 Thermodynamic Exergy

A stationary closed system at a certain state (p, T), under conditions where it can only exchange heat with the environment, can do the maximum useful work when it reversibly changes to an incomplete equilibrium environment state (p_0, T_0). This is called the thermodynamic exergy of the closed system at that state, denoted as $E_{x,\,U}$.

When the closed system undergoes a small reversible change process, according to the first law of thermodynamics, we have

$$dQ = dU + p_0 dV + dW_{max} \tag{3.19}$$

Where dQ is the heat exchanged with the environment during;

dU is the change of thermodynamic energy;

$p_0 dV$ is the volumetric work done by the system against the environmental pressure;

dW_{max} is the maximum useful work done by the system.

Because it is a reversible process, the environmental temperature is T_0, according to the second law of thermodynamics, we have

$$dQ = T_0 dS \tag{3.20}$$

Substituting into Eq. (3.19), we get

$$T_0 dS = dU + p_0 dV + dW_{max} \tag{3.21}$$

That is

$$dW_{max} = -dU - p_0 dV + T_0 dS \tag{3.22}$$

Integrating from any state (p, T) to an incomplete equilibrium state (p_0, T_0), after rearranging, we get

$$E_{x,U} = (U + p_0 V - T_0 S) - (U_0 + p_0 V_0 - T_0 S_0) \tag{3.23}$$

For unit mass working substance, the specific thermodynamic exergy of the closed system is

$$e_{x,u} = (u + p_0 v - T_0 s) - (u_0 + p_0 v_0 - T_0 s_0) \tag{3.24}$$

Thermodynamic energy anergy is

$$A_{n,U} = U - E_{x,U} = (U_0 + p_0 V_0 - T_0 S_0) - (p_0 V - T_0 S) \tag{3.25}$$

Specific thermodynamic energy anergy is

$$a_{n,u} = u - e_{x,u} = (u_0 + p_0 v_0 - T_0 s_0) - (p_0 v - T_0 s) \tag{3.26}$$

Thermodynamic exergy has the following properties:

1. The thermodynamic exergy of a stationary closed system is not only related to the state of the system, but also to the incomplete equilibrium environment state. Only when the environment state is fixed, the thermodynamic exergy depends on the system state and can be considered as a state parameter. The thermodynamic exergy of a stationary closed system at an incomplete equilibrium environment state is zero.
2. The thermodynamic exergy of a stationary closed system is an inherent characteristic of the system's stored energy at a given state, regardless of what kind of reversible process is used to reversibly change from the given state to an incomplete equilibrium environment state.

3. The maximum useful work that a stationary closed system can do when it can only exchange heat with the environment, and reversibly changes from state 1 to another state 2, is equal to the drop in the system's thermodynamic energy.

Example 3.3 The fresh steam state at the inlet of the steam turbine is $p_1 = 1$ MPa and $t_1 = 300$ °C, the steam volume is $V_1 = 0.015$ m^3, after expansion work, the exhaust steam pressure is $p_2 = 0.14$ MPa, and the steam volume is $V_2 = 0.075$ m^3. Assume the environmental state is $p_0 = 0.1$ MPa and $t_0 = 20$ °C, try to calculate:

1. The thermodynamic energy of fresh steam and exhaust steam;
2. The maximum useful work that steam can do inside the steam turbine;
3. If the expansion process dissipates 1.0 kJ of heat to the environment, find the actual work done by the steam turbine.

Solution According to the fresh steam parameters p_1, t_1, look up the steam table, get

$$v_1 = 0.25798\,\mathrm{m^3/kg}, h_1 = 3{,}052.1\,\mathrm{kJ/kg}, s_1 = 7.1251\,\mathrm{kJ/(kg\cdot K)}$$

According to the environmental parameters p_0, t_0, look up the steam table, get

$$v_0 = 0.0010017\,\mathrm{m^3/kg}, h_0 = 83.95\,\mathrm{kJ/kg}, s_0 = 0.2963\,\mathrm{kJ/(kg\cdot K)}$$

The thermodynamic temperature corresponding to the environmental parameters is

$$T_0 = t_0 + 273.15 = 293.15\,\mathrm{K}$$

The mass of steam entering the steam turbine is

$$m = \frac{V_1}{v_1} = \frac{0.015}{0.25798} = 0.05814\,\mathrm{kg}$$

The specific volume of the exhaust steam at the end of expansion is

$$v_2 = \frac{V_2}{m} = \frac{0.075}{0.05814} = 1.2900\,\mathrm{m^3/kg}$$

According to the exhaust steam parameters p_2, v_2, look up the steam table, get

$$t_2 = 124.6\,°\mathrm{C}, h_2 = 2{,}721.6\,\mathrm{kJ/kg}, s_2 = 7.3268\,\mathrm{kJ/(kg\cdot K)}$$

The thermodynamic energy of unit mass steam at each state is respectively

$$u_1 = h_1 - p_1 v_1 = 3{,}052.1 - 1\times10^6 \times 0.25798\times10^{-3} = 2{,}794.12\,\mathrm{kJ/kg}$$

$$u_2 = h_2 - p_2 v_2 = 2{,}721.6 - 0.14\times10^6 \times 1.2900\times10^{-3} = 2{,}541.00\,\mathrm{kJ/kg}$$

$$u_0 = h_0 - p_0 v_0 = 83.95 - 0.1\times10^6 \times 0.0010017\times10^{-3} = 83.85\,\mathrm{kJ/kg}$$

1. The thermodynamic energy of fresh steam is

$$\begin{aligned}E_{x,U_1} &= m\left[(u_1-u_0)+p_0(v_1-v_0)-T_0(s_1-s_0)\right]\\ &= 0.05814\times[(2{,}794.12-83.85)+0.1\times10^3\times(0.25798-0.0010017)-\\ &\qquad 293.15\times(7.1251-0.2963)]\\ &= 0.05814\times734.11=42.68\,\text{kJ}\end{aligned}$$

The thermodynamic energy of exhaust steam is

$$\begin{aligned}E_{x,U_2} &= m\left[(u_2-u_0)+p_0(v_2-v_0)-T_0(s_2-s_0)\right]\\ &= 0.05814\times[(2{,}541.00-83.85)+0.1\times10^3\times(1.2900-0.0010017)-\\ &\qquad 293.15\times(7.3268-0.2963)\\ &= 0.05814\times525.06=30.53\,\text{kJ}\end{aligned}$$

2. The maximum useful work that steam can do inside the steam turbine is

$$W_{\max}=E_{x,U_1}-E_{x,U_2}=42.68-30.53=12.15\,\text{kJ}$$

3. The useful work done in the actual process is

$$\begin{aligned}W &= m[w-p_0(v_2-v_1)=m\left[q-(u_2-u_1)-p_0(v_2-v_1)\right]\\ &= -1.0-0.05814\times\left[(2{,}541.00-2794.12)+0.1\times10^3\times(1.2900-0.25798)\right]\\ &= 7.72\,\text{kJ}\end{aligned}$$

Since the actual process is an irreversible process, the actual useful work is only 63.5% of the maximum useful work.

3.2.5.2 The Physical Energy of Steady Flow Material

In steady flow, the maximum useful work that a material can do when it can only exchange heat with the environment and reversibly change from any state to an incomplete equilibrium environmental state is called the physical energy of the steady flow material.

A material of mass m is in a certain thermodynamic state (p, T, h, z, c), assuming the material flows through a steady flow open system, changing to an incomplete equilibrium environmental state (p_0, T_0, h_0, $z_0 = 0$, $c_0 = 0$). In this process, it absorbs heat dQ from the environment and does the maximum useful work d$W_{\max}$.

According to the steady flow energy equation

$$\mathrm{d}Q=\mathrm{d}H+\frac{1}{2}m\mathrm{d}c^2+mg\mathrm{d}z+\mathrm{d}W_{\max} \tag{3.27}$$

Since the environmental temperature T_0 is constant, we have

$$\mathrm{d}Q = T_0 \mathrm{d}S$$

Where dS is the entropy increase of the steady flow material from the inlet to the outlet.

Substitute the above equation into (3.27), get

$$\mathrm{d}W_{\max} = T_0 \mathrm{d}S - \left(\mathrm{d}H + \frac{1}{2} m\mathrm{d}c^2 + mg\mathrm{d}z \right) \tag{3.28}$$

Integrate the above equation from the given state to the environmental state, considering that $z_0 = 0$, $c_0 = 0$, get

$$E_x = W_{\max} = (H - H_0) - T_0 (S - S_0) + \frac{1}{2} mc^2 + mgz \tag{3.29}$$

For unit mass steady flow material, its specific physical energy is

$$e_x = w_{\max} = (h - h_0) - T_0 (s - s_0) + \frac{1}{2} c^2 + gz \tag{3.30}$$

Then the anergy of the steady flow material is

$$\begin{aligned} A_n &= E - E_x \\ &= \left(H + \frac{1}{2} mc^2 + mgz \right) - \left[(H - H_0) - T_0 (S - S_0) + \frac{1}{2} mc^2 + mgz \right] \\ &= H_0 + T_0 (S - S_0) \end{aligned} \tag{3.31}$$

The specific anergy is

$$a_n = h_0 + T_0 (s - s_0) \tag{3.32}$$

The derivation of this formula does not limit the change from the initial state to the reference state to be a physical change, a chemical change, or both. Therefore, it can be used as the basic calculation formula for physical exergy. As long as the enthalpy and entropy of the system can be determined at a given state and the reference state, the physical exergy of the steady flow system can be calculated from the above formula.

However, it is not always convenient to determine the enthalpy and entropy of various substances at any state, so other methods must be sought to solve the calculation of physical exergy of homogeneous systems at any state.

3.2.5.3 Physical Exergy of Homogeneous Systems at any State

From the thermodynamic relationship, for the isobaric temperature change process, the following relationship exists:

$$\begin{cases} \Delta h = \int_T^{T_0} c_p \mathrm{d}T \\ \Delta s = \int_T^{T_0} \frac{c_p}{T} \mathrm{d}T \end{cases} \tag{3.33}$$

The thermal exergy caused by thermal imbalance is

$$e_{x,t} = T_0 \Delta s - \Delta h = T_0 \int_T^{T_0} \frac{c_p}{T} \mathrm{d}T - \int_T^{T_0} c_p \mathrm{d}T = \int_{T_0}^{T} \left(1 - \frac{T_0}{T}\right) c_p \mathrm{d}T \tag{3.34}$$

For the isothermal pressure change process, the following relationship exists:

$$\begin{cases} \Delta h = \int_p^{p_0} \left[v - T\left(\frac{\partial v}{\partial T}\right)_p \right] \mathrm{d}p \\ \Delta s = \int_p^{p_0} \left[-\left(\frac{\partial v}{\partial T}\right)_p \right] \mathrm{d}p \end{cases} \tag{3.35}$$

The pressure exergy caused by force imbalance is

$$\begin{aligned} e_{x,p} &= T_0 \Delta s - \Delta h \\ &= T_0 \int_p^{p_0} \left[-\left(\frac{\partial v}{\partial T}\right)_p \right] \mathrm{d}p - \int_p^{p_0} \left[v - T\left(\frac{\partial v}{\partial T}\right)_p \right] \mathrm{d}p \\ &= \int_{p_0}^{P} \left[v - (T - T_0)\left(\frac{\partial v}{\partial T}\right)_p \right] \mathrm{d}p \end{aligned} \tag{3.36}$$

Combining the above two formulas, we get the general formula for calculating the physical exergy of a homogeneous system at any state:

$$e_{x,ph} = e_{x,t} + e_{x,p} = \int_{T_0}^{T} \left(1 - \frac{T_0}{T}\right) \mathrm{c}_p \mathrm{d}T + \int_{p_0}^{p} \left[v - (T - T_0)\left(\frac{\partial v}{\partial T}\right)_p \right] \mathrm{d}p \tag{3.37}$$

The above formula shows that the physical exergy of any homogeneous system can be decomposed into two parts: one part is the thermal exergy caused by thermal imbalance under constant pressure; the other part is the pressure exergy caused by force imbalance under ambient temperature.

3.2.5.4 Physical Exergy of Ideal Gas

For an ideal gas, there is

$$pv = RT$$

$$\left(\frac{\partial v}{\partial T}\right)_P = \frac{R}{p}$$

$$\int_{p_0}^{p}\left[v - (T - T_0)\left(\frac{\partial v}{\partial T}\right)_p\right]\mathrm{d}p = RT_0 \ln\frac{p}{p_0}$$

Therefore, the molar physical exergy of pure ideal gas is

$$e_{x,ph} = \int_{T_0}^{T}\left(1 - \frac{T_0}{T}\right)c_p \mathrm{d}T + RT_0 \ln\frac{p}{p_0} \tag{3.38}$$

If $\overline{c}_p$ is taken as the average constant pressure specific heat between T_0 and T, then the above formula can be simplified as

$$e_{x,ph} = \overline{c}_p\left[(T - T_0) - T_0 \ln\frac{T}{T_0}\right] + RT_0 \ln\frac{p}{p_0} \tag{3.39}$$

For ideal gas mixtures, the molar physical exergy is

$$e_{x,ph} = \int_{T_0}^{T}\left(1 - \frac{T_0}{T}\right)c_{pm} \mathrm{d}T + RT_0 \ln\frac{p}{p_0} \tag{3.40}$$

Where c_{pm} is the constant pressure specific heat of the ideal gas mixture, which can be calculated by the following formula:

$$c_{pm} = \sum_{i=1}^{k} x_i c_{pi}$$

Where x_i is the molar fraction of component i in the ideal gas mixture;

c_{pi} is the constant pressure specific heat of component i in the ideal gas mixture.

If $\overline{c}_{pi}$ is taken as the average constant pressure specific heat of component i between temperatures T_0 and T, then the average constant pressure specific heat of the ideal gas mixture is

$$\overline{c}_{pm} = \sum_{i=1}^{k} x_i \overline{c}_{pi} \tag{3.41}$$

The molar physical exergy of the ideal gas mixture is

$$e_{x,\mathrm{ph}} = \overline{c}_{pm}\left[(T - T_0) - T_0 \ln\frac{T}{T_0}\right] + RT_0 \ln\frac{p}{p_0} \tag{3.42}$$

3.2.5.5 Physical Exergy of Real Gas

For real gas, if the appropriate state equation and the change rule of constant pressure specific heat with temperature are known, the formula (3.36) can be used for solution, but the state equation is complex and the calculation is difficult, so the basic formula of steady flow system physical exergy is often used for calculation.

For pure real gas, there is

$$E_{x,\mathrm{ph}} = G\left[h - h_0 - T_0\left(s - s_0\right)\right]$$

For real gas mixtures

$$E_{x,\mathrm{ph}} = G_{\mathrm{m}}\left[\left(h - h_0\right)_{\mathrm{m}} - T_0\left(s - s_0\right)_{\mathrm{m}}\right]$$

3.2.5.6 Physical Exergy of Incompressible Fluid

For incompressible fluids, their specific volume does not change with temperature and pressure, and can be regarded as a constant. The second term in formula (3.37) can be written as

$$\int_{p_0}^{p}\left[v - (T - T_0)\left(\frac{\partial v}{\partial T}\right)_p\right]\mathrm{d}p = \int_{p_0}^{p} v\mathrm{d}p = v\left(p - p_0\right) \tag{3.43}$$

Then formula (3.37) simplifies to

$$e_{x,\mathrm{ph}} = \int_{T_0}^{T}\left(1 - \frac{T_0}{T}\right)c_p\mathrm{d}T + v\left(p - p_0\right) \tag{3.44}$$

If c_p is regarded as a constant, the above formula is further simplified to

$$e_{x,\mathrm{ph}} = c_p\left[\left(T - T_0\right) - T_0\ln\frac{T}{T_0}\right] + v\left(p - p_0\right) \tag{3.45}$$

For pure water, $c_p = 4.1868$ kJ/(kg · K), $v = 10^{-3}$ m^3/kg, then

$$e_{x,\mathrm{ph}} = 4.1868\left[\left(T - T_0\right) - T_0\ln\frac{T}{T_0}\right] + \left(p - p_0\right)\times 10^{-6} \tag{3.46}$$

Example 3.4 Try to calculate the exergy values of superheated steam at three different states: 10 MPa, 500 °C; 4 MPa, 500 °C; 10 MPa, 400 °C. The reference state is 0.101 3 MPa, 25 °C.

Solution The calculation results are listed in Table 3.3.

Table 3.3 Calculation results

Pressure/MPa	Temperature/°C	Enthalpy/(kJ/kg)	Entropy/[kJ/(kg·K)	Exergy/(kJ/kg)
10	500	3372	6.596	11410.08
4	500	3345	7.087	1336.69
10	400	3093	6.207	1247.06
3.171×10^{-3}	25	104.8	0.367 2	0

3.2.6 Chemical Exergy

The maximum useful work provided when the system goes from a constrained equilibrium state to an unconstrained equilibrium state (silent state) is the chemical exergy of the system. That is to say, the exergy possessed by the system due to the difference in composition with the environment is called chemical exergy (standard chemical exergy). Usually, the system goes from a constrained equilibrium state to an unconstrained equilibrium state through two processes: chemical reaction and physical diffusion. The chemical reaction first transforms the substances of the original system into environmental substances (reference substances); physical diffusion "adjusts" (diffuses) the concentration of the original system or the substances produced after the chemical reaction to the concentration of the reference substances in the specified environment. Therefore, chemical exergy can be further divided into diffusion exergy and chemical reaction exergy.

3.2.6.1 Diffusion Exergy

For pure substances that do not react chemically with the environment, such as N_2, O_2, CO_2, Ar, etc., these substances themselves are reference substances, and can be regarded as ideal gases. Their diffusion exergy is equal to the maximum useful work done when this ideal gas reversibly changes to the partial pressure that this substance has in the environment at the environmental state. According to the basic expression of exergy, we have

$$E_{x,\mathrm{ch}} = h - h_0 - T_0\left(s - s_0\right) = T_0 \Delta s - \Delta h$$

For the isothermal process of an ideal gas, $\Delta h = 0$, therefore

$$E_{x,\mathrm{ch}} = T_0 \Delta s = T_0 R \ln \frac{p_0}{p_{0,i}} = RT_0 \ln \frac{p_0}{x_{0,i} p_0} = -RT_0 \ln x_{0,i} \tag{3.47}$$

Where $p_{0,\,i}$ is the partial pressure of component i in the environment.$x_{0,\,i}$ is the molar fraction of component i in the environment.

The diffusion exergy of each pure component in the environmental air calculated according to the above formula is listed in Table 3.4.

Table 3.4 The diffusion exergy of each pure component in the environmental air

Components	N_2	O_2	H_2O	CO_2	Ar	He	Ne
Mole fraction/%	75.60	20.34	3.12	0.03	0.91	0.000 52	0.001 8
Proliferate exergy/(kJ/mol)	693.36	3947.8	8594.9[a]	20,108	11,649	30,159	27,081

[a]For the diffusion exergy of ideal gas steam, the diffusion exergy of liquid water is zero

3.2.6.2 Chemical Reaction Exergy

The chemical reaction exergy of a substance can be calculated through the relevant chemical reaction, for example, any chemical reaction expressed by an algebraic equation can be written as

$$\gamma_1 A_1 + \gamma_2 A_2 + \gamma_3 A_3 + \cdots = \sum_i \gamma_i A_i = 0 \tag{3.48}$$

Where A_i represents the molecular formula of the i-th substance involved in the chemical reaction;

γ_i is the stoichiometric coefficient of substance i.

Assume that the chemical reaction takes place under standard conditions. According to the relationship between exergy and ideal work, the standard chemical exergy change of the reaction is equal to the ideal work of the reaction process:

$$-\sum_i \gamma_i E_{\text{xch},i} = W_{\text{id}} \tag{3.49}$$

For the chemical reaction

$$W_{\text{id}} = T_0 \sum_i \gamma_i s_i^0 - \sum_i \gamma_i \Delta h_{\text{f},i}^0 \tag{3.50}$$

Combine the above two formulas to get

$$\sum_i \gamma_i E_{\text{xch},i} = \sum_i \gamma_i \Delta h_{\text{f},i}^0 - T_0 \sum_i \gamma_i s_i^0 \tag{3.51}$$

Where $\Delta h_{\text{f},i}^0$ is the standard formation enthalpy of substance i;

s_i^0 is the standard entropy of substance i.

As long as the standard formation enthalpy, standard entropy, and standard chemical exergy of other substances except the substance to be calculated are known, the standard chemical exergy of this substance can be obtained from the above formula.

In addition, according to the definition of free enthalpy $G = h - T_s$, for the state change at T_0, we have

$$\begin{aligned} W_{id} &= -\Delta h + T_0 \Delta s = \left(h_1 - T_0 s_1\right) - \left(h_2 - T_0 s_2\right) \\ &= G_1 - G_2 = -\Delta G \end{aligned} \tag{3.52}$$

For the chemical reaction under standard environment

$$W_{id} = -\sum_i \gamma_i \Delta G_{f,i}^0 \tag{3.53}$$

Combine formulas (3.49) and (3.53) to get

$$\sum_i \gamma_i E_{xch,i} = \sum_i \gamma_i \Delta G_{f,i}^0 \tag{3.54}$$

Where $\Delta G_{f,i}^0$ represents the standard free enthalpy of formation of substance i.

3.2.6.3 Fuel Exergy

The fuel exergy based on environmental temperature is called the chemical exergy of the fuel, and the exergy value that the fuel possesses under unordinary pressure is called the physical exergy of the fuel. In energy and energy-saving engineering practice, the chemical exergy of fuel is often used. The chemical exergy of fuel can be expressed as:

$$e_{x,f} = Q_{f,1} + T_0 \Delta S \tag{3.55}$$

Where $Q_{f,1}$ represents the lower heating value of the fuel.

ΔS represents the difference between the entropy of formation and the entropy of reaction.

This formula can be used to calculate the chemical exergy of a single substance fuel.

Chemical Exergy of Gaseous Fuel

When the various components of gaseous fuel are known, the chemical exergy of the fuel can be calculated by the following formula:

$$e_{x,f} = \sum_{i=1}^{n} x_i e_{x,fi}^0 + RT_0 \sum_{i=1}^{n} x_i \ln x_i \tag{3.56}$$

Where x_i represents the volume ratio of the i-th component;

$e_{x,fi}^0$ represents the standard chemical exergy of the i-th component fuel;

R represents the gas constant.

When calculating the exergy of gaseous fuel composed of two or more carbon atoms, the Lange approximation formula can be used:

$$e_{x,f} = 0.95 Q_{g,h} \tag{3.57}$$

Where $Q_{g,h}$ represents the higher heating value of the gaseous fuel.

Chemical Exergy of Liquid Fuel

The calculation of the chemical exergy of liquid fuel can use the following empirical formula:

$$e_{x,f} = Q_{l,l}\left(1.0038 + 0.1365\frac{\beta_H}{\beta_C} + 0.0308\frac{\beta_O}{\beta_C} + 0.0104\frac{\beta_S}{\beta_C}\right) \quad (3.58)$$

Where $Q_{l,l}$ represents the lower heating value of the liquid fuel. β_H, β_C, β_O, β_S represent the weight percentages of hydrogen, carbon, oxygen, and sulfur elements in the liquid fuel, respectively.

The approximate calculation formula proposed by Lange can also be used:

$$e_{x,f} = 0.975Q_{l,h} \quad (3.59)$$

Where $Q_{l,h}$ represents the higher heating value of the liquid fuel.

Chemical Exergy of Solid Fuel

The commonly used calculation formula is

$$e_{x,f} = Q_{s,l}\left(1.0064 + 0.1519\frac{\beta_H}{\beta_C} + 0.0616\frac{\beta_O}{\beta_C} + 0.0429\frac{\beta_N}{\beta_C}\right) \quad (3.60)$$

Where $Q_{s,l}$ represents the lower heating value of the solid fuel;

β_H, β_C, β_O, β_N represent the weight percentages of hydrogen, carbon, oxygen, and nitrogen elements in the solid fuel, respectively, where the weight of the oxygen element does not include the oxygen in the moisture.

The approximate calculation formula proposed by Rant is

$$e_{x,f} = Q_{s,l} + r_0\beta \quad (3.61)$$

Where r_0 represents the latent heat of vaporization of water at environmental temperature. β represents the total moisture weight percentage of the fuel application base.

Table 3.5 lists the standard molar high and low heating values, reaction enthalpy, and fuel enthalpy of some chemically homogeneous fuels for reference.

Table 3.5 The standard molar high and low heating values, reaction enthalpy, and fuel enthalpy of some chemically homogeneous fuels for reference

Fuel formula	Q_h^n	Q_l^n	$-\Delta G^n$	$e_{x,f}^n$	Fuel formula	Q_h^n	Q_l^n	$-\Delta G^n$	$e_{x,f}^n$
C	393.5	393.5	394.4	410.5	S	296.6	296.6	299.8	–
H_2	285.9	241.8	237.3	235.3	CO	283.0	283.0	257.3	275.4
CH_4	890.5	802.3	818.1	830.3	C_2H_2	1300	1256	1235	1265
C_2H_4	1411	1323	1332	1360	C_2H_6	1560	1428	1468	1494
C_3H_6	2059	1926	1970	2012	C_3H_8	2220	2044	2109	2149
n-C_4H_{10}	2879	2658	2748	2803	n-C_5H_{12}	3510	3245	3386	3455
C_6H_6	3268	3135	3202	3293	C_6H_{12}	3920	3656	3817	3902
n-C_6H_{14}	4164	3855	4023	4106	n-C_7H_{14}	3906	3730	3820	3925
n-C_7H_{16}	4817	4465	4660	4757	n-C_8H_{18}	5471	5074	5297	5408
n-C_9H_{20}	6125	5684	5934	6059	n-$C_{10}H_{22}$	6779	6294	6571	6711
CH_2OH	7266	638.4	702.5	716.7	C_2H_5OH	1367.1	1234.8	1325.8	1354.1

3.3 Enthalpy Loss and Enthalpy Balance Equation

The second law of thermodynamics dictates the direction of processes in an energy system, which can be expressed using the concepts of exergy and anergy: in a reversible process, the system's exergy is conserved; in an irreversible process, there will be a transformation from exergy to anergy, reducing the total amount of exergy. Once exergy has transformed into anergy, it is impossible for anergy to revert back to exergy. The decrease in exergy in an irreversible process is referred to as the exergy loss caused by the irreversible process, or simply exergy loss.

All real thermodynamic processes involve irreversible factors, leading to exergy loss. Heat exchange under finite temperature difference, mass exchange under finite concentration difference, and heat generation from friction in viscous flow are all typical irreversible phenomena. Since irreversible processes result in exergy loss, and all exergy loss turns into anergy, there is no conservation of exergy in irreversible processes. To facilitate problem analysis, following the method of establishing the entropy balance equation for irreversible processes, an additional term for irreversible exergy loss is added on the basis of non-conservation of exergy, thus establishing the following quantity balance relationship, namely

$$E_{x,in} = E_{x,out} + \Delta E_x + E_{xl} \tag{3.62}$$

Where $E_{x,\,in}$ is the exergy input into the system;
$E_{x,\,out}$ is the exergy output from the system;
ΔE_x is the change in exergy within the system;
E_{xl} is exergy loss.

The exergy loss in the above formula is actually all anergy, so this formula is not an exergy conservation equation, but is called an exergy balance equation because the quantities on both sides of the equation are equal.

The exergy balance equation and the energy balance equation have essential differences. The energy balance equation is based on the first law of thermodynamics, focusing on the quantity relationship between energies, considering the tangible loss of energy in the thermodynamic system (i.e., external loss); while the exergy balance equation is based on both the first and second laws of thermodynamics, focusing on the change in the ability of energy to do work, considering the loss of energy in terms of mass in the thermodynamic process (i.e., internal loss). Through energy balance, the actual quantity of effective use and loss of energy in the energy system and its various devices and links can be given, which is used to compare the working effects of cycles and thermodynamic devices under the same conditions; through exergy balance, the loss of workability (i.e., exergy loss) caused by irreversibility in the energy system and its various links can be given, which is used to measure the thermodynamic perfection of the thermodynamic system.

3.3.1 Exergy Balance Equation and Exergy Loss of Typical Systems and Processes

3.3.1.1 Steady Flow System

For an open system with a steady flow of working substance in and out, and heat and work exchange with the environment, the exergy input into the system is

$$E_{x,\text{in}} = E_{x,Q} + E_{x,H_1} + \frac{1}{2}mc_1^2 + mgz_1 \tag{3.63}$$

Where $E_{x,Q}$ is the heat exergy input into the system from a heat source at temperature T_H;

E_{x,H_1} is the enthalpy exergy of the working substance at the system inlet;

$\frac{1}{2}mc_1^2$, mgz_1 are the kinetic exergy and potential exergy of the working substance at the system inlet, respectively.

The exergy output from the system is

$$E_{x,\text{out}} = E_{x,W} + E_{x,H_2} + \frac{1}{2}mc_2^2 + mgz_2 \tag{3.64}$$

Where $E_{x,W}$ is the mechanical work output from the system;

E_{x,H_2} is the enthalpy exergy of the working substance at the system outlet;

$\frac{1}{2}mc_2^2$, mgz_2 are the kinetic exergy and potential exergy of the working substance at the system outlet, respectively.

The heat input to the system is

$$E_{x,Q}=\int_1^2\left(1-\frac{T_0}{T_H}\right)\mathrm{d}Q \tag{3.65}$$

For a steady flow system, the change in exergy of the system, $\Delta E_{\mathrm{x},U}=0$, then the exergy balance equation for the steady flow system is

$$E_{x,Q}+E_{\mathrm{x},H_1}+\frac{1}{2}mc^2+mgz_1=E_{\mathrm{x},W}+E_{\mathrm{x},H_2}+\frac{1}{2}mc^2+mgz_2+E_{\mathrm{xl}} \tag{3.66}$$

The exergy loss of the system can be calculated from the exergy balance equation

$$\begin{aligned}E_{\mathrm{xl}}&=\left(E_{\mathrm{x},Q}-E_{\mathrm{x},W}\right)+\left(E_{\mathrm{x},H_1}-E_{\mathrm{x},H_2}\right)+\frac{1}{2}m\left(c_1^2-c_2^2\right)+mg\left(z_1-z_2\right)\\&=\int_1^2\left(1-\frac{T_0}{T_{\mathrm{H}}}\right)\mathrm{d}Q+\left(H_1-H_2\right)-T_0\left(S_1-S_2\right)+\\&\quad\frac{1}{2}m\left(c_1^2-c_2^2\right)+mg\left(z_1-z_2\right)-E_{\mathrm{x},W}\end{aligned} \tag{3.67}$$

3.3.1.2 Static Closed System

For a static closed system that only exchanges energy with the outside world and does not exchange mass, the exergy balance equation is

$$E_{\mathrm{x},Q}=E_{\mathrm{x},W}+\Delta E_{\mathrm{x},U}+E_{\mathrm{xl}} \tag{3.68}$$

The change in exergy of the static closed system is the change in thermodynamic energy exergy, that is

$$\Delta E_{\mathrm{x},U}=E_{\mathrm{x},U_2}-E_{\mathrm{x},U_1}=\left(U_2-U_1\right)+p_0\left(V_2-V_1\right)-T_0\left(S_2-S_1\right) \tag{3.69}$$

The mechanical work output of the system is equal to the difference between the volume work W_{s} done by the system in the process of changing from state 1 to state 2 and the expansion work $p_0(V_2-V_1)$ under environmental pressure:

$$E_{\mathrm{x},W}=W_{\mathrm{s}}-p_0\left(V_2-V_1\right)$$

Therefore, the exergy loss of the system undergoing an irreversible process is

$$E_{\mathrm{xl}}=\int_1^2\left(1-\frac{T_0}{T_{\mathrm{H}}}\right)\mathrm{d}Q+\left(U_1-U_2\right)-T_0\left(S_1-S_2\right)-W_{\mathrm{s}} \tag{3.70}$$

3.3.1.3 System Undergoing Cyclic Process

The closed system or steady flow system can complete a thermal cycle through a series of processes, and the working fluid returns to the initial state after a cycle, and the exergy of the working fluid does not change, so the exergy balance equation is

$$. \left(1-\frac{T_0}{T}\right)\mathrm{d}Q = .\ \mathrm{d}W + E_{\mathrm{xl}} \tag{3.71}$$

The exergy loss of the system after the thermal cycle is

$$E_{\mathrm{xl}} = .\left(1-\frac{T_0}{T}\right)\mathrm{d}Q - .\ \mathrm{d}W = \int_{\mathrm{H}}\left(1-\frac{T_0}{T_{\mathrm{H}}}\right)\mathrm{d}Q_1 - \int_{\mathrm{L}}\left(1-\frac{T_0}{T_{\mathrm{L}}}\right)\mathrm{d}Q_2 - W \tag{3.72}$$

If the thermal cycle is reversible, then $E_{\mathrm{xl}} = 0$, the maximum useful work of the cycle is

$$W_{\max} = \int_{\mathrm{H}}\left(1-\frac{T_0}{T_{\mathrm{H}}}\right)\mathrm{d}Q_1 - \int_{\mathrm{L}}\left(1-\frac{T_0}{T_{\mathrm{L}}}\right)\mathrm{d}Q_2 \tag{3.73}$$

If the environment is used as a cold source, then

$$E_{\mathrm{xl}} = \int_{\mathrm{H}}\left(1-\frac{T_0}{T_{\mathrm{H}}}\right)\mathrm{d}Q_1 - W \tag{3.74}$$

$$W_{\max} = \int_{\mathrm{H}}\left(1-\frac{T_0}{T_{\mathrm{H}}}\right)\mathrm{d}Q_1 \tag{3.75}$$

As can be seen from the above two formulas, the exergy loss caused by irreversible factors in the thermal cycle is

$$E_{\mathrm{xl}} = W_{\max} - W \tag{3.76}$$

That is, exergy loss refers to the part of the exergy converted into anergy caused by the irreversible process.

3.3.1.4 Heat Transfer Process

For the heat transfer process, its exergy balance equation is

$$E_{\mathrm{x,h}} = E_{\mathrm{x,c}} + E_{\mathrm{xl}} \tag{3.77}$$

Where $E_{\mathrm{x,h}}$ is the heat exergy released by the heating medium;
$E_{\mathrm{x,c}}$ is the heat exergy obtained by the heat-absorbing medium.

Taking a surface heat exchanger as an example, under the condition of environmental temperature T_0, the heat release of the hot fluid is Q_h, the heat absorption of the cold fluid is Q_c, if it is assumed that the temperatures of the hot fluid and the cold fluid remain T_h and T_c unchanged during the heat exchange process, then the heat exergy released by the hot fluid and the heat exergy obtained by the cold fluid are

$$E_{x,h} = Q_h\left(1-\frac{T_0}{T_h}\right), \quad E_{x,c} = Q_c\left(1-\frac{T_0}{T_c}\right) \tag{3.78}$$

Then the exergy loss in the heat exchange process is

$$E_{xl} = E_{x,h} - E_{x,c} = Q_h\left(1-\frac{T_0}{T_h}\right) - Q_c\left(1-\frac{T_0}{T_c}\right) \tag{3.79}$$

Considering the heat dissipation to the environment in the heat exchange process $\Delta Q = Q_h - Q_c$, the above formula can be transformed into

$$E_{xl} = \Delta Q\left(1-\frac{T_0}{T_h}\right) + Q_c T_0\left(\frac{1}{T_c}-\frac{1}{T_h}\right) \tag{3.80}$$

In the actual heat exchange process, the temperatures of the cold and hot fluids are usually changing. Assuming that the temperature of the hot fluid drops from $T_{h,2}$ to $T_{h,1}$ during the heat release process, and the temperature of the cold fluid rises from $T_{c,1'}$ to $T_{c,2'}$ during the heat absorption process, as shown in Fig. 3.4b, the heat exergy of the hot and cold fluids in the heat exchange process are

$$\begin{aligned} E_{x,h} &= \int_1^2\left(1-\frac{T_0}{T_h}\right)dQ_h = Q_h - G_h c_{p,h} T_0 \ln\frac{T_{h,2}}{T_{h,1}} \\ E_{x,c} &= \int_{1'}^{2'}\left(1-\frac{T_0}{T_c}\right)dQ_c = Q_c - G_c c_{p,c} T_0 \ln\frac{T_{c,2'}}{T_{c,1'}} \end{aligned} \tag{3.81}$$

Where G_h, G_c are the mass flow rates of the hot and cold fluids in the heat exchanger, respectively;

$c_{p,\,h}$, $c_{p,\,c}$ are the average specific heat at constant pressure of the hot and cold fluids, respectively.

The exergy loss in the heat exchange process is

$$E_{xl} = (Q_h - Q_c) - G_h c_{p,h} T_0 \ln\frac{T_{h,2}}{T_{h,1}} + G_c c_{p,c} T_0 \ln\frac{T_{c,2'}}{T_{c,1'}} \tag{3.82}$$

If the heat dissipation loss in the heat exchange process is ignored, and considering that

$$G_h c_{p,h} \Delta T_h = G_c c_{p,c} \Delta T_c \tag{3.83}$$

Where ΔT_h is the temperature drop of hot fluid in a heat exchanger, $\Delta T_h = T_{h,2} - T_{h,1}$;

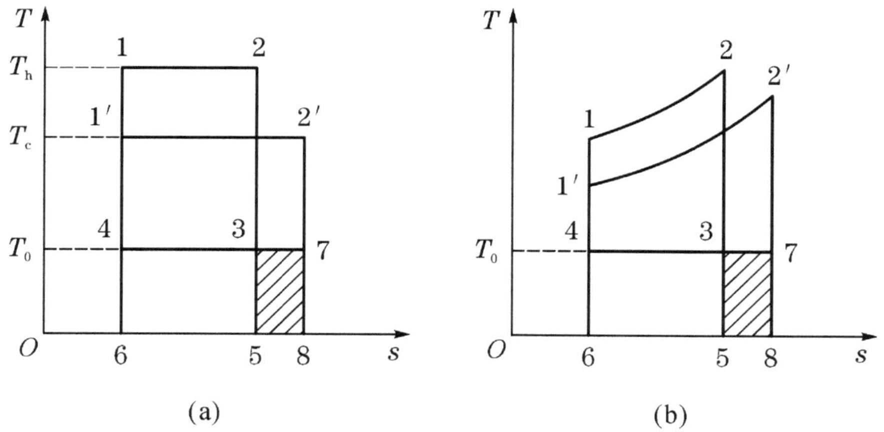

Fig. 3.4 The exergy loss of temperature difference heat exchange: (**a**) Constant heat transfer temperature; (**b**) Varied heat transfer temperature

ΔT_c is the temperature drop of cold fluid in a heat exchanger, $\Delta T_c = T_{c,2'} - T_{c,1'}$. Then the above equation can be transformed into

$$E_{xl} = G_h c_{p,h} T_0 \left(\frac{\Delta T_h}{\Delta T_c} \ln \frac{T_{c,2'}}{T_{c,1'}} - \ln \frac{T_{h,2}}{T_{h,1}} \right) \tag{3.84}$$

Figure 3.4 represents the exergy loss of temperature difference heat exchange on the *T*-*s* diagram, where (a) represents the situation where both the heat release temperature and the heat absorption temperature are constant, and (b) represents the situation where both the heat release temperature and the heat absorption temperature vary.

In the figure, 1-2 is the heat release process, and 1′-2′ is the heat absorption process. If the heat dissipation loss in the heat transfer process is not considered, then the area of region 1-2-5-6-1 is the same as the area of region 1′-2′-8-6-1′, and the exergy loss in the heat exchange process is the area enclosed by region 3-7-8-5-3.

As can be seen from Eq. (3.78), the actual exergy loss in the heat exchange process consists of two parts: one part is the external exergy loss caused by heat dissipation, and the other part is the internal exergy loss caused by the existence of a heat exchange temperature difference. If good thermal insulation measures are taken to eliminate the heat dissipation loss in the heat exchange process, i.e., $Q_h = Q_c$, then the external exergy loss can be reduced to zero. However, in order to ensure the normal progress of the heat transfer process, the heat transfer temperature difference between the hot fluid and the cold fluid must exist, so the internal exergy loss cannot be eliminated.

3.3.1.5 Adiabatic Throttling Process

When a fluid flows through a throttle valve under adiabatic conditions, assuming the ambient temperature is T_0, the exergy value of the fluid before the throttle valve is

$$E_{x,1} = (H_1 - H_0) - T_0(S_1 - S_0)$$

Where H_0, S_0 are the enthalpy and entropy of the fluid under ambient conditions.

The exergy value of the fluid after the throttle valve is

$$E_{x,2} = (H_2 - H_0) - T_0(S_2 - S_0)$$

The exergy balance equation for the adiabatic throttling process is

$$E_{x,1} = E_{x,2} + E_{xl}$$

Since the enthalpy before and after the adiabatic throttling process is unchanged, i.e., $H_1 = H_2$, the exergy loss in the adiabatic throttling process is

$$E_{xl} = E_{x,1} - E_{x,2} = T_0(S_2 - S_1) = T_0 \Delta S_{12} \tag{3.85}$$

3.3.2 Classification and Properties of Exergy Loss

All actual processes are irreversible processes, which are inevitably accompanied by various exergy losses. The size of the exergy loss reveals the degree of irreversibility of the process and reflects the perfection of energy conversion and utilization in the process. According to the location of exergy loss, it can be divided into internal exergy loss and external exergy loss. This division method has a certain relativity. For example, for an isolated system, since all irreversible factors are included within the system, the exergy loss of the system is all internal exergy loss; for a non-isolated system, irreversible factors may exist both inside and outside the system, so the exergy loss of the system consists of internal exergy loss and external exergy loss.

3.3.2.1 Internal Exergy Loss and External Exergy Loss

Internal exergy loss is the exergy loss caused by the irreversible factors of the internal process of the thermodynamic system, including throttling, fluid resistance, mechanical friction, finite temperature difference heat transfer, and concentration difference. Internal exergy loss is also known as exergy dissipation, which refers to the part of exergy that is dissipated as anergy in the irreversible process within the thermodynamic system.

External exergy loss is the exergy loss caused by the irreversible factors of the external environment of the thermodynamic system, which is related to both the system and the environmental medium. Typical external exergy losses include: exergy loss caused by poor thermal insulation of the system and heat dissipation to the environment (also known as exergy dissipation), exergy loss caused by discharging waste heat to the environment without utilization (also known as exergy discharge), etc.

3.3.2.2 The Nature of Exergy Loss

Whether it's exergy dissipation or exergy emission, no actual benefits are produced, and they are simply discarded into the environment. Therefore, from the perspective of energy conservation, under reasonable technical and economic conditions, they should be reduced to the minimum as much as possible.

Energy loss and exergy loss, in principle, have different contents. Energy loss is unscientific in nature, because according to the principle of energy conservation, energy will not disappear. However, for the loss of a certain system, or for the loss of a certain purpose, if a part of the energy is unfavorable to it in terms of its form or parameters, it can be said so. On the contrary, exergy loss can completely disappear, it is associated with energy dissipation and disappears.

In summary, exergy is a reflection of one aspect of energy, that is, its convertibility under certain environmental conditions.

The following conclusions can be drawn from the above analysis:

1. Exergy dissipation is caused by the irreversibility within the system, and the magnitude of its value depends on the degree of irreversibility of the process. The greater the degree of irreversibility of the process, the greater the exergy dissipation.
2. Since all actual processes are irreversible processes, exergy dissipation is actually unavoidable. It can only be reduced, but not completely eliminated.
3. In most cases, the dissipated exergy is high-quality energy, such as mechanical exergy dissipated due to friction, kinetic exergy dissipated due to flow resistance, and electrical exergy dissipated due to resistance.

However, internal exergy loss is not always the case. Although the internal irreversibility causes part of the exergy to transform into anergy, this transformation is closely related to the driving force of the process. In order for the process to proceed at a certain rate, there must be various corresponding potential differences as driving forces to overcome certain resistances. For example, the flow of fluid requires a pressure difference as a driving force to overcome flow resistance; heat transfer requires a temperature difference as a driving force to overcome thermal resistance; and the migration of mass and the progress of chemical reactions require a chemical potential difference as a driving force. The existence of various potential differences will inevitably lead to internal exergy loss, the greater the potential difference, the greater the exergy loss. If you want to reduce exergy loss, you must reduce the potential difference, which will also reduce the driving force of the process. In this sense, the internal exergy loss in the process is unavoidable, and can be understood as the price of exergy paid to promote the process.

Example 3.5 The inlet steam parameters of the steam turbine are 3.5 MPa and 435 °C, with a flow rate of 11 kg/s; steam is extracted at a flow rate of 2 kg/s when it expands to 0.5 MPa and 180 °C in the steam turbine, and the remaining steam continues to expand to 0.005 MPa and a dryness of 0.82 when it is discharged from the steam turbine. Assuming the environmental state is 0.1 MPa and 20 °C, ignore the kinetic exergy and potential exergy of the steam. Try to calculate:

1. The exergy values of the fresh steam, extraction steam, and exhaust steam;
2. The maximum useful power output that the steam turbine could possibly have;
3. The actual power output of the steam turbine;
4. The exergy loss caused by the irreversibility of the expansion process in the steam turbine.

Solution According to the fresh steam parameters $p_1 = 3.5$ MPa, $t_1 = 435$ °C, look up the steam table

$$h_1 = 3304.1\,\text{kJ/kg}, s_1 = 6.9599\,\text{kJ/(kg}\cdot\text{K)}$$

According to the extraction steam parameters $p_2 = 0.5$ MPa, $t_2 = 180$ °C, look up the steam table

$$h_2 = 2811.4\,\text{kJ/kg}, s_1 = 6.9647\,\text{kJ/(kg}\cdot\text{K)}$$

According to the exhaust steam parameters $p_3 = 0.005$ MPa, $x_3 = 0.82$, the steam table is consulted

$$h_3 = 137.77 + 0.82 \times (2561.6 \quad 137.77) = 2125.3\,\text{kJ/kg}$$
$$s_3 = 0.4763 + 0.82 \times (8.3960 \quad 0.4763) = 6.9705\,\text{kJ/(kg}\cdot\text{K)}$$

The exhaust steam flow rate is

$$G_3 = G_1 - G_2 = 11 - 2 = 9\,\text{kg/s}$$

According to the environmental conditions $p_0 = 0.1$ MPa, $t_0 = 20$ °C, the water and steam tables are consulted

$$h_0 = 84.0\,\text{kJ/kg}, s_0 = 0.2963\,\text{kJ/(kg}\cdot\text{K)}$$

1. Calculate the exergy values of each steam flow

   The exergy value of the new steam:

$$\begin{aligned}E_{x,1} &= G_1\left[(h_1 - h_0) - T_0(s_1 - s_0)\right]\\ &= 11 \times \left[(3304.1 \quad 84.0) \quad (20 + 273) \times (6.9599 \quad 0.2963)\right]\\ &= 13944.3\,\text{kW}\end{aligned}$$

The exergy value of the extracted steam:

$$\begin{aligned}E_{x,2} &= G_2\left[(h_2 - h_0) - T_0(s_2 - s_0)\right]\\ &= 2 \times \left[(2811.4 \quad 84.0) \quad (20 + 273) \times (6.9647 \quad 0.2963)\right]\\ &= 1547.1\,\text{kW}\end{aligned}$$

The exergy value of the exhaust steam:

$$\begin{aligned} E_{x,3} &= G_3\left[(h_3 - h_0) - T_0(s_3 - s_0)\right] \\ &= 9\times\left[(2125.3 \quad 84.0) \quad (20+273)\times(6.9705 \quad 0.2963)\right] \\ &= 771.8\,\text{kW} \end{aligned}$$

2. Calculate the maximum useful power output of the steam turbine
   The exergy balance equation under reversible conditions can be obtained

$$N_{max} = E_{x,1} - E_{x,2} - E_{x,3} = 13944.3 \quad 1547.1 \quad 771.8 = 11625.4\,\text{kW}$$

3. Calculate the actual power output of the steam turbine
   The energy equation of the steam turbine can be obtained

$$\begin{aligned} N_t &= G_1h_1 - G_2h_2 - G_3h_3 = 11\times3304.1 \quad 2\times2811.4 \quad 9\times2125.3 \\ &= 11594.6\,\text{kW} \end{aligned}$$

4. Calculate the exergy loss caused by the irreversibility of the expansion process
   The exergy balance equation of the actual process can be obtained

$$E_{xl} = E_{x,1} - E_{x,2} - E_{x,3} - N_t = N_{max} - N_t = 11625.4 \quad 11594.6 = 30.8\,\text{kW}$$

3.3.3 Exergy Efficiency

Through exergy balance, the exergy loss of each thermodynamic process and the entire system or equipment can be calculated, which only indicates the absolute amount of functional loss. If we want to fully analyze the energy conversion and utilization of each process and system or equipment, it is obviously not enough to only determine the exergy loss, because we cannot directly analyze, compare and judge the exergy utilization of the process and system or equipment, and the impact of each thermodynamic process on the entire system or equipment. Therefore, it is very necessary to determine the relative degree of energy loss of the process and system or equipment. To evaluate the thermodynamic perfection of the thermodynamic process and system or equipment, the concept of exergy efficiency was introduced.

Exergy efficiency is a measure of the thermodynamic perfection of a process, device, or system based on the analysis of the second law of thermodynamics. It is related to the thermal efficiency proposed by the analysis of the first law of thermodynamics, but it can more profoundly reveal the essence of energy conversion, utilization, and loss. The general definition of exergy efficiency is

$$\eta_{\mathrm{ex}} = \frac{\text{exergy efficiently utilized by the system}}{\text{exergy supplied to the system from outside}} = \frac{E_{\mathrm{x,eff}}}{E_{\mathrm{x,sup}}} \tag{3.86}$$

For different thermodynamic systems or equipment, the determination of the exergy effectively utilized and the exergy supplied by the outside world may vary. Even for the same system or equipment, different exergy efficiencies may be determined due to different understandings, especially when there are more exergy flows entering and leaving the system or equipment. Therefore, the definition of exergy efficiency has a certain uncertainty.

3.3.3.1 Transfer Exergy Efficiency

Transfer exergy efficiency is a commonly used exergy efficiency, defined as the ratio of the total output exergy to the total input exergy of a thermodynamic system or device, that is

$$\eta_{\mathrm{ex}} = \frac{\text{Total exergy output of the system}}{\text{Total exergy input of the system}} = \frac{\sum_{i=1}^{n} \left(E_{x,i}\right)_{\mathrm{out}}}{\sum_{j=1}^{n} \left(E_{x,j}\right)_{\mathrm{in}}} \tag{3.87}$$

The total input exergy and total output exergy in the above formula are the total exergy brought into and out of the system in the form of heat, work, material flow, and other possible forms.

Transfer exergy efficiency is actually the ratio of the exergy obtained through a system or device to the exergy entrusted to the system or device for transfer. Its definition is based on the exergy balance equation and has the following characteristics:

1. Generally applicable to all processes;
2. Only characterizes the internal thermodynamic perfection of the system or device;
3. Does not reflect the nature and purpose of the process in the system or device.

The shortcoming of exergy efficiency is that it does not distinguish which parts of the system output exergy are useful and which are not, therefore it cannot differentiate various situations, truly reflecting the thermodynamic perfection of the system or device. Essentially, exergy efficiency only reflects the single impact of the irreversibility of the process in the system or device on the thermodynamic perfection, and is more suitable for energy transfer processes such as heat transfer, fluid transmission, and shaft power transfer.

For example, for the condenser in the thermal system of a NPP, if the heat dissipation during operation is ignored, the exergy efficiency of the condenser is

$$\eta_{ex} = \frac{\text{the heat exergy absorbed by the circulating cooling water}}{\text{the heat exergy released from condensation of turbine exhaust steam}}$$

In fact, the heat absorbed by the circulating cooling water passing through the condenser is ultimately brought into the environment by the circulating cooling water and cannot be utilized at all. Therefore, according to the definition of formula (3.80), the exergy efficiency of the condenser should be equal to zero.

From this, it can be seen that the output exergy of the thermal system or device is not all effectively utilized. If this fact is ignored, the economy of the thermal system cannot be correctly reflected.

3.3.3.2 Purpose Exergy Efficiency

From the perspective of energy utilization, the processes carried out in the thermal system or device are all for the purpose of energy conversion or transmission. According to this purpose, it can be determined which parts of the output exergy are effectively utilized and which are not, thereby determining the exergy efficiency of the thermal system or device. The ratio of the exergy obtained for the purpose to the exergy supplied by the outside world is usually called the purpose exergy efficiency, and the definition is the same as formula (3.80).

Still taking the condenser as an example, its main function is to serve as a heat trap in the thermal system of the power plant, realizing the constant temperature heat release process in the thermal cycle. The exergy supplied by the outside world is equal to the heat released by the complete condensation of the turbine exhaust steam in the condenser, and the output exergy is discharged into the environment, so the exergy that can be effectively utilized in the heat transfer process inside the condenser is zero.

The exergy efficiency mentioned later in this book, if not specifically stated, refers to the purpose exergy efficiency.

Table 3.6 lists the expressions of the supplied exergy and effectively utilized exergy of common thermal engineering equipment or processes.

3.3.3.3 Relationship Between Exergy Efficiency and Thermal Efficiency

There are both differences and connections between exergy efficiency and thermal efficiency, mainly manifested in the following aspects:

1. Thermal efficiency is based on the first law of thermodynamics, representing the ratio of various forms of energy obtained from the process to the various forms of energy consumed. Although both the numerator and the denominator involve energy, their grades may be different, so this efficiency inevitably leads to the equal view of high-grade energy and low-grade energy. For this reason, thermal efficiency is not an ideal, unified scientific measure to measure and evaluate the rationality of energy use in the process. Since energy has different qualities, in

Table 3.6 The supplied exergy and effectively utilized exergy of common thermal engineering equipment or processes

No.	Thermal equipment	Supplied exergy	Effectively utilized exergy
1	Stove	Fuels exergy $E_{x,f}$	Vapor absorption exergy$G_s(e_{x,fh} - e_{x,fw})$
2	Combustion chamber	$E_{x,f}$	$G_g e_{x,g} - G_a e_{x,a}$
3	Steam turbine	$G_s(e_{x,1} - e_{x,2})$	N_i
4	Compressor or pump	Engine power N_p	Fluid absorption exergy$G_f(e_{x,2} - e_{x,1})$
5	Throttle	Pre-valve fluid exergy $G_f e_{x,1}$	Post valve fluid $G_f e_{x,2}$
6	Compression chillers	N_p	$\left(1-\frac{T_0}{T_1}\right)Q_1$
7	Absorption chillers	$\int_1^2\left(1-\frac{T_0}{T_h}\right)dQ_h$	$\left(1-\frac{T_0}{T_1}\right)Q_1$
8	Vapor injection chillers	$G_s(e_{x,1} - e_{x,2})$	$\left(1-\frac{T_0}{T_1}\right)Q$
9	Vapor compression to heat	N_p	$\left(1-\frac{T_0}{T_h}\right)Q$
10	Hybrid heat exchanger	$G_h(e_{x,h1} - e_{x,12})$	$G_l(e_{x,12} - e_{x,11})$
11	Surface heat exchanger	$G_h(e_{x,h1} - e_{x,h2})$	$G_l(e_{x,12} - e_{x,11})$
12	Warm gas Heating	Steam gives off exergy $G_s(e_{x,1} - e_{x,2})$	$\left(1-\frac{T_0}{T_h}\right)Q$
13	Electric heating	Power consumption Ne	$\left(1-\frac{T_0}{T_h}\right)Q$

order to obtain the true efficiency of the process, the numerator and denominator must belong to the true same category, that is, they must be compared with equivalent energy, which is exergy efficiency. Because the same amount of exergy, regardless of which kind of material flow or energy flow it possesses, theoretically they are all the same grade, the same use value, the same amount of energy, they are all strictly the same category.

2. The conclusions they analyze vary greatly. For example, for boilers, the thermal efficiency of large industrial boilers can now reach over 90%. From the perspective of heat balance, it is quite perfect. However, from the perspective of exergy balance, the boiler is very imperfect. This is because the heat loss of the boiler is not large in quantity, but in terms of energy quality, there is a large temperature difference between the flue gas and the working fluid, and this high temperature difference heat transfer causes a large exergy loss. Also, in the analysis of the condenser, from the perspective of heat balance, it is believed that the amount of heat dissipation is too much, the heat loss is large, and it is the focus of energy saving. However, from the perspective of exergy balance, although the heat dissipation is large, its quality is very low, so the proportion of exergy loss is very small, and the real potential for energy saving is not here.

3. Thermal efficiency is not equal to exergy efficiency, and it is not reasonable to use it to measure the pros and cons of the process or system; exergy efficiency truly and essentially reflects the quality and quantity of energy conversion and utilization.

However, there is a certain connection between them, because both are studying the same object, so there must be an inherent connection. In addition, the connection between heat balance and exergy balance methods is also reflected in that their analysis results can be mutually checked, and there is a certain quantitative relationship between thermal efficiency and exergy efficiency.

Exergy efficiency is simpler and clearer than using the size of power loss or entropy generation to evaluate the process, and it can more accurately reflect the irreversibility of the process. It has become a major indicator of process thermodynamic analysis.

Therefore, exergy efficiency can accurately and quantitatively reflect the irreversibility of the process, and is a unified measure of the thermodynamic perfection of various thermal processes and energy-using equipment.

Using exergy efficiency to evaluate and measure the perfection of energy utilization of various thermal engineering and energy-using equipment is a very meaningful job. This is because the average value of the exergy efficiency of a certain type of thermal engineering and energy-using equipment directly reflects the pros and cons of the energy conversion of this type of equipment. By analyzing the composition of exergy loss of various equipment and comparing the average level of exergy, it can reveal the weak link in the energy conversion process, that is, where the process is more irreversible, thereby pointing out the target and direction of energy-saving transformation.

Because exergy efficiency reflects the thermodynamic irreversibility of the process and system or equipment from both quality and quantity, it is currently widely used in various thermal analyses in energy and energy-saving engineering. There are mainly the following aspects:

Determine the impact of exergy loss of various thermodynamic processes on exergy efficiency, so as to determine their proportion and status in the entire system, and point out the main contradictions for improving thermodynamic processes and increasing energy utilization rate; analyze the conditions of each unit process on exergy loss The impact, in order to determine the measures to be taken to reduce exergy loss; as a unified measurement indicator, used to evaluate the pros and cons of thermodynamic processes or systems and equipment.

Exergy efficiency points out how much potential there is to improve the utilization of exergy in a certain system or equipment, but exergy efficiency cannot directly point out the distribution of exergy loss in the entire system or equipment and the proportion of exergy loss in each link. Therefore, it cannot directly reveal the weak link. By defining the exergy loss coefficient or exergy loss rate, it can reveal the location and degree of exergy transformation in the process, and it can complement exergy efficiency.

3.3.4 Exergy Loss Rate and Exergy Loss Coefficient

3.3.4.1 Exergy Loss Rate

The total exergy loss in the system is $\sum E_{xl,\,i}$, where the proportion of the local exergy loss $E_{xl,\,i}$ of a certain link or process is called the exergy loss rate of this link or process, that is

$$d_i = \frac{E_{xl,i}}{\Sigma E_{xl,i}} \tag{3.88}$$

The exergy loss rate represents the share of local exergy loss relative to the total exergy loss. It has many advantages in comparing and evaluating the exergy loss of each link or process in the thermodynamic system and its impact on the total exergy loss. However, when comparing and evaluating the exergy utilization and loss of different systems, due to the different total exergy losses of each system, the exergy loss rate based on the total exergy loss as a comparison benchmark has no comparability between different systems and cannot represent the size of exergy loss and the degree of exergy utilization of different systems. Moreover, there is no clear and direct relationship between the exergy loss rate and exergy efficiency, and the exergy efficiency cannot be directly obtained from the exergy loss rate.

3.3.4.2 Exergy Loss Coefficient

The proportion of the local exergy loss of the *j*th process or link in the thermodynamic system or equipment to the total exergy supplied by the entire system or equipment is called the exergy loss coefficient of this process or link, that is

$$\xi_j = \frac{\text{Localized exergy loss from process or link j}}{\text{Total exergy supplied to the system or equipment}} = \frac{E_{xl,j}}{E_{x,sup}} \tag{3.89}$$

The total exergy loss coefficient of the system or equipment is

$$\xi = \frac{E_{xl}}{E_{x,sup}} = \sum_{j=1}^{n} \frac{E_{xl,j}}{E_{x,sup}} = \sum_{j=1}^{n} \xi_j \tag{3.90}$$

According to the exergy balance equation of the system

$$E_{x,sup} = E_{x,eff} + \sum_{j=1}^{n} E_{xl,j}$$

The relationship between system exergy efficiency and process exergy loss coefficient can be obtained

$$1 = \frac{E_{x,\mathrm{eff}}}{E_{x,\mathrm{sup}}} + \sum_{j=1}^{n} \frac{E_{xl,j}}{E_{x,\mathrm{sup}}} = \eta_{\mathrm{ex}} + \sum_{j=1}^{n} \xi_j = \eta_{\mathrm{ex}} + \xi \tag{3.91}$$

This is another form of the exergy balance equation.

This formula provides a method for analyzing thermodynamic systems. By analyzing the exergy loss of each process in the thermodynamic system, it can reflect the weak links in the energy use process of the thermodynamic system. At the same time, the rationality and effectiveness of system energy use can be conveniently determined using the above formula.

3.4 Exergy Analysis Method [24–26]

The exergy analysis method is a second type of thermodynamic analysis method based on the first law of thermodynamics and the second law of thermodynamics. It takes the thermodynamic system or process as the analysis object, calculates the exergy values of various material flows and energy flows on the basis of mass balance and energy balance, determines the exergy loss and its distribution through exergy balance, finds out the causes of exergy loss and the weak links in energy utilization, thereby pointing out the correct direction for energy saving and consumption reduction, and improving energy utilization rate.

The main contents of the exergy analysis method are:

1. Carry out mass balance and energy balance calculations, determine the input and output system's working fluid flow, energy flow and the state parameters of the working fluid (such as temperature, pressure, etc.);
2. Calculate the exergy of material flow and heat flow;
3. Determine the exergy loss of each process by the exergy balance equation;
4. Determine the exergy efficiency of the system.

3.4.1 Unit Method of Exergy Analysis

When it is necessary to conduct a detailed analysis of the energy system, it is far from enough to simply evaluate the system as a whole. It is necessary to analyze in detail the size of the exergy loss of all links that make up the energy system, and their status in the total exergy loss, so as to determine the location of the largest loss and find out the direction of efforts to improve the thermal economy of the system.

In the actual analysis process, a complete energy system is always divided into several constituent units. By analyzing each constituent unit one by one, the overall thermal economic perfection of the system is mastered. This method of dividing an independent energy system into several units and calculating and analyzing the exergy loss of each unit one by one is called the unit method of exergy analysis, which includes the following three stages:

1. Divide the energy system into units as needed;
2. Perform exergy analysis calculations on each divided unit;
3. Represent the calculation results in the form of tables or line graphs.

The division of units in the energy system is an important part of unit method analysis and is the basis of the entire calculation analysis. How to divide the units depends on the different requirements of different problems, that is, it depends solely on the tasks and purposes of system analysis. Generally speaking, the principle of dividing units in the energy system is:

1. When dividing a given system into units, the number of units divided depends on the level of detail of the research problem. The more detailed the research requirements, the more units are divided, and the more detailed the analysis results given.
2. The units that make up the system can be main and auxiliary equipment, or they can be processes involved in the system's connection channels. Therefore, the specific units that make up the system can be divided into two categories: equipment units and process units, the former can include the latter.

3.4.2 *Exergy Analysis of Typical Units*

3.4.2.1 Expansion Process

In the energy system, expansion and compression are common processes. The expansion process is generally carried out under conditions where its starting point is higher than the ambient temperature, with the aim of converting the exergy of the flow into mechanical work. In the thermal system of a power plant, this conversion process takes place in a steam turbine.

Taking the single-stage expansion of unit mass steam as an example, if the steam turbine is adiabatic to the environment, the exergy balance equation is

$$e_{x,1} = e_{x,2} + w + e_{xl} \tag{3.92}$$

Where $e_{x,1}$, $e_{x,2}$ are the specific exergy of the steam before and after expansion. w is the work done by the unit mass steam during the expansion process, $w = h_1 - h_2.e_{xl}$ is the specific exergy loss during the expansion process.

The efficiency of this steam turbine is

$$\eta_{ex} = \frac{w}{e_{x,1} - e_{x,2}} = 1 - \frac{e_{xl}}{e_{x,1} - e_{x,2}}$$

Under adiabatic conditions, there is no external exergy loss in the system, then

$$e_{xl} = T_0\left(s_2 - s_1\right)$$

If mechanical friction is not considered, considering that $e_{x,1} - e_{x,2} = (h_1 - h_2) - T_0(s_1 - s_2)$, the exergy efficiency can be expressed as

$$\eta_{ex} = \frac{h_1 - h_2}{\left(h_1 - h_2\right) - T_0\left(s_1 - s_2\right)} = \frac{h_1 - h_2}{\left(h_1 - h_2\right) + e_{xl}} \tag{3.93}$$

3.4.2.2 Throttling Process

In the thermal system of a power plant, the throttling process is ubiquitous. Intentionally caused throttling processes are often used for two purposes: one is to reduce the pressure of the working fluid with higher pressure through throttling to meet the process requirements of the production process; the other is to reduce the quantity of working fluid exergy by reducing the pressure of the working fluid through throttling, thereby achieving the purpose of reducing the output power of the steam turbine. From a thermal economic point of view, this is a waste of exergy, but because it is a simple and easy power regulation method, it is still widely used in engineering.

For the adiabatic throttling process of unit mass working fluid, its exergy balance equation is

$$e_{x,1} = e_{x,2} + e_{xl} \tag{3.94}$$

Where $e_{x,1}$, $e_{x,2}$ are the specific exergy of the working fluid before and after adiabatic throttling;

e_{xl} is the specific exergy loss during the throttling process.

The exergy loss of the throttling process can be expressed as

$$e_{xl} = \left(h_1 - h_2\right) - T_0\left(s_1 - s_2\right)$$

Where h_1, h_2 are respectively the enthalpy of the unit mass working fluid before and after throttling, respectively, kJ/kg;

s_1, s_2 are respectively represent the entropy of the working substance per unit mass before and after throttling, respectively, kJ/(kg·K).

Since the enthalpy of the working substance remains constant during the adiabatic throttling process, i.e., $h_1 = h_2$, the above formula can be transformed into

$$e_{xl} = T_0\left(s_2 - s_1\right)$$

The entropy loss during the throttling process is caused by viscous friction, and the entropy efficiency of this process is

$$\eta_{ex} = \frac{e_{x,2}}{e_{x,1}} \tag{3.95}$$

3.4.2.3 Compression Process

The starting temperature of the compression process is generally higher than the ambient temperature. Like the expansion process, compression also has single-stage compression and multi-stage compression.

Taking the single-stage compression of the working substance per unit mass as an example, its entropy balance equation is

$$e_{x,1} + w_c = e_{x,2} + e_{x,q} + e_{xl} \tag{3.96}$$

Where $e_{x,1}$, $e_{x,2}$ are respectively represent the specific entropy of the working substance before and after compression, respectively;

w_c represents the external work consumed by compressing the working substance per unit mass;

$e_{x,q}$ represents the specific heat entropy transferred to the outside world during the compression process;

e_{xl} represents the specific entropy loss during the compression process.

If the compression process is adiabatic, then $e_{x,q} = 0$, the above formula becomes

$$w_c = e_{x,2} - e_{x,1} + e_{xl}$$

Then the entropy efficiency of the adiabatic compression process is

$$\eta_{ex} = \frac{e_{x,2} - e_{x,1}}{w_c} = 1 - \frac{e_{xl}}{w_c} \tag{3.97}$$

3.4.2.4 Heat Transfer Process

The heat transfer process can generally be divided into two types. The first type is to transfer heat from one heat-carrying medium to another through conduction, convection heat transfer or radiation heat transfer, thereby realizing energy exchange. This is the most common process in energy systems; the second type of heat transfer process is characterized by transferring heat from the system to the environment, or from the environment into the system.

The irreversibility of the heat transfer process caused by finite temperature difference heat exchange is the most basic form of irreversible factors causing entropy loss in heat exchangers. This is the result of the decrease in thermal energy quality caused by heat transfer from high temperature to low temperature. In actual heat exchangers, in order to maintain a certain heat transfer rate, a sufficient temperature difference between the heat transfer media must be ensured, so this form of irreversibility is inevitable.

The entropy balance equation for the internal heat transfer process of the heat exchanger is

$$E_{x,Q_1} = E_{x,Q_2} + E_{x,\Delta Q} + E_{xl} \tag{3.98}$$

Where E_{x,Q_1}, E_{x,Q_2} are respectively represent the heat entropy released by the hot fluid and the heat entropy absorbed by the cold fluid, respectively;

$E_{x,\Delta Q}$ is represents the heat entropy lost to the environment;

E_{xl} is represents the entropy loss during the heat transfer process.

Since the heat entropy lost to the environment cannot be effectively utilized, it should also be included in the entropy loss, then the entropy efficiency of the heat exchanger is

$$\eta_{ex} = \frac{E_{x,Q_2}}{E_{x,Q_1}} = \frac{\int \left(1 - \frac{T_0}{T_2}\right) dQ_2}{\int \left(1 - \frac{T_0}{T_1}\right) dQ_1} \tag{3.99}$$

Example 3.6 A well-insulated shell-and-tube heat exchanger, the flow rate of the heating medium on the tube side is 0.05 kg/s, the inlet and outlet temperatures are respectively 150 °C and 35 °C, the inlet and outlet temperatures of the heat-absorbing medium on the shell side are respectively 25 °C and 110 °C. The average specific heat at constant pressure of the heating medium and the heat-absorbing medium are respectively 4.35 kJ/(kg·K) and 4.69 kJ/(kg·K), and the ambient temperature is 25 °C, try to determine the energy loss and efficiency of the heat exchanger during the heat transfer process.

Solution During the heat transfer process, the heat released by the heating medium is

$$Q_1 = G_1 c_{p1} \left(t_{1i} - t_{1o}\right) = 0.05 \times 4.35 \times (150 - 35) = 25\,\text{kW}$$

Ignoring the heat dissipation loss during the heat transfer process, the heat absorbed by the heat-absorbing medium equals the heat released by the heat-releasing medium, that is, $Q_2 = Q_1$.

According to the heat absorption equation of the heat-absorbing medium

$$Q_2 = G_2 c_{p2} \left(t_{2o} - t_{2i}\right)$$

The mass flow rate of the heat-absorbing medium is obtained

$$G_2 = \frac{Q_2}{c_{p2}\left(t_{2o} - t_{2i}\right)} = \frac{25}{4.69 \times (110 - 25)} = 0.0627\,\text{kg/s}$$

The exergy released by the heating medium is

$$E_{x1} = \int_{T_{1o}}^{T_{1i}} \left(1 - \frac{T_0}{T}\right) G_1 c_{p1} \mathrm{d}T = Q_1 - G_1 c_{p1} T_0 \ln \frac{T_{1i}}{T_{1o}}$$
$$= 25 - 0.05 \times 4.35 \times (25 + 273.15) \ln \frac{150 + 273.15}{35 + 273.15}$$
$$= 4.43\,\mathrm{kW}$$

The exergy obtained by the heat-absorbing medium is

$$E_{x2} = \int_{T_{2i}}^{T_{2o}} \left(1 - \frac{T_0}{T}\right) G_2 c_{p2} \mathrm{d}T = Q_2 - G_1 c_{p2} T_0 \ln \frac{T_{2o}}{T_{2i}}$$
$$= 25 - 0.0627 \times 4.69 \times (25 + 273.15) \ln \frac{110 + 273.15}{25 + 273.15}$$
$$= 3\,\mathrm{kW}$$

The exergy loss during the heat transfer process is

$$E_{xl} = E_{x1} - E_{x2} = 4.43 - 3 = 1.43\,\mathrm{kW}$$

The exergy efficiency of the heat exchanger is

$$\eta_{ex} = \frac{E_{x2}}{E_{x1}} \times 100\% = \frac{3}{4.43} \times 100\% = 67.72\%$$

3.4.3 Series and Parallel Models of Units

Any complex system can be considered as composed of several subsystems, and each subsystem can be divided into several subsystems, each of which can be considered as composed of several basic units (or devices). The combination of basic units is nothing more than the two most basic forms of parallel and series.

If the input exergy of the latter unit is the effective output of the previous unit, or the exergy of the latter unit as a cost is the exergy of the previous unit as a benefit, it is called series. If the inputs and effective outputs of several units are parallel to each other, or the exergy of several units as costs and benefits are parallel to each other, it is called parallel.

The so-called parallel or series here is not divided according to the direction or combination of the working substance flow, so it may be consistent with the natural flow direction of the working substance, or it may not be consistent. In addition, in the parallel combination, the direction of the cost exergy and benefit exergy of several units can be exactly the same, or it can be opposite, the former is called the same direction parallel, and the latter is called the reverse direction parallel.

According to the different combination methods of parallel and series, energy systems can generally be divided into "parallel-series combination" and "series-parallel combination" two different basic combination models. Complex systems are often seen as a recombination of these two basic combination models in parallel or series.

In the exergy analysis of the thermodynamic system, in order to simplify the problem, the concept of "black box model" is often used. The black box model of the device is to regard the device as a system surrounded by opaque boundaries. The boundary is represented by a solid line, the input and output are represented by solid lines with arrows, and the irreversible loss inside the device is represented by a dotted line with arrows. This is a method to study the macro characteristics of the device's energy use process with the help of input and output energy flow information, but the exergy loss coefficient of each specific process inside the device cannot be determined.

The concept opposite to the "black box model" is the "white box model", which regards the analysis object as a system surrounded by transparent boundaries, and can analyze each process in the system one by one, and calculate the exergy loss coefficient of each process.

If all the devices in the thermodynamic system are regarded as black boxes, and the black boxes are connected by exergy flow lines to form a network, this analysis model is called the black box network model or the gray box model.

3.4.3.1 Black Box Series Model

As shown in Fig. 3.5, the entire system is composed of several units connected in series, and the effective output exergy of the previous unit is supplied to the next unit, that is, $E_{x,\text{eff}}(i) = E_{x,\text{sup}}(i + 1)$.

The exergy balance equation and exergy efficiency of each unit are

$$E_{x,\text{sup}}(1) = E_{x,\text{eff}}(1) + E_{xl}(1), \eta_{ex}(1) = \frac{E_{x,\text{eff}}(1)}{E_{x,\text{sup}}(1)}$$

$$E_{x,\text{sup}}(2) = E_{x,\text{eff}}(2) + E_{xl}(2), \eta_{ex}(2) = \frac{E_{x,\text{eff}}(2)}{E_{x,\text{sup}}(2)}$$

$$\cdots$$

$$E_{x,\text{sup}}(n) = E_{x,\text{eff}}(n) + E_{xl}(n), \eta_{ex}(n) = \frac{E_{x,\text{eff}}(n)}{E_{x,\text{sup}}(n)}$$

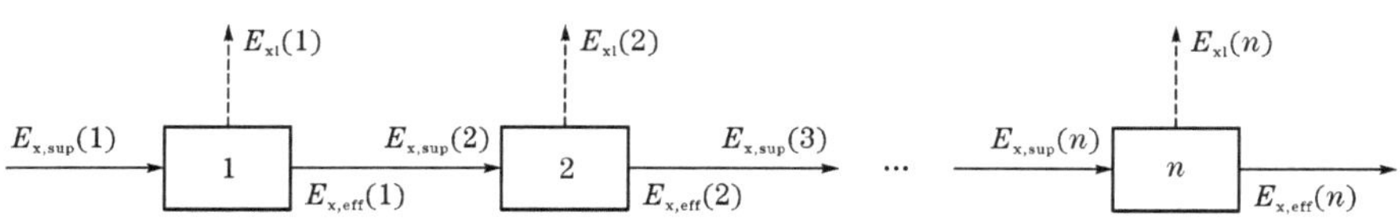

Fig. 3.5 Series unit model

Except for the first unit, no other units have received any supply energy from outside the system, so the total supply energy of the system is the supply energy of the first unit, i.e., $E_{x,sup} = E_{x,sup}(1)$; the total effective energy used by the system is the effective output energy of the last unit, i.e., $E_{x,eff} = E_{x,eff}(n)$.

Therefore, the energy balance equation of the entire system is

$$E_{x,sup}(1) = E_{x,eff}(n) + \sum_{i=1}^{n} E_{xl}(i) \tag{3.100}$$

The energy efficiency of the entire system is

$$\eta_{ex} = \frac{E_{x,eff}(n)}{E_{x,sup}(1)} = \frac{E_{x,eff}(1)}{E_{x,sup}(1)} \frac{E_{x,eff}(2)}{E_{x,sup}(2)} \cdots \frac{E_{x,eff}(n)}{E_{x,sup}(n)} = \eta_{ex}(1)\eta_{ex}(2)\cdots\eta_{ex}(n) \tag{3.101}$$

The energy loss coefficient of the ith unit in the system is

$$\xi_i = \frac{E_{xl}(i)}{E_{x,sup}(1)} = \frac{E_{x,sup}(i) - E_{x,eff}(i)}{E_{x,sup}(1)} \tag{3.102}$$

The relationship between the energy efficiency of the entire system and the energy loss coefficients of each unit is

$$\eta_{ex} = 1 - \sum_{i=1}^{n} \xi_i \tag{3.103}$$

3.4.3.2 Black Box Parallel Model

As shown in Fig. 3.6, the entire system is composed of several parallel units, each of which receives supply energy from the outside world. The sum of the effective output energy of all units is the effective output energy of the entire system.

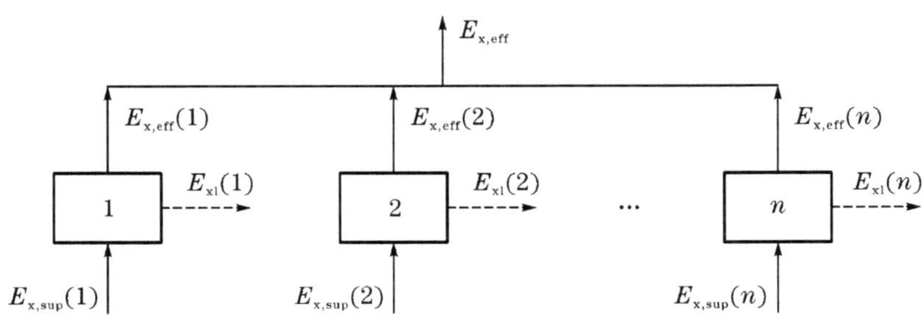

Fig. 3.6 Parallel unit model

The energy balance equation and energy efficiency of each unit are

$$E_{x,sup}(1)=E_{x,eff}(1)+E_{xl}(1),\eta_{ex}(1)=\frac{E_{x,eff}(1)}{E_{x,sup}(1)}$$

$$E_{x,sup}(2)=E_{x,eff}(2)+E_{xl}(2),\eta_{ex}(2)=\frac{E_{x,eff}(2)}{E_{x,sup}(2)}$$

$$\cdots$$

$$E_{x,sup}(n)=E_{x,eff}(n)+E_{xl}(n),\eta_{ex}(n)=\frac{E_{x,eff}(n)}{E_{x,sup}(n)}$$

The energy balance equation and energy efficiency of the entire system are

$$\sum_{i=1}^{n}E_{x,sup}(i)=\sum_{i=1}^{n}E_{x,eff}(i)+\sum_{i=1}^{n}E_{xl}(i) \tag{3.104}$$

$$\eta_{ex}=\frac{\sum_{i=1}^{n}E_{x,eff}(i)}{\sum_{i=1}^{n}E_{x,sup}(i)} \tag{3.105}$$

The energy loss coefficient of the *i*th device in the system is

$$\xi_i=\frac{E_{xl}(i)}{\sum_{i=1}^{n}E_{x,sup}(i)} \tag{3.106}$$

The relationship between the energy efficiency of the entire system and the energy loss coefficients of each unit is

$$\eta_{ex}=1-\sum_{i=1}^{n}\xi_i \tag{3.107}$$

3.5 Principle of Energy Level Analysis

The concept of energy level is based on energy analysis, used to characterize the quality of energy. It means the proportion of energy value in the unit energy provided to the system, i.e.,

$$\lambda=\frac{E_x}{E} \tag{3.108}$$

Where λ is energy level coefficient. E is energy provided to the system;

E_x is the energy value contained in the energy provided to the system.

The energy level coefficient is a measure of energy quality, and its size represents the level of energy. Obviously, for high-level energy such as mechanical energy, electrical energy, nuclear energy, $\lambda = 1$; for low-level energy such as thermal energy, $0 < \lambda < 1$; for heat in the natural environment (atmosphere, natural water, earth), $\lambda = 0$, which is rigid energy or silent energy.

The energy level is defined as the ratio of energy value to energy. It should be noted that this formula only holds when energy and energy take the same calculation basis. Energy is a relative value calculated based on a given environment, and the basis for energy calculation has become a convention with relative invariance. Any system in equilibrium with the given environment has zero energy. Energy is also a relative value calculated based on a human-selected state, but the basis for energy calculation is different from the basis for energy calculation. The basis for energy calculation is more arbitrary. In order to make the energy level truly represent the quality of energy, the calculation of energy and energy should have the same starting point (basis). When the calculation basis of energy and energy is different, the energy under any basis should be converted to the energy when it is consistent with the energy calculation basis, so that the definition of energy level equal to the ratio of energy value to energy can be used to calculate the energy level. Usually, the environmental state is taken as the basis for energy and energy calculation. According to the definition of energy level, when the system is in the environmental state, its energy level is 0.

Since energy has both quantity and quality characteristics, it is necessary to pay attention to the full utilization of energy in quantity and the reasonable matching of quality in the process of energy use. Energy analysis reveals the thermodynamic perfection of the energy use process, and the rationality of energy use in large systems can only be correctly evaluated through energy level balance analysis. The correct meaning of energy level matching refers to the matching between the energy level of the supplied energy (energy source) and the required energy (user), not the matching between other energy levels.

Taking a surface heat exchanger as an example, if the heat and exergy values released by the hot fluid are Q_1 and $E_{x,1}$, the heat and exergy values absorbed by the cold fluid are Q_2 and $E_{x,2}$, and the heat and exergy values dissipated by the heat exchanger to the environment are Q_3 and $E_{x,3}$, then the energy balance equation of this heat exchanger is

$$Q_1 = Q_2 + Q_3 \tag{3.109}$$

The exergy balance equation of this device is

$$E_{x,1} = E_{x,2} + E_{x,3} + E_{xl,in} \tag{3.110}$$

Where E_{xl}, in is the internal exergy loss of the heat exchanger during the heat transfer process.

Divide Eq. (3.110) by Eq. (3.109) to get the energy level balance equation

$$\frac{E_{x,1}}{Q_1} = \frac{E_{x,2} + E_{x,3} + E_{xl,in}}{Q_2 + Q_3} \tag{3.111}$$

The exergy essentially discharged by the heat exchanger to the environment is the external exergy loss during the heat transfer process, let $E_{xl,\ out} = E_{x,\ 3}$, then Eq. (3.111) can be expressed as

$$\frac{E_{x,1}}{Q_1} = \frac{E_{x,2} + E_{xl,out} + E_{xl,in}}{Q_2 + Q_3} = \frac{E_{x,2}}{Q_2 + Q_3} + \frac{E_{xl}}{Q_2 + Q_3} \tag{3.112}$$

Where E_{xl} is the total exergy loss of the heat exchanger during the heat transfer process, $E_{xl} = E_{xl,\ out} + E_{xl,\ in}$.

According to the definition of energy level, the energy level of the heat released by the hot fluid is $\lambda_1 = E_{x1}/Q_1$, and the energy level of the heat absorbed by the cold fluid is $\lambda_2 = E_{x2}/Q_2$, substitute into Eq. (3.112) to get

$$\lambda_1 = \lambda_2 \frac{Q_2}{Q_2 + Q_3} + \frac{E_{xl}}{Q_2 + Q_3} = \lambda_2 \eta_e + \frac{E_{xl}}{Q_1} \tag{3.113}$$

Where $\eta_e = Q_2/Q_1$ is the thermal efficiency of the heat exchanger.

The energy level difference between the hot and cold fluids of the heat exchanger is

$$\Delta\lambda = \lambda_1 - \lambda_2 = \frac{Q_2 E_{xl} - Q_3 E_{x,2}}{Q_2 (Q_2 + Q_3)} \tag{3.114}$$

The energy level efficiency of the heat exchanger is

$$\eta_\lambda = \frac{\lambda_2}{\lambda_1} \tag{3.115}$$

The energy level loss coefficient is

$$\xi_\lambda = \frac{\Delta\lambda}{\lambda_1} \tag{3.116}$$

Combine Eq. (3.113) and Eq. (3.115), and considering that the exergy loss coefficient of the heat exchanger is $\xi = E_{xl}/E_{x1}$, we have

$$\eta_\lambda \eta_e + \xi = 1 \tag{3.117}$$

Example 3.7 Air with a mass flow rate of 1.20 kg/s passes through a counterflow heat exchanger, and is heated from a state of 0.28 MPa, 25 °C by the combustion gas to 125 °C; the combustion gas with a mass flow rate of 1.04 kg/s is cooled from a state of 0.13 MPa, 250–98 °C, and then discharged into the environment. Assuming the average specific heat at constant pressure of the air is 1.01 kJ/(kg·K), the average specific heat at constant pressure and constant volume of the combustion gas are 0.84 and 0.63 kJ/(kg·K) respectively, and the environmental state is 0.1 MPa, 20 °C, use energy analysis, exergy analysis and energy level analysis to determine the efficiency of the heat exchanger.

Solution In the calculation process, the pressure and kinetic energy changes of the air and combustion gas passing through the heat exchanger are not considered.

I. Energy Analysis

In the heat exchanger, the heat released by the combustion gas is

$$Q_{\text{gas},1} = 1.04 \times 0.84 \times (250 - 98) = 132.79\,\text{kW}$$

The heat absorbed by the air is

$$Q_{\text{air},1} = 1.20 \times 1.01 \times (125 - 25) = 121.20\,\text{kW}$$

The heat loss of the heat exchanger is

$$\Delta Q = Q_{\text{gas},1} - Q_{\text{air}} = 132.79 \quad 121.20 = 11.59\,\text{kW}$$

The heat released by the combustion gas discharged from the heat exchanger into the environment is

$$Q_{\text{gas},2} = 1.04 \times 0.84 \times (98 - 20) = 68.14\,\text{kW}$$

In the process of using combustion gas to heat the air, the total heat released by the combustion gas is

$$Q_{\text{gas}} = Q_{\text{gas},1} + Q_{\text{gas},2} = 132.79 + 68.14 = 200.93\,\text{kW}$$

Therefore, the thermal efficiency and energy loss coefficient of this thermodynamic process are

$$\eta_Q = \frac{Q_{\text{air}}}{Q_{\text{gas}}} = \frac{121.20}{200.93} = 0.60$$

$$\xi_Q = \frac{\Delta Q + Q_{\text{gas},2}}{Q_{\text{gas}}} = \frac{11.59 + 68.14}{200.93} = 0.40$$

II. Exergy Analysis

In the heat exchanger, the exergy of the heat released by the combustion gas is

$$E_{\mathrm{x1,gas}} = G_{\mathrm{gas}} c_p \left[(T_1 - T_2) - T_0 \ln \frac{T_1}{T_2} \right]$$
$$= 1.04 \times 0.84 \times \left[(250 - 98) - (20 + 273.15) \times \ln \frac{250 + 273.15}{98 + 273.15} \right]$$
$$= 44.88\,\mathrm{kW}$$

The exergy of the heat brought into the environment by the combustion gas is calculated by the following formula

$$E_{\mathrm{x2,gas}} = G_{\mathrm{gas}} \left[c_p (T_2 - T_0) - c_p T_0 \ln \frac{T_2}{T_0} + R T_0 \ln \frac{p}{p_0} \right]$$

Where, the gas constant of the combustion gas is

$$R = c_p - c_v = 0.84 - 0.63 = 0.21\,\mathrm{kJ/(kg \cdot K)}$$

Then we have

$$E_{\mathrm{x2,gas}} = 1.04 \times [0.84 \times (98 - 20) - 0.84 \times (20 + 273.15) \times \ln \frac{98 + 273.15}{20 + 273.15} +$$
$$0.21 \times (20 + 273.15) \ln \frac{0.13}{0.1}] = 24.52\,\mathrm{kW}$$

The exergy of the heat released by the combustion gas in the entire heat release process is

$$E_{\mathrm{x,gas}} = E_{\mathrm{x1,gas}} + E_{\mathrm{x2,gas}} = 44.88 + 24.52 = 69.40\,\mathrm{kW}$$

The exergy of the heat absorbed by the air in the heat exchanger is

$$E_{\mathrm{x,air}} = G_{\mathrm{air}} c_p \left[(T_1 - T_2) - T_0 \ln \frac{T_1}{T_2} \right]$$
$$= 1.2 \times 1.01 \times \left[(125 - 25) - (20 + 273.15) \times \ln \frac{125 + 273.15}{25 + 273.15} \right]$$
$$= 18.44\,\mathrm{kW}$$

The exergy loss in the heat exchanger is

$$E_{\mathrm{xl}} = E_{\mathrm{x1,gas}} - E_{\mathrm{x,air}} = 44.88 - 18.44 = 26.44\,\mathrm{kW}$$

The exergy efficiency and exergy loss coefficient of the entire heat release process are respectively

$$\eta_{ex} = \frac{E_{x,air}}{E_{x,gas}} = \frac{18.44}{69.40} = 0.266$$

$$\xi_{ex} = \frac{E_{xl} + E_{x2,gas}}{E_{x,gas}} = \frac{26.44 + 24.52}{69.40} = 0.734$$

III. Energy level analysis

In the entire heat release process, the energy level of the gas heat release is

$$\lambda_{gas} = \frac{E_{x,gas}}{Q_{gas}} = \frac{69.40}{200.93} = 0.3454$$

The energy level of air heat absorption is

$$\lambda_{air} = \frac{E_{x,air}}{Q_{air}} = \frac{18.44}{121.2} = 0.1521$$

The energy level difference between gas and air is

$$\Delta\lambda = \lambda_{gas} - \lambda_{air} = 0.3454 \quad 0.1521 = 0.1933$$

The energy level efficiency and energy level loss coefficient are respectively

$$\eta_{\lambda} = \frac{\lambda_{air}}{\lambda_{gas}} = \frac{0.1521}{0.3454} = 0.44$$

$$\xi_{\lambda} = \frac{\Delta\lambda}{\lambda_{gas}} = \frac{0.1933}{0.3454} = 0.56$$

For comparison, the calculation results of the three methods are listed in Table 3.7.

Example 3.8 In winter room heating, the outdoor ambient temperature is −5 °C, to maintain the room temperature at 20 °C, 100 kW of heat power should be provided to the room. There are four heating schemes:

1. Electric furnace heating. The electric furnace is placed indoors, and the working temperature of the resistance wire is 1200 °C.
2. Coal-fired heating. The working temperature of the stove is 800 °C.

Table 3.7 Comparison of heat exchanger thermal analysis results

No.	Energy analysis		Exergy analysis		Energy level analysis	
1	Gas gives off heat/kW	200.93	Heat exergy of gas	69.40	Input energy level	0.345 4
2	Air absorbs heat/kW	121.20	Air absorbs heat exergy/kW	18.44	Output energy level	0.152 1
3	Bringing in ambient heat/kW	68.14	Bringing in ambient exergy/kW	24.52	Loss energy level	0.193 3
4	Heat loss/kW	11.59	Exergy loss	26.44	–	–
5	Thermal efficiency	0.60	Exergy efficiency	0.266	Energy level efficiency	0.44
6	Energy loss factor	0.40	Exergy loss coefficient	0.734	Energy level loss coefficient	0.56

Table 3.8 Results of thermal analysis of the room heating process

Analytical methods	Computational Items		Electric furnace heating	Coal-fired heating	Warm air heating	Heat pump heating
Energy analysis	Energy input	Formality	Electricity	Chemical energy	Heat energy	Mechanical energy
		Value/kJ	100	115	100	10
	Energized	Formality	Heat energy	Heat energy	Heat energy	Heat energy
		Value/kJ	100	100	100	100
	Thermo-Economic Indicators	Formality	Furnace efficiency	Combustion efficiency	Heat transfer efficiency	Heating factor
		Value/kJ	100%	87%	100%	10
Analytical Methods	Computational items		Electric Furnace Heating	Coal-fired heating	Warm Air Heating	Heat Pump Heating
Exergy analysis	Heat supplied by a heat source/kW		100	110.4	34.4	10
	Effectively utilized exergy/kW		8.5	8.5	8.5	8.5
	Exergy loss/kW		91.5	101.9	25.9	1.5
	Exergy efficiency/%		8.5	7.7	24.7	85.0
Energy level analysis	Input energy level		1	0.96	0.344	0.1
	Access to energy levels		0.085	0.085	0.085	0.085
	Energy level difference		0.915	0.875	0.249	0.015
	Energy efficiency		8.5	8.85	24.71	85

3. Steam heating. The heating steam is 0.5 MPa saturated steam, which releases heat to the room through the steam pipeline, the outlet water temperature is 20 °C and is recycled for reuse.
4. Heat pump heating.

The results of energy analysis, entropy analysis and energy level analysis are shown in Table 3.8.

Thought Questions and Exercises

1. Briefly describe the basic concept and properties of entropy.
2. A stream of hot water with a temperature of 90 °C and a flow rate of 20 kg/s is adiabatically mixed with another stream of hot water with a temperature of 50 °C and a flow rate of 30 kg/s. Assuming the ambient temperature is 20 °C, try to calculate the entropy loss during the mixing process.
3. The parameters of the new steam at the outlet of the steam generator in the NPP are p_1 = 5.4 MPa, x_1 = 0.9975. When the steam enters the turbine through the main steam pipeline, the steam parameters change to p_2 = 5.2 MPa, x_2 = 0.9925. Assuming the environmental parameters are p_0 = 0.1 MPa, t_0 = 20 °C, try to calculate the entropy loss during the steam flow process in the main steam pipeline.
4. The parameters of the new steam at the turbine inlet are 3.5 MPa, 450 °C, and the parameters of the exhaust steam at the outlet are 0.2 MPa, 160 °C, with the environmental state being 0.1 MPa, 20 °C. Ignoring the changes in macroscopic kinetic energy and potential energy, try to calculate:

    1. The actual output power of the turbine;
    2. The maximum useful power that the turbine could produce;
    3. The entropy loss during the steam expansion process in the turbine.

5. The steam parameters at the turbine inlet are p_1 = 6.4 MPa, x_1 = 0.9975, the exhaust pressure is p_2 = 0.004 MPa, and the relative internal efficiency of the turbine is η_{oi} = 0.82. If the environmental parameters are p_0 = 0.1 MPa, t_0 = 20 °C, try to calculate the entropy efficiency of the turbine.
6. Use an electric heater to heat 5 kg of water under environmental conditions (0.1 MPa, 15 °C) to 90 °C. Assuming the average specific heat at constant pressure of water is 4.19 kJ/(kg·K), try to find the entropy loss and energy level difference during the heating process.
7. A room is heated by electricity, and when the indoor temperature is 25 °C, the electricity consumption is 5 kW, when the ambient temperature is 0 °C, try to calculate the efficiency of the electric heating device.

Chapter 4
Thermal Cycle of NPPs

NPPs use fissionable materials such as uranium and plutonium as nuclear fuel, converting the energy produced in a controllable chain fission reaction into electrical energy. At the current level of technology, it is not yet possible to directly convert nuclear fission energy into electrical energy on a large scale, but it needs to go through several forms of energy conversion, as shown in Fig. 4.1.

The process of converting nuclear fission energy into thermal energy is realized in the reactor, and the heat released by the reactor core nuclear fuel is transferred to the thermal power conversion system through the reactor coolant system, thereby realizing the process of converting thermal energy into mechanical energy, and finally the generator converts mechanical energy into electrical energy. The form of the thermal power conversion system depends on the thermal cycle used. For water-cooled reactors (such as PWRs, boiling water reactors, heavy water reactors, graphite water-cooled reactors, etc.) NPPs, liquid metal cooled reactor NPPs, and gas-cooled reactor NPPs using steam cycles, their thermal power conversion systems use water/steam as the working fluid and use the Rankine cycle of steam power devices; for gas-cooled reactor NPPs using direct cycles, their thermal power conversion systems use high-temperature gas as the working fluid and use the Brayton cycle of gas.

The perfection of the thermal cycle has a significant impact on the thermal economic indicators of NPPs. Therefore, analyzing and evaluating the perfection of the thermal power conversion process, and studying ways to improve the thermal economic indicators of NPPs, is an important topic in the design and operation of NPPs. The study of the thermal cycle is based on the first and second laws of thermodynamics. The study of the actual thermal cycle is usually divided into two steps:

1. Simplify the actual cycle into an ideal reversible cycle, that is, temporarily ignore the impact of irreversible factors, study the main factors affecting the efficiency of this thermal cycle and the measures that may be taken to improve thermal efficiency;

M. Peng et al., *Thermal Analysis of Nuclear Power Plants*, Nuclear Science and Technology, https://doi.org/10.1007/978-981-96-8983-5_4

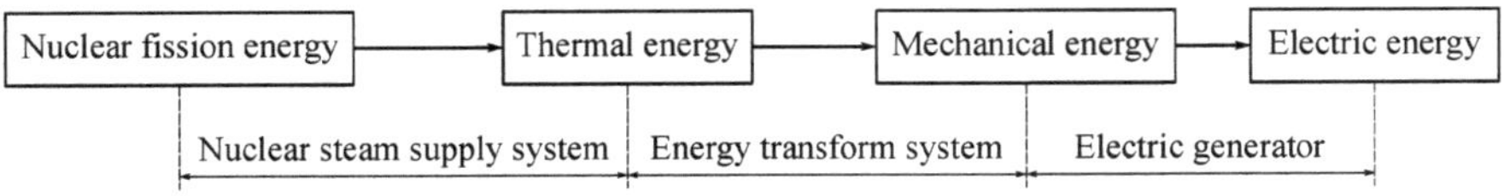

Fig. 4.1 Energy conversion process in a NPP

2. Based on the study of the ideal reversible cycle, further study the irreversible losses in the actual cycle, find out the location, size, cause and relationship of the losses, study the methods to reduce irreversible losses, and analyze the possible degree of improving thermal economy.

The following takes the PWR NPP and the high-temperature gas-cooled reactor NPP using a direct cycle as examples, and introduces the basic principles of the Rankine cycle and the Brayton cycle and the basic methods of thermal cycle analysis.

4.1 Rankine Cycle

The thermal system of the PWR nuclear power device is generally divided into the PWR and the primary loop system, and the secondary loop system, which belong to the nuclear island and the conventional island in the NPP, respectively. Pressurized water reactor NPPs generally use natural circulation steam generators that can only produce saturated steam, so their secondary loop system uses a saturated steam Rankine cycle [27, 28].

According to the second law of thermodynamics, the thermal efficiency of the Carnot cycle is the highest between the same boundary temperatures, and Fig. 4.2 shows the saturated steam Carnot cycle on the *T-s* diagram.

Where 1-2 is an adiabatic expansion process, 2-*c* is an isothermal (constant pressure) heat release process, *c-5* is an adiabatic compression process, and 5-1 is an isothermal (constant pressure) heat absorption process. The thermal efficiency of the Carnot cycle can be expressed as

$$\eta_t = 1 - \frac{T_2}{T_1} \tag{4.1}$$

Despite the highest thermal efficiency of the Carnot cycle between the heat absorption temperature T_1 and the heat release temperature T_2, the Carnot cycle is not used in actual nuclear power devices. Although process 1-2 can be approximately realized in a steam turbine, process 2-*c* can be approximately realized in a condenser, and process 5-1 can be approximately realized in a steam generator, the state c is in the wet steam area where saturated water and saturated steam coexist, and the specific volume of wet steam is 1000–2000 times larger than that of water, making it extremely difficult to use a pump to achieve the adiabatic compression process *c*-5.

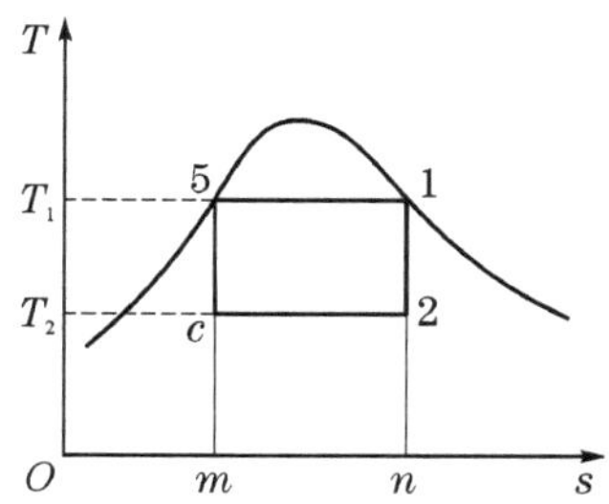

Fig. 4.2 Carnot cycle with saturated steam

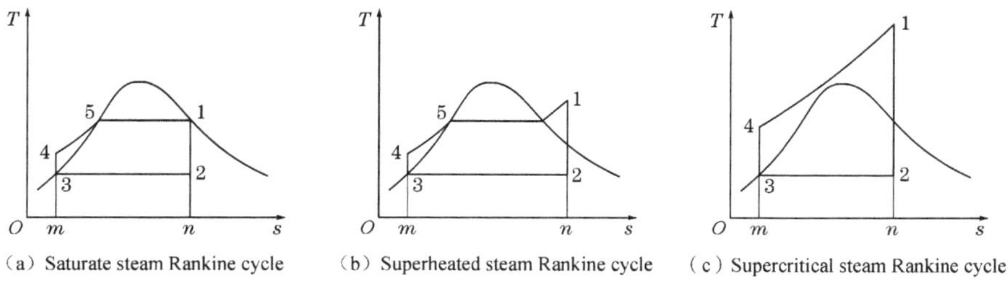

Fig. 4.3 Ideal rankine cycle on the *T-s* diagram: (**a**) Rankine cycle of saturated steam; (**b**) Rankine cycle of superheated steam; (**c**) Supercritical steam Rankine cycle

4.1.1 Ideal Rankine Cycle

To avoid the difficulties brought about by the implementation of the steam Carnot cycle, the Rankine cycle is commonly used in actual steam power devices, as shown in Fig. 4.3 for the *T-s* diagram of the ideal Rankine cycle.

The main difference between the Rankine cycle and the steam Carnot cycle is that the condensation of the exhaust steam is complete, that is, it does not stop at state c, but continues to state 3 to completely liquefy the exhaust steam; complete condensation adds a heating process of unsaturated water 4-5 in the cycle, reducing the average temperature difference of the cycle, lowering the thermal efficiency, but compressing water is much easier than compressing steam-water mixture, thus simplifying the equipment.

NPPs, thermal power plants, and marine steam power devices all use steam power cycles that are improved based on the Rankine cycle. The Rankine cycle is the basic cycle of various complex steam power devices.

The difference between the Rankine cycle and the Carnot cycle is that the heating of water does not occur at a constant temperature. If superheated steam is used, heating in the superheated area is not under constant temperature conditions. Therefore, the thermal efficiency of the Rankine cycle is lower than that of the Carnot cycle operating between the same temperature limits.

Figure 4.4 is the basic thermal cycle of a PWR nuclear power device.

Corresponding to Fig. 4.3, 4-5-1 is the constant pressure heat absorption process of the secondary circuit working fluid in the steam generator, changing from

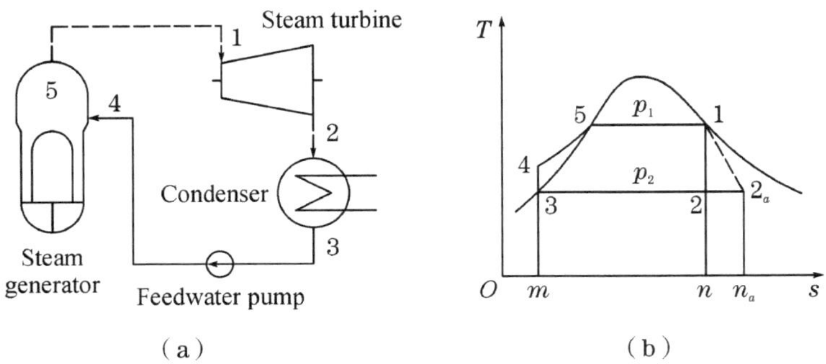

Fig. 4.4 Basic thermodynamic cycle of PWR NPPs: (**a**) Thermodynamic system principle flow; (**b**) *T*-*s* diagram

unsaturated water to saturated steam under the heating of the primary circuit coolant; 1-2 is the adiabatic expansion process of steam in the steam turbine, the end state 2 is low-pressure wet steam, referred to as exhaust steam or waste steam; 2-3 is the constant pressure (constant temperature) heat release process of the exhaust steam in the condenser, cooled and condensed into saturated water by circulating water; 3-4 is the adiabatic compression process of the condensed water in the feedwater pump, the pressure increases, re-enters the steam generator, and starts the next cycle.

When the nuclear power device is operating normally under stable conditions, the working fluid is in a stable flow state. If based on per kilogram of working fluid, in the constant pressure heat absorption process 4-5-1 in the steam generator, the heat absorbed by the secondary circuit working fluid from the primary side coolant is

$$q_1 = h_1 - h_4 \tag{4.2}$$

It is equal to the area enclosed by the m-4-5-1-n-m on the *T*-*s* diagram.

In the adiabatic expansion process 1-2 in the steam turbine, the theoretical work done by the steam is

$$w_T = h_1 - h_2 \tag{4.3}$$

In the constant pressure (constant temperature) heat release process 2-3 in the condenser, the heat released by the exhaust steam to the circulating cooling water is

$$q_2 = h_2 - h_3 \tag{4.4}$$

It is equal to the area enclosed by the *m*-3-2-*n*-*m* on the *T*-*s* diagram.

In the adiabatic compression process 3-4, the work done by the water pump on the condensed water is

$$w_P = h_4 - h_3 \tag{4.5}$$

The net work done by the working substance in the entire cycle is

$$w_0 = w_T - w_p = (h_1 - h_2) - (h_4 - h_3) \tag{4.6}$$

It is equal to the area enclosed by the 1-2-3-4-5-1 on the *T-s* diagram.

The effective heat of the cycle is

$$q_0 = q_1 - q_2 = (h_1 - h_4) - (h_2 - h_3) = (h_1 - h_2) - (h_4 - h_3) \tag{4.7}$$

It is equal to the area enclosed by the 1-2-3-4-5-1 on the *T-s* diagram.

It can be seen that

$$q_0 = w_0$$

The thermal efficiency of the cycle is

$$\eta_t = \frac{w_0}{q_1} = \frac{q_1 - q_2}{q_1} = \frac{(h_1 - h_2) - (h_4 - h_3)}{h_1 - h_4} = \frac{w_T - w_P}{q_1} \tag{4.8}$$

The energy consumed by the feedwater pump is usually very small compared to the work done by the main steam turbine, not exceeding 1%. Neglecting the pump work has little effect on the calculation accuracy, but it is more convenient for analyzing the general trend of the change in cycle thermal efficiency. At this time, there is

$$h_4 \oplus h_3$$

The approximate expression of cycle thermal efficiency is

$$\eta_t = \frac{w_T}{q_1} = \frac{h_1 - h_2}{h_1 - h_3} \tag{4.9}$$

The steam consumption rate of the ideal Rankine cycle is

$$d_0 = \frac{3600}{w_T - w_P} \approx \frac{3600}{h_1 - h_2} \quad \text{kg}/(\text{kW}\cdot\text{h}) \tag{4.10}$$

4.1.2 Actual Rankine Cycle with Friction

In fact, all processes of steam in power devices are irreversible. For example, fresh steam has to go through a series of pipes and valves from the outlet of the steam generator to the inlet of the main steam turbine. Due to flow friction and heat dissipation, the pressure and temperature of the fresh steam are reduced; the adiabatic expansion process of the steam passing through the turbine is significant due to the high speed of the steam flow, the internal friction loss of the steam flow, and the friction loss of the nozzle inner wall and blades to the steam flow. Therefore, there is a significant difference between the actual cycle and the ideal cycle.

For the actual steam power cycle, the losses of the thermal cycle itself mainly include heat dissipation and throttling losses in high-pressure steam pipes, and internal losses in turbines.

4.1.2.1 Irreversible Losses in the Internal Expansion Process of the Turbine

When steam expands and does work in the turbine, it will generate various internal losses such as throttle losses in the steam inlet mechanism, nozzle losses, moving blade losses, residual speed losses, wet steam losses, stage internal steam leakage losses, and blower friction losses. As a result, only a part of the available enthalpy drop of steam in the turbine can be converted into mechanical work. In Fig. 4.4, the ideal adiabatic expansion process is 1-2. After considering the various losses inside the steam in the turbine, the actual expansion process becomes 1-2-*a*. Under the condition that the heat absorption q_1 of the thermal cycle remains unchanged, the heat dissipation q_2 of the thermal cycle increases, and the increased part is the area $2 - 2_a - n_a - n - 2$.

The actual internal work done by the steam passing through the turbine is

$$w_{T_a} = h_1 - h_{2_a} = \left(h_1 - h_{2_a}\right) - \left(h_{2_a} - h_2\right) \tag{4.11}$$

Obviously, the less work done equals the more heat discharged.

The degree of irreversibility in the internal expansion process of the turbine is described by the relative internal efficiency of the turbine, which is defined as the ratio of the actual work done by the internal steam of the turbine to the theoretical work, also known as the turbine efficiency, that is

$$\eta_{oi} = \frac{w_{T_a}}{w_T} = \frac{h_1 - h_{2_a}}{h_1 - h_2} \tag{4.12}$$

The relative internal efficiency of the turbine is generally 0.8–0.88, and for large power turbines, it can reach 0.92.

4.1.2.2 Heat Dissipation and Throttling Losses in Steam Pipes

The heat dissipation loss in the steam pipe is due to the difference in temperature between the steam and the environment, but the steam pipe has a good thermal insulation layer, so the heat dissipation loss is generally small, less than 1%. The result of heat dissipation loss is that the temperature of the steam is reduced, which reduces the ability of the steam to do work.

Throttling losses are caused by the resistance when steam flows through pipes, valves, and throttling valves, with the pressure drop reaching up to 3% or more of the initial pressure. The enthalpy of the steam before and after throttling remains unchanged, so throttling losses are different from heat dissipation losses. They are not losses of heat, but losses of work capacity.

The larger the pipe diameter, the smaller the resistance, and the smaller the throttling loss and pressure drop. However, this requires a larger equipment investment and also increases heat dissipation losses. Therefore, there is an optimal size for the pipe, and thus a certain value for the throttling loss.

The heat dissipation and throttling losses in steam pipes are described by the pipe loss coefficient, which is defined as

$$\eta_{\mathrm{mp}} = \frac{\left(h_1 - h_{2_\mathrm{a}}\right) - \Delta h_{\mathrm{mp}}}{h_1 - h_{2_\mathrm{a}}} = 1 - \frac{\Delta h_{\mathrm{mp}}}{h_1 - h_{2_\mathrm{a}}} \tag{4.13}$$

Where Δh_{mp} is the decrease in steam enthalpy due to heat dissipation and throttling.

Usually, $\eta_{\mathrm{mp}} = 0.95\sim0.98$.

4.1.2.3 Losses Outside the Thermodynamic Cycle

Figure 4.5 shows the main energy transfer process in a PWR nuclear power device.

The primary circuit coolant flows through the reactor core, carrying out the heat released by the nuclear fuel, and then transfers it to the secondary circuit feedwater in the steam generator, causing it to boil and produce saturated steam. Due to the heat dissipation loss in the primary circuit main pipe, if the energy transferred to the coolant by the operation of the main pump is not considered, the energy utilization coefficient η_1 of the primary circuit can be used to measure the perfection of the energy transfer of the primary circuit system:

$$\eta_1 = \frac{Q_{\mathrm{s}}}{Q_{\mathrm{R}}} \tag{4.14}$$

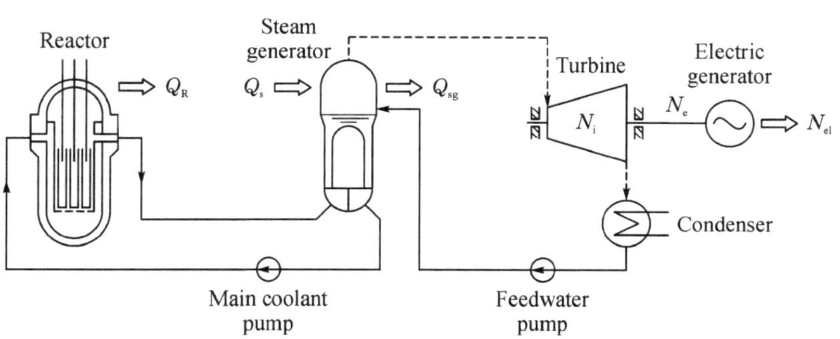

Fig. 4.5 Main processes of energy transfer in pressurized water reactor NPP

Where Q_s is the thermal power of the nuclear steam supply system;

Q_R is the thermal power of the reactor.

There are heat dissipation and pollution discharge losses in the steam generator, usually measured by the steam generator efficiency η_{sg}, which measures the perfection of energy transfer between the primary and secondary sides of the steam generator:

$$\eta_{sg} = \frac{Q_{sg}}{Q_s} \tag{4.15}$$

Where Q_{sg} is the thermal power of the secondary side of the steam generator.

Considering the internal losses of the turbine and the flow losses in the pipe, the actual work done by the steam in the turbine is

$$N_i = Q_{sg}\eta_t\eta_{oi}\eta_{mp} \tag{4.16}$$

The mechanical work produced by the steam in the turbine is not all transferred to the generator. Some of it is used to overcome bearing friction, drive the main oil pump, and consume energy in the regulation system. Therefore, the effective power N_e obtained on the input shaft of the generator is always less than the internal power N_i of the turbine, and their ratio is called the mechanical efficiency of the turbine set:

$$\eta_m = \frac{N_e}{N_i} \tag{4.17}$$

Considering the electrical losses and mechanical friction of the generator, the output electric power Nel of the generator will be less than the effective power Ne of the turbine, and their ratio is called the efficiency of the generator:

$$\eta_{ge} = \frac{N_{el}}{N_e} \tag{4.18}$$

Considering all the above losses, the efficiency of the NPP can be expressed as

$$\eta_{el} = \frac{N_{el}}{Q_R} = \frac{Q_s}{Q_R}\frac{Q_{sg}}{Q_s}\frac{N_i}{Q_{sg}}\frac{N_e}{N_i}\frac{N_{el}}{N_e} = \eta_1\eta_{sg}\eta_t\eta_{oi}\eta_{mp}\eta_m\eta_{ge} \tag{4.19}$$

Example 4.1 For the PWR nuclear power device shown in Fig. 4.4, the fresh steam pressure at the outlet of the steam generator is 4 MPa, the dryness is 0.9975, and the exhaust steam pressure of the turbine is 4 kPa. Considering the ideal Rankine cycle and the actual Rankine cycle (the relative internal efficiency of the turbine $\eta_{oi} = 0.8$), with the unit mass flow rate (1 kg/s) of the working fluid as the calculation basis, try to determine:

1. The heat absorbed by the fresh steam in the steam generator, the heat released by the exhaust steam in the condenser, and the humidity of the exhaust steam;

2. The theoretical power output of the turbine and the theoretical power consumption of the water pump;
3. The thermal efficiency and steam consumption rate of the cycle.

Solution (I) Calculation according to the ideal Rankine cycle

First, check the water and steam parameter table to get the parameters of the steam at each state point.

State 1: New steam at the outlet of the steam generator.

Known $p_1 = 4$ MPa, $x_1 = 0.9975$, the enthalpy and entropy of the new steam are

$$h_1 = h_1^{'} + x_1\left(h_1^{''} - h_1^{'}\right)$$
$$= 1087.4 + 0.9975 \times (2800.3 - 1087.4) = 2796.02 \mathrm{kJ/kg}$$
$$s_1 = s_1^{'} + x_1\left(s_1^{''} - s_1^{'}\right)$$
$$= 2.7965 + 0.9975 \times (6.0685 - 2.7965) = 6.0603 \mathrm{kJ/(kg \cdot K)}$$

State 2: Exhaust steam at the outlet of the steam turbine.

Known $p_2 = 4$ kPa, the steam expands adiabatically inside the steam turbine, i.e.$s_2 = s_1 = 6.0603$ kJ/(kg · K).

The dryness and enthalpy of the exhaust steam are

$$x_2 = \frac{s_2 - s_2^{'}}{s_2^{''} - s_2^{'}} = \frac{6.0603 - 0.4225}{8.4755 - 0.4225} = 0.70$$
$$h_2 = h_2^{'} + x_2\left(h_2^{''} - h_2^{'}\right)$$
$$= 121.41 + 0.70 \times (2554.5 - 121.41) = 1824.57 \mathrm{kJ/kg}$$

State 3: Condensate at the outlet of the condenser

The exhaust steam releases heat at constant pressure (constant temperature) in the condenser, and the condensate is saturated water under operating pressure, so

$$p_3 = p_2 = 4\mathrm{kPa}, h_3 = h_2^{'} = 121.41 \mathrm{kJ/kg}, v_3 = 1.0041 \times 10^{-3} \mathrm{m^3/kg}$$

State 4: Feedwater at the inlet of the steam generator

The specific volume of water changes very little under the compression of the pump, ignoring its change, we have

$$p_4 = p_1 = 4\mathrm{MPa}, v_4 = v_3 = 1.0041 \times 10^{-3} \mathrm{m^3/kg}$$
$$h_4 = h_3 + w_p = h_3 + v_4\left(p_4 - p_3\right) = 121.42 + 1.0041 \times 10^{-3} \times \left(4 \times 10^3 - 4\right)$$
$$= 125.42 \mathrm{~kJ/kg}$$

1. The heat absorbed by the new steam in the steam generator is

$$q_1 = G_s\left(h_1 - h_4\right) = 1 \times (2796.02 - 125.42) = 2670.6 \mathrm{kW}$$

The heat released by the exhaust steam in the condenser is

$$q_2 = G_s\left(h_2 - h_3\right) = 1 \times \left(1824.57 - 121.41\right) = 1703.16\text{kW}$$

2. The theoretical power of the steam turbine is

$$w_T = G_s\left(h_1 - h_2\right) = 1 \times \left(2796.02 - 1824.57\right) = 971.45\text{kW}$$

The theoretical power consumed by the water pump is

$$w_p = G_s\left(h_4 - h_3\right) = 1 \times \left(125.42 - 121.41\right) = 4.01\text{kW}$$

The ratio of the power consumed by the water pump to the theoretical power of the steam turbine is

$$\frac{w_P}{w_T} = \frac{4.01}{971.45} = 0.411\%$$

3. Cycle thermal efficiency

$$\eta_t = \frac{w_T - w_P}{q_1} = \frac{971.45 - 4.01}{2.670.6} = 0.3622$$

The steam consumption rate of the cycle is

$$d_0 = \frac{G_s}{w_T - w_p} = \frac{1 \times 3 \cdot 600}{971.45 - 4.01} = 3.72\text{kg}/\left(\text{kW} \cdot \text{h}\right)$$

(II) Calculation according to the actual Rankine cycle.

The parameters of the working fluid at state points 1, 3, and 4 are the same as in the ideal Rankine cycle. The actual expansion end point of the steam in the steam turbine is 2_a. Since $\eta_{oi} = \frac{\left(h_1 - h_{2_a}\right)}{\left(h_1 - h_2\right)}$, we have

$$h_{2_a} = h_1 - \eta_{oi}\left(h_1 - h_2\right)$$
$$= 2796.02 - 0.80 \times \left(2796.02 - 1824.57\right) = 2018.86\text{kJ/kg}$$

$$x_{2_a} = \frac{h_{2_a} - h_2'}{h_2'' - h_2'} = \frac{2018.86 - 121.41}{2.554.5 - 121.41} = 0.78$$

1. The heat absorbed by the new steam in the steam generator is

$$q_1 = G_s\left(h_1 - h_4\right) = 2670.6\text{kW}$$

The heat released by the exhaust steam in the condenser is

$$q_2 = G_s\left(h_{2_a} - h_3\right) = 1 \times \left(2018.86 - 121.41\right) = 1897.45\text{kW}$$

Compared with the ideal Rankine cycle, the heat released to the condenser has increased

$$\Delta q_2 = 1897.45 - 1703.16 = 194.29\text{kW}$$

2. The actual power of the steam turbine is

$$w_{\text{T}} = G_{\text{s}}\left(h_1 - h_{2_{\text{a}}}\right) = 1 \times (2796.02 - 2018.86) = 777.16\text{kW}$$

Compared with the ideal Rankine cycle, the power of the steam turbine has decreased

$$\Delta w_{\text{T}} = 971.45 - 777.16 = 194.29\text{kW}$$

The power consumed by the water pump is

$$w_{\text{p}} = G_{\text{s}}\left(h_4 - h_3\right) = 4.01\text{kW}$$

The ratio of the power consumed by the water pump to the actual power of the steam turbine is

$$\frac{w_{\text{P}}}{w_{\text{T}}} = \frac{4.01}{777.16} = 0.516\%$$

3. Cycle thermal efficiency

$$\eta_{\text{t}} = \frac{w_{\text{T}} - w_{\text{P}}}{q_1} = \frac{777.16 - 4.01}{2.670.6} = 0.2895$$

The steam consumption rate of the cycle is

$$d_0 = \frac{G_{\text{s}}}{w_{\text{r}} - w_{\text{p}}} = \frac{1 \times 3600}{777.16 - 4.01} = 4.66\text{kg}/(\text{kW} \cdot \text{h})$$

From the above calculation results, it can be seen that:

① The power consumed by the water pump is very small compared to the output power of the steam turbine, and is usually ignored.
② Due to the irreversible loss in the steam expansion process, the actual power of the steam turbine is reduced, the thermal efficiency of the cycle is reduced, and the steam consumption rate is increased.
③ The decrease in the actual power of the steam turbine compared to the ideal power is equal to the additional heat discharged into the environment through the condenser in the actual cycle.
④ The steam dryness at the end point of the actual expansion process is less than that of the adiabatic expansion process.

4.2 The Impact of Steam Parameters on Cycle Efficiency

From Eq. (4.8), it is known that the thermal efficiency η_t of the Rankine cycle depends on the enthalpy h_1 of the new steam at the turbine inlet, the enthalpy h_2 of the exhausted steam at the turbine outlet, and the enthalpy h_3 of the condensate. Among them, h_1 is determined by the pressure p_1 and temperature T_1 of the new steam, and h_2 and h_3 are determined by the pressure p_2 and temperature T_2 of the exhausted steam. Therefore, the initial parameters (p_1, T_1) and final parameters (p_2, T_2) of the steam are important factors affecting the thermal efficiency of the Rankine cycle.

Pressurized water reactor nuclear power units often use the Rankine cycle of saturated steam, and the pressure and temperature of the new steam and exhausted steam correspond one-to-one. Changing the steam parameters will cause the pressure and temperature of the steam to change simultaneously.

4.2.1 The Impact of Initial Steam Parameters

If the final steam parameters are kept constant and the initial steam parameters are increased, as shown in Fig. 4.6, the thermodynamic cycle changes from 1-2-3-4-5-1 to $1_a - 2_a - 3 - 4_a - 5_a - 1_a$. Since the average heat release temperature of the cycle remains unchanged and the average heat absorption temperature increases, the cycle thermal efficiency η_t increases.

Figure 4.7 shows the change in the efficiency of the Rankine cycle in a thermal power plant with the initial steam temperature and pressure. From the curve in the figure, it can be seen that as the initial steam parameters increase, the cycle thermal efficiency also increases correspondingly.

For PWR NPPs that use the saturated steam Rankine cycle, although increasing the initial steam parameters can improve the cycle thermal efficiency, it also brings some problems.

1. The humidity of the steam at the turbine outlet increases. If the humidity of the turbine exhaust steam is too high, it will cause erosion to the last few stages of the turbine blades, affecting the safe operation of the turbine and also reducing

Fig. 4.6 Impact of initial steam parameters

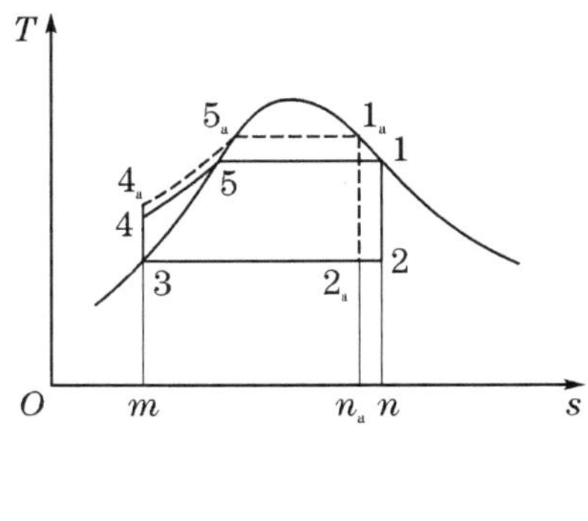

Fig. 4.7 Relationship between Rankine cycle efficiency and initial steam parameters

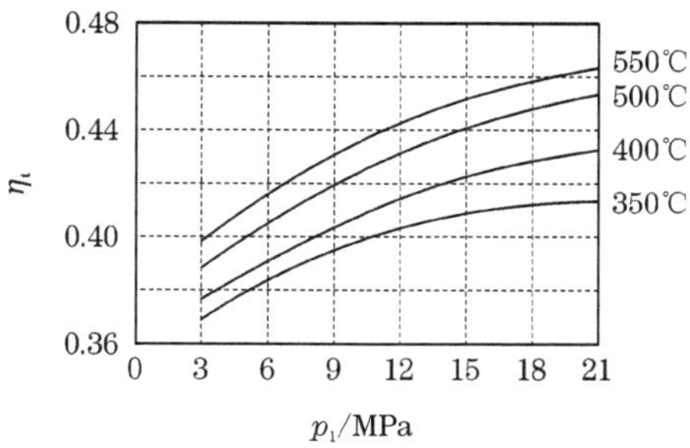

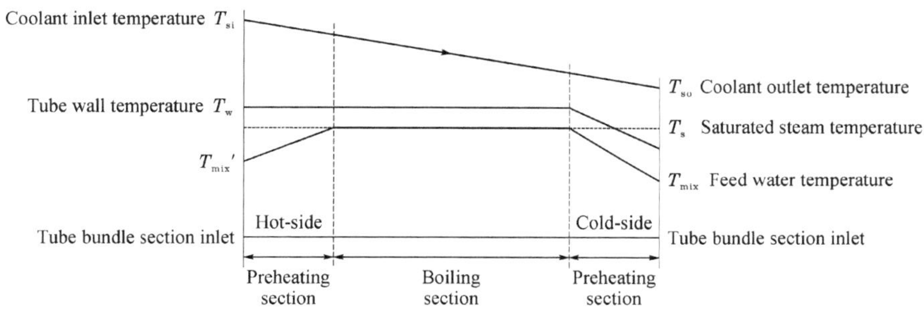

Fig. 4.8 Temperature distribution of working substance on the primary and secondary sides of the steam generator

the internal efficiency of the turbine. In engineering, the humidity of the turbine exhausted steam is generally required to be no more than 12%.

2. The operating pressure of the steam generator increases, requiring an increase in the lift of the feedwater pump, which not only increases the axial thrust of the feedwater pump impeller, but also increases the power consumed by the feedwater pump for the working fluid of unit mass flow.
3. The strength and temperature resistance requirements of the steam generator, turbine, steam pipes, and valves are increased, and the equipment investment is increased.

Increasing the initial steam parameters is also affected by the operating temperature of the coolant on the primary side of the steam generator. Figure 4.8 shows the temperature distribution of the working fluid on the primary and secondary sides of the steam generator. In order to increase the steam temperature under the condition that the structural parameters of the steam generator and the heat transfer tube material remain unchanged, in addition to adopting enhanced heat exchange measures and increasing the heat transfer coefficients on the primary and secondary sides of the steam generator, it is also necessary to increase the temperature of the coolant on the primary side, but it is restricted by the thermal safety regulations of the reactor.

In recent years, there has been a clear trend of increasing the secondary loop steam parameters in PWR NPPs. For example, the steam pressure of early American NPPs was 4.2 MPa, and it has now been increased to 7.5–7.7 MPa, which has played an important role in improving the economy of NPPs. However, the further increase of the initial parameters of the steam in PWR NPPs is strictly restricted by the primary side parameters, and there will be no significant increase.

4.2.2 The Impact of Final Steam Parameters

If the initial parameters of the steam are kept constant and the final parameters of the steam are reduced, as shown in Fig. 4.9, the thermodynamic cycle changes from $1-2-3-4-5-1$ to $1-2_a-3_a-4_a-5-1$. Since the average heat absorption temperature of the cycle remains unchanged, and the average heat dissipation temperature decreases, the thermal efficiency η_t of the cycle increases. Figure 4.10 shows the influence curve of the final pressure of the steam on the efficiency of the Rankine cycle. It can be seen from the figure that if the exhaust pressure of the steam turbine is reduced, the thermal efficiency of the cycle will correspondingly increase, and it is more sensitive to lower initial steam parameters. Therefore, reducing the exhaust pressure is one of the main methods to improve the thermal economy of the power plant. Reducing the final parameters of the steam, while improving the thermal efficiency of the cycle, also brings some disadvantages.

(1) It will increase the humidity of the steam turbine exhaust, which is not conducive to the safe and effective operation of the steam turbine. (2) Excessive reduction of the final pressure of the steam will cause the steam to expand in the oblique cut at the outlet of the steam turbine nozzle, and when the expansion capacity in the oblique cut is exhausted, the steam will expand outside the stage, and the useful

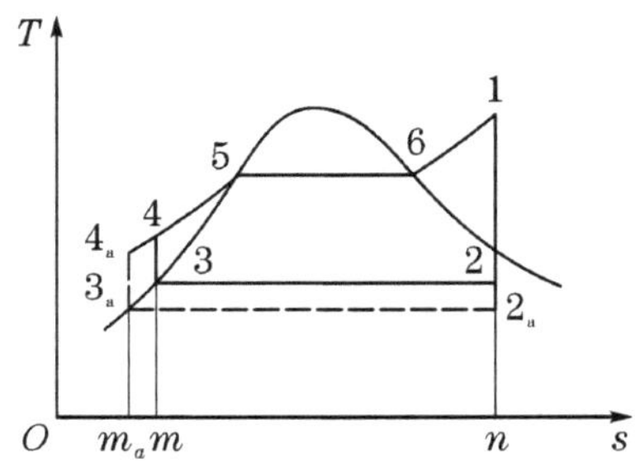

Fig. 4.9 Impact of final steam parameters

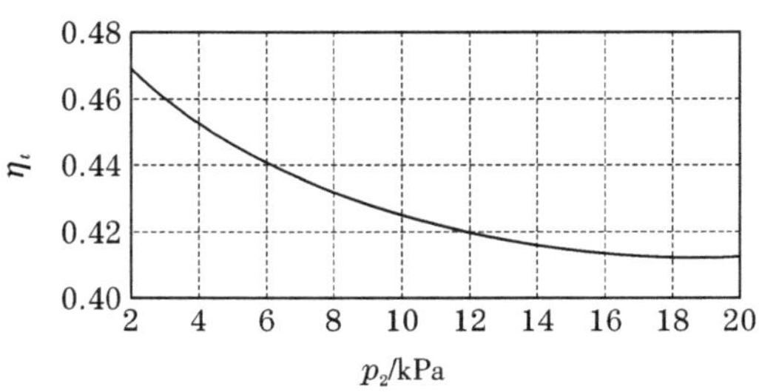

Fig. 4.10 Relationship between Rankine Cycle efficiency and final steam parameters

enthalpy drop will no longer change. (3) The outlet water temperature of the condenser decreases, requiring an increase in the amount of heating steam in the first-stage feedwater heater, resulting in a decrease in the steam flow through the last few stages of the low-pressure cylinder of the steam turbine, and a decrease in the power output of the steam turbine. Reducing the final parameters of the steam is mainly restricted by natural conditions and technical conditions. (1) The temperature of the condensate in the condenser cannot be lower than the local cooling water temperature. The circulating cooling water required by the condenser of the NPP comes from the sea or rivers, and its temperature T_{sw} is affected by local geographical conditions. For the thermal system of the power plant, the circulating cooling water plays the role of a cold source. The exhaust steam of the steam turbine condenses into water in the condenser, and its condensation temperature must be higher than the temperature of the circulating cooling water to ensure a sufficiently large heat transfer temperature difference. Therefore, it can be said that the temperature of the circulating cooling water is the theoretical limit of the exhaust temperature in the steam cycle. In the design of thermal power plants, different countries adopt different exhaust pressures according to their natural conditions. In China, it is generally taken as 4.4–5.4 kPa. The back pressure of modern large-scale thermal power plants is usually 4 kPa, corresponding to a condensation temperature of 28.98 °C. (2) The temperature rise ΔT_{sw} of circulating cooling water and the terminal difference δT of the condenser The temperature rise ΔT_{sw} of the inlet and outlet of the circulating cooling water and the terminal difference δT between the condensing steam and the outlet temperature of the circulating cooling water constitute the technical limit of the exhaust temperature, as shown in Fig. 4.11.

The condensation temperature of the steam turbine exhaust in the condenser can be expressed as

$$T_{cd} = T_{sw,1} + \frac{h_{cs} - h_{cd}}{c_p m} + \delta T = T_{sw,1} + \Delta T_{sw} + \delta T \tag{4.20}$$

Where $T_{sw,1}$ is the inlet temperature of circulating cooling water in the condenser;

h_{cs}, h_{cd} are the enthalpy of the steam turbine exhaust and the condensate, respectively;

c_p is the specific heat at constant pressure of the circulating cooling water;

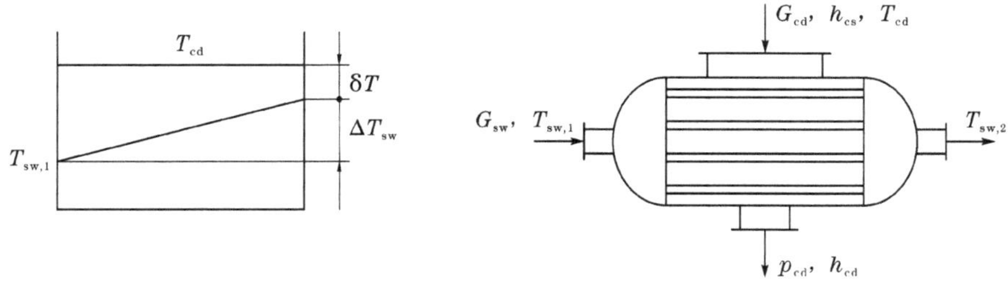

Fig. 4.11 Schematic diagram of condenser operation process

m is the cooling multiple of the condenser, defined as

$$m = \frac{G_{sw}}{G_{cd}} \tag{4.21}$$

Where G_{sw} is the flow rate of circulating cooling water entering the condenser;
G_{cd} is the flow rate of steam entering the condenser.

Obviously, to reduce the exhaust temperature of the steam turbine, we can only reduce ΔT_{sw} and δT.

If we reduce the temperature rise ΔT_{sw} of the circulating cooling water, it requires increasing the flow of circulating cooling water and the corresponding power of the circulating pump, which will increase the operating cost. On the other hand, reducing ΔT_{sw} can reduce the heat transfer area of the condenser, which will reduce the equipment investment cost. Calculations for a certain NPP show that when ΔT_{sw} increases from 7 to 8 °C, the heat transfer area of the condenser increases by about 10%, while the amount of circulating water decreases by 14%. According to a preliminary analysis of electricity generation costs, the increased equipment costs can be recovered in about 7 years through savings in operating costs.

In order to ensure the smooth progress of the heat transfer process in the condenser, it is required that the terminal difference δT takes a sufficiently large value, therefore, the reduction of δT is extremely limited.

In addition to the impact on thermal economy, the back pressure of the condenser has a significant impact on the length of the last few stages of turbine blades and the size of the exhaust port. Therefore, the selection of steam final parameters should be considered comprehensively based on various factors.

4.2.3 *Mutual Constraints of the First and Second Circuit Parameters*

Figure 4.12 shows the mutual relationship between the coolant temperature of the first circuit and the working fluid temperature of the second circuit in a PWR NPP.

The working pressure of the reactor and the primary loop system is about 15.5 MPa, the core outlet temperature is about 320 °C, and the boiling margin (underheating) of the coolant is about 25 °C; the steam inlet pressure of the turbine is between 5.5–7 MPa, and the corresponding saturation temperature is 270–286 °C. To ensure smooth heat transfer, there must be a sufficiently large temperature difference between the primary and secondary sides of the steam generator.

The heat transfer equation of the steam generator is

$$q_s = FK\Delta T_m \tag{4.22}$$

Where F is the effective heat transfer area of the steam generator;

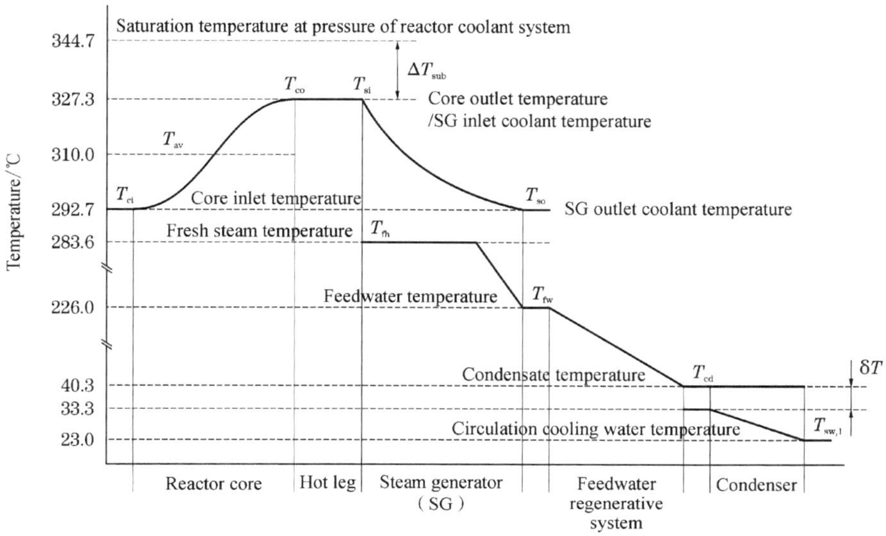

Fig. 4.12 The mutual relationship between the coolant temperature of the primary loop and the working fluid temperature of the secondary loop in a PWR NPPs

K is the effective heat transfer coefficient of the steam generator;

ΔT_{m} is the average heat transfer temperature difference of the steam generator, its calculation formula is

$$\Delta T_{\mathrm{m}} = \frac{T_{\mathrm{si}} - T_{\mathrm{so}}}{\ln \dfrac{T_{\mathrm{si}} - T_{\mathrm{fh}}}{T_{\mathrm{so}} - T_{\mathrm{fh}}}} \tag{4.23}$$

According to the heat transfer equation of the steam generator, there are mainly the following two ways to improve the initial parameters of the second circuit steam:

1. Increase the temperature of the primary side coolant, and under the condition that the structural parameters and heat transfer conditions of the steam generator remain unchanged, the parameters of the secondary side steam are correspondingly increased.

   At present, the fuel element cladding of the reactor core is widely made of zirconium-4 alloy. Zirconium alloy has good nuclear performance and mechanical properties. Compared with zirconium-2 alloy, zirconium-4 alloy (1.55Sn, 0.20Fe, 0.10Cr, Ni < 0.007) has reduced nickel content and added an equivalent amount of iron to increase its oxidation resistance and mechanical properties at high temperatures. Therefore, the mechanical properties of zirconium-4 alloy are comparable to those of zirconium-2 alloy, but its hydrogen absorption performance is only 1/3–1/2 of that of zirconium-2 alloy.

Zirconium alloy reacts with high-temperature water to produce a metal-water chemical reaction and releases reaction heat. The reaction equation is

$$Zr + 2H_2O \rightarrow ZrO_2 + 2H_2 + \text{heat} \tag{4.24}$$

This process leads to the oxidation of the zirconium cladding, and 1 kg of zirconium oxidation can produce 0.5 m^3 of hydrogen gas and 6500 kJ of heat, the generated hydrogen and heat will cause the casing to become brittle, eventually leading to the component bursting or melting. Therefore, in the thermal design of the reactor, the maximum temperature of the outer surface of the core fuel element casing in contact with the coolant is usually limited to below 350 °C.

From this, it can be seen that the potential to increase the initial parameters of steam by increasing the temperature of the primary side coolant in a PWR NPP is extremely limited.

2. Reduce the average heat exchange temperature difference ΔT_m between the primary and secondary sides of the steam generator. With the parameters of the primary side coolant unchanged, the parameters of the secondary side steam are correspondingly increased.

   Enhanced heat exchange measures can be adopted to reduce the average heat exchange temperature difference ΔT_m when the effective heat transfer area of the steam generator remains unchanged; or ΔT_m can be reduced by increasing the effective heat transfer area F of the steam generator, but both require an increase in equipment investment costs. Properly balancing the parameters of the primary and secondary circuits can minimize the cost of electricity generation. There is a choice of optimal values here, at which point the benefits brought by the increased cycle thermal efficiency just balance the increased investment and power plant operating costs.

4.3 Feedwater Regenerative Cycle

In the Rankine cycle of steam power devices, the heat obtained by the feedwater from the heat source, about 60% is discharged to the cold source in the constant pressure (constant temperature) heat release process 2-3, which is the basic reason for the low economy of the thermal cycle; on the other hand, the constant pressure heat absorption process 4-5-1, the section 4-5 is the process of the feedwater absorbing heat from unsaturated water to become saturated water, the temperature is relatively low, which reduces the average heat absorption temperature of the thermal cycle and affects the economy of the thermal cycle.

The regenerative cycle is to extract steam from the intermediate stage of the turbine, heat the feedwater of the steam generator, raise the temperature of the feedwater before it enters the steam generator, increase the average heat absorption temperature of the secondary circuit working fluid in the steam generator, reduce the temperature difference with the primary side coolant, and reduce the amount of exhaust steam discharged by the turbine and the heat discharged to the cold source, all of which are beneficial to improving the cycle thermal efficiency. Practice has shown that the use of the regenerative cycle can improve the economy of the power plant thermal system by 10–15%, so almost all modern large and medium-sized thermal power plants and NPPs use the regenerative cycle.

Figure 4.13 shows the principle thermal system diagram and *T-s* diagram of the ideal single-stage regenerative cycle.

4.3.1 Feedwater Regenerative Cycle Thermal Efficiency

4.3.1.1 Thermal Efficiency of the Single-Stage Regenerative Cycle

As shown in Fig. 4.13, the unit mass (1 kg) of fresh steam expands adiabatically from state 1 to state 2_e in the turbine, extracts αkg of steam to heat the feedwater, the steam releases heat to raise the temperature of the feedwater, its condensate and feedwater fully mix to become saturated water at state 4_b; the remaining (1-α) kg of steam continues to expand adiabatically to state 2 in the turbine, enters the condenser for constant pressure heat release to become condensate, is pumped into the feedwater heater by the condensate pump to absorb heat, and then is pumped into the steam generator by the feedwater pump.

According to the energy balance equation

$$\alpha\left(h_{2_e}-h_{4_b}\right)=(1-\alpha)\left(h_{4_b}-h_{4_a}\right) \tag{4.25}$$

The expression for the extraction steam quantity α is

$$\alpha=\frac{h_{4b}-h_{4a}}{h_{2e}-h_{4a}} \tag{4.26}$$

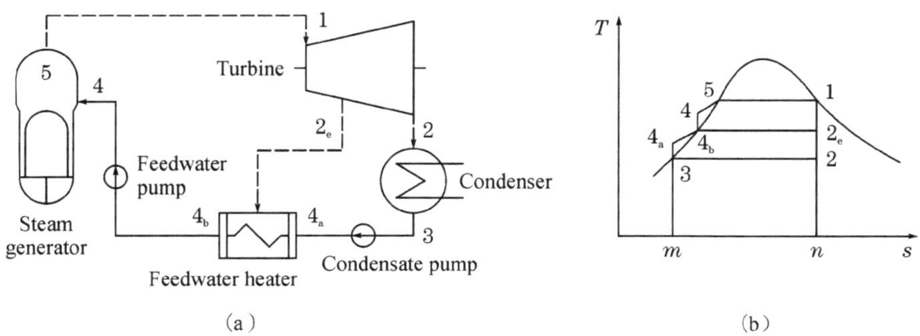

Fig. 4.13 Ideal single-stage feedwater regenerative cycle: (**a**) Thermodynamic system principle flow; (**b**) *T-s* diagram of regenerative cycle

Ignoring the pump work, then the cycle work is

$$w_{\mathrm{T}} \approx \left(h_1 - h_{2_e}\right) + (1-\alpha)\left(h_{2_e} - h_2\right) \tag{4.27}$$

The cycle heat absorption is

$$q_1 = h_1 - h_{4_b} \tag{4.28}$$

The thermal efficiency of the cycle is

$$\eta_t = 1 - \frac{h_2 - h_3}{\left(h_1 - h_1\right) + \frac{\alpha}{1-\alpha}\left(h_1 - h_{2_e}\right)} \tag{4.29}$$

Since $\frac{\alpha}{(1-\alpha)\left(h_1 - h_{2_e}\right)} > 0$, the thermal efficiency of the extraction steam regenerative cycle is greater than that of the Rankine cycle.

The steam consumption rate of the ideal single-stage regenerative cycle is

$$d_0 = \frac{3600}{w_{\mathrm{T}}} = \frac{3600}{\left(h_1 - h_{2_e}\right) + (1-\alpha)\left(h_{2_e} - h_2\right)} \tag{4.30}$$

The above formula can be transformed to obtain

$$d_0 = \frac{3600}{\left(h_1 - h_2\right) - \alpha\left(h_{2_e} - h_2\right)} > \frac{3600}{h_1 - h_2} \tag{4.31}$$

That is, the steam consumption rate of the ideal single-stage regenerative cycle is greater than that of the ideal Rankine cycle.

4.3.1.2 Thermal Efficiency of Multi-Stage Regenerative Cycle

According to thermodynamic principles, when saturated steam is used for work, the corresponding limit regenerative cycle has the same thermal efficiency as the Carnot cycle. However, the limit regenerative cycle requires an infinite number of heating stages, which is unrealistic in system design. In engineering practice, a finite-stage feedwater regenerative system is generally adopted to achieve design optimization.

Figure 4.14 shows a Z-stage feedwater regenerative heating system, with the working fluid based on a unit mass flow rate (1 kg/s). The meanings of the variables in the figure are as follows:

α_i is the amount of extraction steam consumed by the i-th stage heater;
$\alpha_{\mathrm{f},i}$ is the feedwater flow rate at the inlet of the i-th stage heater;
$q_{\mathrm{h},i}$ is the heat released by the condensation of steam in the i-th stage heater;

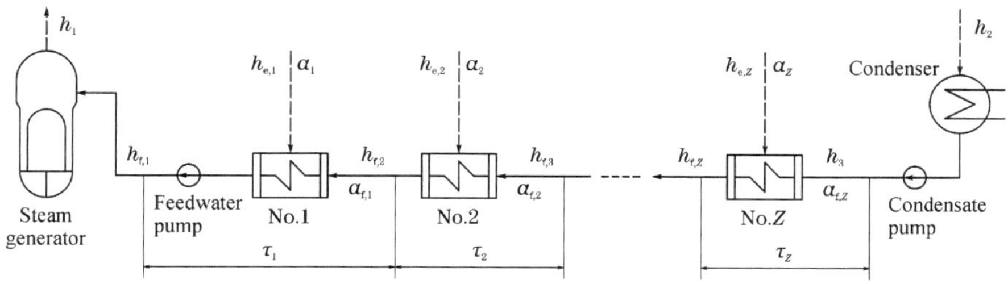

Fig. 4.14 Multi-stage feedwater regenerative cycle

Table 4.1 Interrelationship of main parameters in the regenerative system

Number of reheat stages	Extraction steam flow rate	Feedwater flow rate	Heat extracted by extraction steam	Enthalpy rise of feedwater
1	α_1	$\alpha_{f,1} = 1 - \alpha_1$	$q_{h,1} = h_{e,1} - h_{f,1}$	$\tau_1 = h_{f,1} - h_{f,2}$
2	α_2	$\alpha_{f,2} = 1-\alpha_1-\alpha_2$	$q_{h,2} = h_{e,2} - h_{f,2}$	$\tau_2 = h_{f,2} - h_{f,3}$
⋮	⋮	⋮	⋮	⋮
Z-1	α_{Z-1}	$\alpha_{f,Z\ 1} = 1-\sum_{i=1}^{Z-1}\alpha_i$	$q_{h,Z-1} = h_{e,Z-1} - h_{f,Z-1}$	$\tau_{z-1} = h_{f,Z\ 1} - h_{f,Z}$
Z	α_Z	$\alpha_{f,Z} = 1-\sum_{i=1}^{Z}\alpha_i$	$q_{h,Z} = h_{e,Z} - h_{f,Z}$	$\tau_z = h_{f,Z} - h_3$

τ_i is the enthalpy rise of the feedwater in the i-th stage heater;
$h_{e,i}$ is the enthalpy value of the extraction steam at the i-th stage;
$h_{f,i}$ is the enthalpy value of the feedwater at the outlet of the i-th stage heater;
h_{s0} is the enthalpy of saturated water under the operating pressure of the steam generator.

The relationships between the main parameters are shown in Table 4.1.

Considering the ideal regenerative cycle, it is assumed that all feedwater heaters are of the mixing type, with zero terminal difference, ignoring the pressure loss of extraction steam and the power consumption of the feedwater pump, and neglecting the heat loss of the feedwater regenerative system.

According to the heat balance equation of the first stage heater

$$\alpha_1\left(h_{e,1} - h_{f,1}\right) = \left(1-\alpha_1\right)\left(h_{f,1} - h_{f,2}\right) \tag{4.32}$$

Substituting $q_{h,1}$, τ_1 into the above equation, after rearranging, we get

$$\begin{cases} \alpha_1 = \dfrac{\tau_1}{q_{h,1} + \tau_1} \\ \alpha_{f,1} = 1 - \alpha_1 = \dfrac{q_{h,1}}{q_{h,1} + \tau_1} \end{cases} \tag{4.33}$$

From the heat balance equation of the second stage heater

$$\alpha_2 q_{\text{h},2} = (1-\alpha_1-\alpha_2)\tau_2 \tag{4.34}$$

We can get

$$\begin{cases} \alpha_2 = (1-\alpha_1)\dfrac{\tau_2}{q_{\text{h},2}+\tau_2} = \left(\dfrac{q_{\text{h},1}}{q_{\text{h},1}+\tau_1}\right)\left(\dfrac{\tau_2}{q_{\text{h},2}+\tau_2}\right) \\ \alpha_{\text{f},2} = 1-\alpha_1-\alpha_2 = \left(\dfrac{q_{\text{h},1}}{q_{\text{h},1}+\tau_1}\right)\left(\dfrac{q_{\text{h},2}}{q_{\text{h},2}+\tau_2}\right) \end{cases} \tag{4.35}$$

In this way, we can get the corresponding coefficients of the Z-stage heater

$$\begin{cases} \alpha_Z = \left(\dfrac{q_{\text{h},1}}{q_{\text{h},1}+\tau_1}\right)\left(\dfrac{q_{\text{h},2}}{q_{\text{h},2}+\tau_2}\right)\cdots\left(\dfrac{q_{\text{h},Z\ 1}}{q_{\text{h},Z\ 1}+\tau_{Z-1}}\right)\left(\dfrac{\tau_Z}{q_{\text{h},Z}+\tau_Z}\right) \\ \alpha_{f,Z} = \left(\dfrac{q_{\text{h},1}}{q_{\text{h},1}+\tau_1}\right)\left(\dfrac{q_{\text{h},2}}{q_{\text{h},2}+\tau_2}\right)\cdots\left(\dfrac{q_{\text{h},Z}}{q_{\text{h},Z}+\tau_Z}\right) = \displaystyle\prod_{i=1}^{Z}\dfrac{1}{1+\dfrac{\tau_i}{q_{\text{h},i}}} \end{cases} \tag{4.36}$$

In the thermal cycle, the heat q1 absorbed by the working fluid from the heat source and the heat q_2 discharged to the cold source are

$$\begin{cases} q_1 = h_1 - h_{\text{f},1} \\ q_2 = (1-\alpha_1-\alpha_2-\cdots-\alpha_z)(h_2-h_3) = \alpha_{\text{i},z}(h_2-h_3) \end{cases} \tag{4.37}$$

Then the efficiency of the regenerative cycle is

$$\eta_\text{t} = 1-\frac{q_2}{q_1} = 1-\frac{h_2-h_3}{h_1-h_{\text{f},1}}\prod_{i=1}^{z}\frac{1}{1+\dfrac{\tau_i}{q_{\text{h},i}}} \tag{4.38}$$

Let $q_{\text{h},0} = h_1 - h_{\text{s},0}$, $q_\text{c} = h_2 - h_3$, $\tau_0 = h_{\text{s},0} - h_{\text{f},1}$, then the above equation becomes

$$\eta_\text{t} = 1-\frac{q_\text{c}}{q_{\text{h},0}+\tau_0}\prod_{i=1}^{z}\frac{1}{1+\dfrac{\tau_i}{q_{\text{h},i}}} \tag{4.39}$$

In the case of multi-stage extraction regeneration, the theoretical power of the steam turbine unit is

$$w_T = (h_1 - h_{e,1}) + (1-\alpha_1)(h_{e,1} - h_{e,2}) + (1-\alpha_1-\alpha_2)(h_{e,2} - h_{e,3}) + \cdots + \left(1-\sum_{i=1}^{z}\alpha_i\right)(h_{e,Z} - h_2) \tag{4.40}$$
$$= (h_1 - h_2) - \alpha_1(h_{e,1} - h_2) - \alpha_2(h_{e,2} - h_2) - \cdots - \alpha_z(h_{e,Z} - h_2)$$

Let $Y_i = \dfrac{h_{e,i} - h_2}{h_1 - h_2}$, defined as the coefficient of insufficient work done by regenerative extraction steam. Then the above equation becomes

$$w_T = (h_1 - h_2)(1 - \alpha_1 Y_1 - \alpha_2 Y_2 - \cdots - \alpha_Z Y_Z) = (h_1 - h_2)\left(1 - \sum_{i=1}^{z}\alpha_i Y_i\right) \tag{4.41}$$

Ignoring the power consumption of the feedwater pump, the steam consumption rate of the multi-stage regenerative cycle is

$$d_0 = \frac{3600}{w_T} = \frac{3600}{(h_1 - h_2)\left(1 - \sum_{i=1}^{z}\alpha_i Y_i\right)} \quad \mathrm{kg/(kW\cdot h)} \tag{4.42}$$

Let $\beta = \dfrac{1}{1 - \sum_{i=1}^{z}\alpha_i Y_i}$, defined as the coefficient of increased steam consumption due to regenerative extraction, then the above equation becomes

$$d_0 = \frac{3600}{h_1 - h_2}\beta > \frac{3600}{h_1 - h_2} \tag{4.43}$$

It can be seen that after using extraction regeneration, the steam consumption rate increases compared with the ideal Rankine cycle.

Under the condition of limited extraction and regenerative heating, the irreversible loss of the entire heat absorption process (including feedwater heating, heat absorption and evaporation in the steam generator) of the reheat cycle is equal to the sum of the following two losses:

1. The loss of work ability caused by a certain temperature difference between the extraction steam and the feedwater in each heater, that is, the loss of work ability caused by the irreversibility of the internal process of each feedwater heater.
2. Compared with the Carnot cycle at the same temperature limit, it is the heat exchange between the heat source at temperature T_0 and the working substance (feedwater or steam), which causes a certain temperature difference and results in the loss of work ability, that is, the loss of work ability caused by the external irreversibility of regeneration.

4.3.2 *Main Factors Affecting the Thermal Economy of the Regenerative Cycle*

The main parameters affecting the thermal economy of the feedwater regenerative cycle are the regenerative heating distribution ratio τ, the corresponding optimal feedwater temperature $T_{\mathrm{fw,\,opt}}$, and the number of regeneration stages Z. The three are closely related and influence each other.

4.3.2.1 Optimal Regenerative Distribution

For the Z-stage ideal reheat cycle, the optimal regenerative distribution is obtained by seeking the maximum thermal efficiency. The optimal regenerative distribution should make η_{t} maximum, that is, seek the extreme value under the following conditions:

$$\frac{\partial \eta_{\mathrm{t}}}{\partial h_{\mathrm{f},1}} = 0, \frac{\partial \eta_{\mathrm{t}}}{\partial h_{\mathrm{f},2}} = 0, \cdots, \frac{\partial \eta_{\mathrm{t}}}{\partial h_{\mathrm{f},\mathrm{Z}}} = 0$$

Equation (4.38) is written in the following form:

$$\eta_{\mathrm{t}} = 1 - \frac{q_{\mathrm{c}} q_{\mathrm{h},1} q_{\mathrm{h},2} \cdots q_{\mathrm{h},\mathrm{Z}}}{\left(q_{\mathrm{h},0} + \tau_0\right)\left(q_{\mathrm{h},1} + \tau_1\right)\left(q_{\mathrm{h},2} + \tau_2\right) \cdots \left(q_{\mathrm{h},\mathrm{Z}} + \tau_{\mathrm{Z}}\right)} \tag{4.44}$$

When the initial and final parameters of the steam in the thermodynamic cycle are given, h_1, h_2, $q_{\mathrm{h},\,0}$, q_{c} are all constants.

First, find the partial derivative of η_{t} with respect to $h_{\mathrm{f},\,1}$, that is, $\frac{\partial \eta_{\mathrm{t}}}{\partial h_{\mathrm{f},1}} = 0$, considering that

$$\frac{\partial \tau_0}{\partial h_{\mathrm{f},1}} = -1, \frac{\partial \tau_1}{\partial h_{\mathrm{f},1}} = 1, \frac{\partial q_{\mathrm{h},1}}{\partial h_{\mathrm{f},1}} = q_{\mathrm{h},1}'$$

And the rest of τ, q is irrelevant to $h_{\mathrm{f},\,1}$, that is,

$$\frac{q_{\mathrm{h},2} \cdots q_{\mathrm{h},\mathrm{Z}}}{\left(q_{\mathrm{h},2} + \tau_2\right) \cdots \left(q_{\mathrm{h},\mathrm{Z}} + \tau_{\mathrm{Z}}\right)} = \text{constant}$$

Therefore, from $\frac{\partial \eta_{\mathrm{t}}}{\partial h_{\mathrm{f},1}} = 0$ we can get

$$\frac{\partial}{\partial h_{\mathrm{f},1}} \left[\frac{q_{\mathrm{h},1}}{\left(q_{\mathrm{h},2} + \tau_0\right)\left(q_{\mathrm{h},\mathrm{Z}} + \tau_{\mathrm{Z}}\right)} \right] = 0 \tag{4.45}$$

After expanding, we can get

$$\left(q_{\text{h},0}+\tau_0\right)-\left(q_{\text{h},1}+\tau_1\right)-\left(q_{\text{h},0}+\tau_0\right)\tau_1\frac{q'_{\text{h},1}}{q_{\text{h},1}}=0 \tag{4.46}$$

That is

$$\tau_1=\frac{q_{\text{h},0}+\tau_0-q_{\text{h},1}}{1+\left(q_{\text{h},0}+\tau_0\right)\dfrac{q'_{\text{h},1}}{q_{\text{h},1}}} \tag{4.47}$$

Then find the partial derivative of η_t with respect to $h_{\text{f},2}$ similarly we can get

$$\frac{\partial}{\partial h_{\text{f},2}}\left[\frac{q_{\text{h},2}}{\left(q_{\text{h},1}+\tau_1\right)\left(q_{\text{h},2}+\tau_2\right)}\right]=0 \tag{4.48}$$

After expanding, we can get

$$\left(q_{\text{h},1}+\tau_1\right)-\left(q_{\text{h},2}+\tau_2\right)-\left(q_{\text{h},1}+\tau_1\right)\tau_2\frac{q'_{\text{h},2}}{q_{\text{h},2}}=0 \tag{4.49}$$

That is

$$\tau_2=\frac{q_{\text{h},1}+\tau_1-q_{\text{h},2}}{1+\left(q_{\text{h},1}+\tau_1\right)\dfrac{q'_{\text{h},2}}{q_{\text{h},2}}} \tag{4.50}$$

In this way, we can get

$$\tau_\text{Z}=\frac{q_{\text{h},\text{Z}\;1}+\tau_{\text{Z}-1}-q_{\text{h},\text{Z}}}{1+\left(q_{\text{h},\text{Z}\;1}+\tau_{\text{Z}-1}\right)\dfrac{q'_{\text{h},z}}{q_{\text{h},\text{Z}}}} \tag{4.51}$$

Equation (4.51) is the general formula for the optimal reheat distribution of the ideal reheat cycle.

If further simplified, ignoring some minor factors, other more approximate optimal regenerative distribution formulas can be obtained. For example, if we ignore the change of q with τ, that is $\dfrac{\partial q_{\text{h},\text{Z}}}{\partial h_{\text{f},z}}=q'_{\text{h},z}=0$, then Eq. (4.51) can be simplified as

$$\begin{aligned}\tau_\text{Z}&=q_{\text{h},\text{Z}\;1}+\tau_{\text{Z}-1}-q_{\text{h},\text{Z}}\\&=\left(h_{\text{e},\text{Z}\;1}-h_{\text{f},\text{Z}\;1}\right)+\left(h_{\text{f},\text{Z}\;1}-h_{\text{f},\text{Z}}\right)-\left(h_{\text{e},\text{Z}}-h_{\text{f},\text{Z}}\right)\\&=h_{\text{e},\text{Z}\;1}-h_{\text{e},\text{Z}}\end{aligned} \tag{4.52}$$

This regenerative distribution method is to make the enthalpy increase of the feedwater in each heater equal to the enthalpy decrease of the steam from the previous stage to this stage in the turbine, which is simply called the "enthalpy drop distribution method.

There are also other regenerative distribution methods, the common ones are as follows:

1. Average distribution method

   This regenerative distribution method is to make the enthalpy increase of each heater equal, that is,

$$\tau_Z = \tau_{Z\,1} = \cdots = \tau_2 = \tau_1 = \tau_0 = \frac{h_{s,0} - h_3}{Z+1} \tag{4.53}$$

2. Equal enthalpy drop distribution method

   This regenerative distribution method is to make the enthalpy increase of the feedwater in each heater equal to the enthalpy drop of each stage of the turbine, that is,

$$h_1 - h_{e,1} = h_{e,1} - h_{e,2} = h_{e,2} - h_{e,3} = \cdots = h_{z-1} - h_z \tag{4.54}$$

3. Geometric series distribution method

   This regenerative distribution method is to distribute the enthalpy increase of the feedwater in each heater according to the geometric series, that is,

$$\frac{\tau_0}{\tau_1} = \frac{\tau_1}{\tau_2} = \frac{\tau_2}{\tau_3} = \cdots = \frac{\tau_{Z-1}}{\tau_Z} = m \tag{4.55}$$

Generally, $m = 1.01$–1.04.

The distribution of extraction points determined by the best regeneration distribution formula is often difficult to strictly adhere to in actual power plants. The number of stages of the steam turbine is limited, the pressure in the extraction chamber is also fixed, and may be different from the pressure that ensures the best thermal efficiency; in addition, the setting of the deaerator and the need to extract steam from the steam turbine exhaust pipe will also make some changes to the calculated distribution of extraction points. However, minor changes usually do not cause significant changes in thermal efficiency.

When finally determining the distribution of extraction points, it is usually necessary to compare the device efficiency calculated based on the selected system with the efficiency obtained from the optimal distribution of extraction points in each specific situation.

4.3.2.2 Optimal Feedwater Temperature

Feedwater regenerative improves the average temperature of the heat absorption process in the cycle, reduces the loss of the cold source, and improves the thermal economy of the cycle. The feedwater temperature that maximizes the thermal economy of the cycle is called the optimal feedwater temperature, which is closely related to the number of regeneration stages and the distribution method of feedwater regeneration.

Figure 4.15 shows the relationship curve between feedwater temperature T_{fw} and cycle efficiency η_t, heat consumption rate q_0, steam consumption rate d_0, and cycle heat absorption q_1 when single-stage regeneration.

From the figure, it can be seen that if the feedwater temperature T_{fw} is equal to the condensate temperature T_{cd} at the outlet of the condenser, there is no regeneration at this time, and the cycle thermal efficiency η_t is equal to the Rankine cycle thermal efficiency $\eta_{t,R}$. When using extraction steam regeneration to heat the feedwater, the feedwater temperature increases with the increase of extraction pressure, and the thermal efficiency also increases. When the extraction pressure reaches a certain value, the cycle efficiency reaches the maximum value, and the corresponding feedwater temperature is called the optimal feedwater temperature $T_{fw,opt}$.

If the extraction pressure continues to increase, although the feedwater temperature increases accordingly, the cycle thermal efficiency begins to decrease. The reason is that although the cycle heat absorption q_1 continues to decrease, the work done by the unit mass of steam in the steam turbine decreases accordingly, which inevitably increases the steam consumption rate d_0, increases the loss of the cold source, and when the increase is faster, the heat consumption rate q_0 continues to increase, and the cycle thermal efficiency η_t decreases accordingly. When the extraction pressure is increased to the new steam pressure, the cycle thermal efficiency η_t

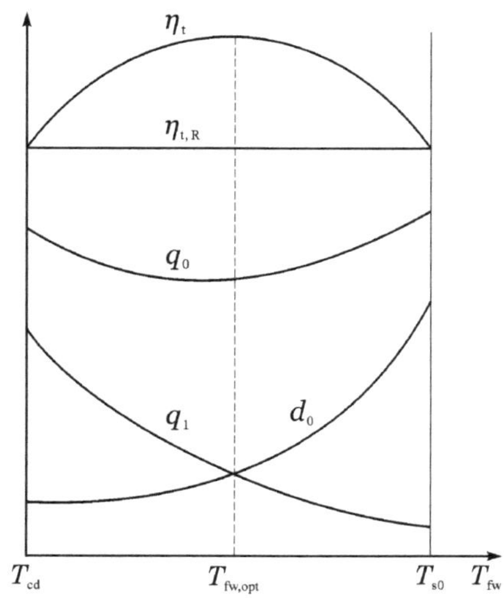

Fig. 4.15 The relationship curve between feedwater temperature

is equal to the Rankine cycle thermal efficiency $\eta_{t,R}$, that is, the feedwater regeneration at this time is no longer beneficial to improve the cycle thermal efficiency.

Theoretical analysis results show that the optimal feedwater temperature for single-stage regeneration is

$$T_{fw,opt} = \frac{T_{s,0} - T_{cd}}{2} \tag{4.56}$$

Where $T_{s,0}$ is the saturation water temperature under the operating pressure of the steam generator.

For multi-stage extraction steam regenerative cycle, there is also an optimal feedwater temperature, which is related to the number of regenerative stages and the regenerative distribution method.

Using different regenerative distribution methods, the expressions for the optimal feedwater temperature obtained are also different. If the regenerative distribution is carried out according to the average distribution method, the enthalpy value corresponding to the optimal feedwater temperature is

$$h_{fw,opt} = h_3 + Z\tau_Z = h_3 + \frac{Z}{Z+1}\left(h_{s,0} - h_3\right) \tag{4.57}$$

When the regenerative distribution is carried out according to the enthalpy drop distribution method, the enthalpy value corresponding to the optimal feedwater temperature is

$$h_{fw,opt} = h_3 + \sum_{i=1}^{Z}\tau_i = h_3 + \left(h_1 - h_{e,Z}\right) \tag{4.58}$$

When the regenerative distribution is carried out according to the geometric series distribution method, the enthalpy value corresponding to the optimal feedwater temperature is

$$h_{fw,opt} = h_s + \sum_{i=1}^{z}\tau_i = h_3 + \tau_z\left(m^{z-1} + \cdots + m + 1\right) = h_3 + \tau_Z\frac{m^z - 1}{m-1} \tag{4.59}$$

When considering the economic benefits of the feedwater regenerative system, one cannot simply pursue thermal economy, but must also comprehensively consider the technical and economic indicators of the power plant, including system and equipment investment, operation and maintenance costs, depreciation costs, and fuel prices. While increasing the feedwater temperature is beneficial for improving the thermal efficiency of the NPP cycle and reducing the consumption of nuclear fuel, adding a feedwater regenerative system will inevitably complicate the thermal system of the NPP, increasing equipment investment and operating costs.

The optimal feedwater temperature determined through technical and economic comparison, referred to as the economically optimal feedwater temperature, is obviously lower than the theoretically optimal feedwater temperature.

4.3.2.3 Optimal Number of Regeneration Stages

If the feedwater regenerative distribution adopts the average distribution method, then q and τ are constants, and Eq. (4.44) can be written as

$$\eta_t = 1 - \left(\frac{q}{q+\tau}\right)^{Z+1} = 1 - \frac{1}{\left(1+\frac{\tau}{q}\right)^{Z+1}} = 1 - \frac{1}{\left[1+\frac{h_{s,0}-h_s}{(Z+1)q}\right]^{Z+1}} \tag{4.60}$$

Let $\frac{h_{s,0}-h_3}{q} = M$, when the initial and final parameters of the thermal cycle are given, M is a constant.

When $Z = \infty$, it can be obtained from the above formula

$$\eta_t = 1 - \frac{1}{e^M} \tag{4.61}$$

From Eq. (4.60), it can be seen that η_t is a function that increases with Z, that is, the larger Z is, the higher η_t is, but $\left(1+\frac{M}{Z+1}\right)^{Z+1}$ is a convergent series, and the growth rate is decreasing, with the extreme value being e^M.

Figure 4.16 shows the relationship between the optimal feedwater temperature and the number of regeneration stages and the thermal efficiency of the cycle. The coordinates in the figure are defined as

$$\Phi = \frac{\Delta\eta_{t,Z}}{\Delta\eta_{t,\infty}}$$

$$\mu = \frac{T_{fw}-T_{cd}}{T_{s,0}-T_{cd}}$$

Where $\Delta\eta_{t,Z}$, $\Delta\eta_{t,\infty}$ the increase in cycle thermal efficiency when the number of regeneration stages is Z and ∞ respectively.

From the curve in the figure, it can be seen that the basic rule of the regenerative cycle is:

1. The cycle thermal efficiency η_t is a function that increases with the number of regeneration stages Z, that is, the thermal efficiency of the regenerative cycle increases with the increase of the number of regeneration stages; at the same time, it is a convergent series, and the increase decreases with the increase of the series, as shown in Table 4.2.
2. When the feedwater temperature is constant, the thermal economy also increases with the increase of the number of regeneration stages, but its growth rate is also decreasing.

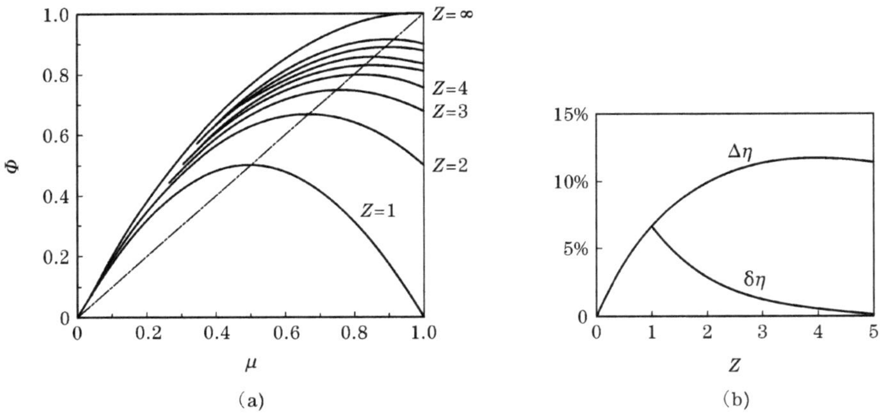

Fig. 4.16 The relationship curve between feedwater temperature

Table 4.2 Thermal efficiency of reheat stages

Item	Number of reheat stages Z								
	0	1	2	3	4	5	6	···	Z
Φ	0	$\frac{1}{2}$	$\frac{2}{3}$	$\frac{3}{4}$	$\frac{4}{5}$	$\frac{5}{6}$	$\frac{6}{7}$	···	$\frac{z}{z+1}$
η_t diminishing marginal returns	$\frac{1}{2}$	$\frac{1}{6}$	$\frac{1}{12}$	$\frac{1}{20}$	$\frac{1}{30}$	$\frac{1}{42}$	···	$\frac{z}{z(z+1)}$	

3. For any number of regeneration stages, there is a corresponding optimal feedwater temperature, which increases with the increase of the series.
4. For the actual feedwater temperature of any number of regeneration stages, although it deviates from the optimal value, it has little effect on the thermal economy.

As we have seen before, for the same number of feedwater regenerative heaters, the amount of heating to the feedwater that is economically reasonable is not equal to the optimal value from the perspective of thermal efficiency, the former is always less than the latter. For a given Z number of regenerative heaters, there exists a specific feedwater temperature T_{fw}, at which the unit discount cost $z_{e,\,opt}$ is the smallest. By comparing the $z_{e,\,opt}$ calculated for different numbers of heaters Z, it is easy to find the optimal number of regenerative heaters z_{opt} and the optimal feedwater temperature $T_{fw,\,opt}$ when the best overall economy ($z_{e,\,opt}$ is the smallest) can be achieved.

The optimal number of feedwater regeneration stages also needs to be determined based on a comprehensive technical and economic analysis, so the actual optimal number of regeneration stages is always less than the theoretical optimal number of regeneration stages.

Table 4.3 gives the forms of the feedwater regenerative system under different new steam parameters and their impact on cycle efficiency, and Table 4.4 lists the

Table 4.3 Effects of feedwater regeneration at various steam conditions

Steam pressure/MPa	Steam temperature/°C	Regeneration stages	Feedwater temperature/°C	Cycle efficiency increment $\Delta\eta_t$/%
3.5	435	3	145	7–8
9	535	5	215	11–13
13	535	7	230	15–16
24	560	8–9	245	17–18

Table 4.4 Feedwater temperature and number of regeneration stages adopted by large power plants in the United States

Steam pressure/MPa	Steam temperature/°C	Feedwater temperature/°C	Number of regeneration stages
10.0–12.7	538–566	230–240	5–6
14–18	538–566	240–260	6–7
⩾24	538–566	260–280	7–9

Table 4.5 Key parameters of feedwater regenerative systems in different nuclear power units

Key parameters	unit	Marble Hill	Tricastin	Standard	Qinshan Phase I	Daya Bay
Electric Power	MWe	1150	920	1000	300	925
New Steam Pressure	MPa	6.82	5.66	6.45	5.34	6.1
New Steam Temperature	°C	Saturated	273	280	268	278
Number of Reheat Stages	–	5 + deaerator	6	6 + deaerator	6 + deaerator	6 + deaerator
Feedwater Temperature	°C	226.7	219.4	218	221.5	226

feedwater temperatures and number of regeneration stages used by large American units under different new steam parameters.

The saturation temperature corresponding to the initial pressure of steam in modern PWR NPPs is about 280–290 °C, and the feedwater temperature used is almost all within the range of 20–240 °C. Table 4.5 shows the data of the feedwater regenerative system of some typical NPPs.

Example 4.2 The nuclear power unit shown in Fig. 4.13 uses a reheating cycle, the fresh steam pressure at the outlet of the steam generator is 4 MPa, the dryness is 0.9975, the extraction pressure of the steam turbine is 0.2 MPa, and the exhaust steam condensation pressure is 4 kPa. Considering the ideal reheating cycle, try to calculate the thermal efficiency and steam consumption rate.

Solution First, check the water and steam parameter table to determine the thermodynamic parameters of each state point.

State 1: Fresh steam at the outlet of the steam generator.

Known $p_1 = 4$ MPa, $x_1 = 0.9975$, the enthalpy and entropy of steam are respectively

$$h_1 = 2796.02\text{kJ/kg}, s_1 = 6.0603\text{kJ/(kg}\cdot\text{K)}$$

State 2_e: Parameters at the extraction point of the steam turbine.
Known $p_{2_e} = 0.2$ MPa, since $1-2-e$ is an adiabatic expansion process, we have

$$s_{2_e} = s_1 = 6.0603\,\text{kJ/(kg}\cdot\text{K)}.$$

The dryness and enthalpy of the extracted steam are respectively

$$x_{2_e} = \frac{s_{2_e} - s'_{2_e}}{s''_{2_e} - s'_{2_e}} = \frac{6.0603 - 1.5301}{7.1268 - 1.5301} = 0.809$$

$$h_{2_e} = h'_{2_e} + x_{2_e}\left(h''_{2_e} - h'_{2_e}\right) = 504.7 + 0.809 \times (2706.3 - 504.7)$$
$$= 2285.79\text{kJ/kg}$$

State 2: Exhaust steam at the outlet of the steam turbine.
Known $p_2 = 4$ kPa, 1—2 is an adiabatic expansion process, we have $s_2 = s_1 = 6.0603$ kJ/(kg · K). The dryness and enthalpy of the exhaust steam are respectively

$$x_2 = 0.70, h_2 = 1824.57\text{kJ/kg}$$

State 3: Condensed water at the outlet of the condenser

$$p_3 = 4\text{kPa}, h_3 = 121.41\text{kJ/kg}, v_3 = 1.0041 \times 10^{-3}\text{m}^3/\text{kg}$$

State 4_a: Feedwater at the inlet of the feedwater heater

$$p_{4_a} \approx p_{2_e} = 0.2\,\text{MPa}$$

$$h_{4_a} = h_3 + v_3\left(p_{4_a} - p_3\right) = 121.41 + 1.0041 \times 10^{-3} \times (0.2 - 0.004) \times 10^3$$
$$= 121.61\text{kJ/kg}$$

State 4b: Feedwater at the outlet of the feedwater heater

$$p_{4_b} \approx p_{2_e} = 0.2\text{MPa}, h_{4_b} = h'_{2_e} = 504.7\text{kJ/kg}, v_{4_b} = 1.0608 \times 10^{-3}\text{m}^3/\text{kg}$$

State 4: Feedwater at the inlet of the steam generator

$$p_4 \approx p_1 = 4\text{MPa}$$
$$h_4 = h_{4_b} + v_{4_b}\left(p_4 - p_{4_b}\right) = 504.7 + 1.0608 \times 10^{-3} \times (4 - 0.2) \times 10^3$$
$$= 508.73\text{kJ/kg}$$

The amount of steam extracted is

$$\alpha = \frac{h_{4_b} - h_{4_a}}{h_{2_e} - h_{4_a}} = \frac{504.7 - 121.61}{2285.79 - 121.61} = 0.177$$

The work done by the steam turbine is

$$\begin{aligned} w_T &= \left(h_1 - h_{2_e}\right) + (1-\alpha)\left(h_{2_e} - h_2\right) \\ &= (2796.02 - 2285.79) + (1 - 0.177)(2285.79 - 1824.57) \\ &= 889.81\,\text{kJ / kg} \end{aligned}$$

The power consumed by the water pump is

$$\begin{aligned} w_P &= (1-\alpha)\left(h_{4_a} - h_3\right) + \left(h_4 - h_{4_b}\right) \\ &= (1 - 0.177)(121.61 - 121.42) + (515.2 - 504.8) \\ &= 4.19\,\text{kJ / kg} \end{aligned}$$

The net work of the cycle is

$$w_s - w_p = 889.81 - 4.19 = 885.62\,\text{kJ / kg}$$

The heat absorbed by the working fluid

$$q_1 = h_1 - h_4 = 2796.02 - 508.73 = 2287.29\,\text{kJ / kg}$$

The thermal efficiency of the cycle

$$\eta_1 = \frac{w_T - w_p}{q_1} = \frac{885.62}{2287.29} = 0.3872$$

The steam consumption rate of the cycle is

$$d_0 = \frac{1}{w_T - w_P} = \frac{3600}{885.62} = 4.065\,\text{kg} / (\text{kW} \cdot \text{h})$$

Compared with Example 4.1, due to the use of steam extraction for reheating, the net work of the cycle is reduced

$$\Delta w_0 = 971.45 - 885.62 = 85.83\,\text{kJ / kg}$$

The heat load of the steam generator is reduced

$$\Delta q_1 = 2670.6 - 2287.29 = 383.31\,\text{kJ / kg}$$

The thermal efficiency is improved

$$\Delta\eta_t = 0.3872 - 0.3622 = 0.025$$

The steam consumption rate increases

$$\Delta d_0 = 4.065 - 3.72 = 0.345\text{kg}/(\text{kW}\cdot\text{h})$$

4.4 Steam Reheating Cycle

4.4.1 Ideal Steam Reheating Cycle

For the Rankine cycle of saturated steam, increasing the initial parameters of steam can improve the thermal efficiency of the cycle, but it will cause the dryness of the exhaust steam to decrease, which will have a negative impact on the operation of the steam turbine. In order to improve the thermal efficiency of the cycle and solve the problem of excessive humidity of the exhaust steam, the steam reheating cycle has been studied and developed. Figure 4.17a is the principle thermodynamic system diagram and Fig. 4.17b is the *T-s* diagram of a single-stage reheating cycle.

Currently, most large-scale thermal power plants adopt a steam reheat system, the main purpose of which is to improve the steam parameters at the inlet of the low-pressure cylinder, thereby enhancing the thermal economy of large-capacity units. However, for PWR NPPs, the main purpose of reheating is to improve the dryness of the steam at the expansion endpoint in the turbine. Taking a saturated steam nuclear turbine as an example, if no measures are taken, when the steam expands to 5 kPa, its steam humidity will approach 30%. In order to ensure the safe operation of the low-pressure cylinder of the turbine unit, an intermediate steam-water separator and a low-pressure cylinder inter-stage dehumidification structure are set up, but the final stage blade humidity is still close to 20%; on this basis, the steam is further reheated, and the steam is heated to superheat, then the final stage blade humidity is about 11%. Therefore, after the nuclear turbine adopts the steam

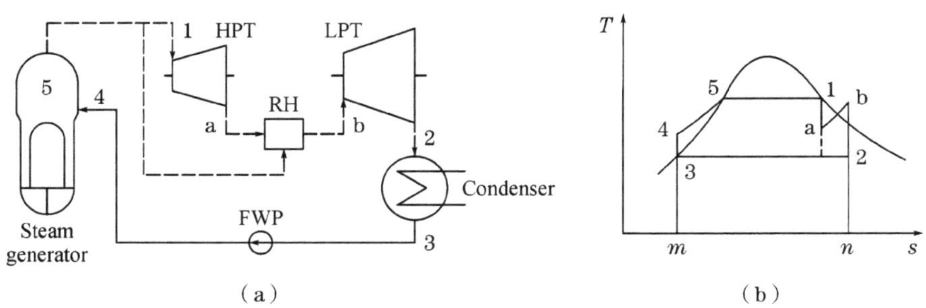

Fig. 4.17 Schematic thermodynamic system and *T-s* diagram for single-reheat cycle

reheat measures, the final stage humidity is similar to that of conventional power plant units. The ideal steam reheat cycle can be seen as a cycle composed of a basic cycle without reheating (i.e., the Rankine cycle) and an additional cycle constituted by the reheating process. Whether the use of steam intermediate reheating can improve the thermal efficiency of the entire reheat cycle depends on whether the average heat absorption temperature of the additional cycle is higher than the corresponding value of the basic cycle. If the pump work is ignored, the thermal efficiency of the ideal reheat cycle is

$$\eta_t = \frac{w_0}{q_1} = \frac{(h_1 - h_a) + (h_b - h_2)}{(h_1 - h_4) + (h_b - h_a)} = \frac{(h_1 - h_2) + \Delta q_{ch}}{(h_1 - h_4) + \Delta q_{ch}} \tag{4.62}$$

Where Δq_{rh} is the heat absorbed by the intermediate reheat, $\Delta q_{rh} = h_b - h_a$.

The steam consumption rate of the ideal reheat cycle is

$$d_0 = \frac{3600}{w_0} = \frac{3600}{(h_1 - h_2) + \Delta q_{rh}} \tag{4.63}$$

Since $\Delta q_{rh} > 0$, under the same initial and final steam parameters, the steam consumption rate of the ideal reheat cycle is lower than that of the ideal Rankine cycle. From the above two formulas, it can be seen that the thermal efficiency of the reheat cycle is related to the intermediate reheat pressure p_a. If the steam temperature Tb is the same after reheating, increasing the intermediate reheat pressure can improve the cycle thermal efficiency, but it will reduce the improvement degree of the dryness of the expansion endpoint exhaust steam. When selecting the intermediate reheat pressure, it is necessary to improve the dryness of the exhaust steam and strive to achieve the purpose of improving the cycle thermal efficiency as much as possible. Therefore, the value of the optimal reheat pressure needs to be determined based on a comprehensive technical and economic analysis of the given cycle conditions. After adopting the reheat cycle, the work done by the unit mass of steam increases, so the steam consumption rate can be reduced, which is beneficial to reduce the burden on the feedwater pump and condenser. However, due to the increase in pipelines, valves, and related equipment, investment costs increase, and operation management costs also increase.

4.4.2 *Optimal Reheat Pressure*

The optimal reheat pressure involves many factors, such as the initial pressure and temperature of the steam, the internal efficiency of the turbine before and after the intermediate reheat, the temperature and pressure after the intermediate reheat, the feedwater temperature, and the distribution of the regeneration. The steam returning from the high-pressure cylinder of the steam turbine to the low-pressure cylinder, due to flow resistance, causes a pressure drop, known as reheat pressure loss. Reheat pressure loss reduces the thermal economy of the unit. Although reducing reheat pressure loss can improve the thermal economy of the unit, it requires increasing the pipe diameter, and the metal consumption and investment costs will increase accordingly. Generally, the reheat pressure loss is taken as 8–12% of the exhaust pressure of the high-pressure cylinder. The currently used intermediate reheat pressure is 20–30% of the initial pressure, at which time the cycle thermal efficiency can be increased by 2–5%. Multiple reheat cycles can also be used, as long as it is economically and technically reasonable.

Example 4.3 shows a nuclear power device with intermediate reheating, as shown in Fig. 4.17. The new steam pressure at the outlet of the steam generator is 4 MPa, with a dryness of 0.9975. The steam is adiabatically expanded in the high-pressure cylinder and then heated to 230 °C in the steam reheater before entering the low-pressure cylinder for adiabatic expansion. The condensation pressure of the spent steam is 4 kPa. Assuming that the pressurization process of the condensate in the pump is also isentropic, when the exhaust pressure of the high-pressure cylinder is 0.8, 0.9, and 1.0 MPa, try to calculate the corresponding cycle thermal efficiency, spent steam dryness, and steam consumption rate.

The calculation process and results are listed in the Table 4.6, referring to Example 4.1.

Table 4.6 Calculation process and results

Parameters	Unit	Calculation formula or source	Computational values		
High pressure cylinder exhaust pressure	MPa	Given	0.8	0.9	1
High pressure cylinder exhaust entropy	kJ/(kg·K)	$s_a = s_1$	6.0603	6.0603	6.0603
High pressure cylinder exhaust steam dryness		$x_a = \frac{s_a - s_a^{'}}{s_a^{''} - s_a^{'}}$	0.87	0.8765	0.882
High pressure cylinder exhaust enthalpy	kJ/kg	$h_a = h_a^{'} + x_a\left(h_a^{'} - h_a^{'}\right)$	2501.45	2521.46	2538.6
Reheat steam pressure	MPa	$p_b = p_a$	0.8	0.9	1.0
Reheat steam temperature		Given	230	230	230
Reheat steam specific enthalpy	kJ/kg	$h_b = f(p_b, T_b)$	2906.6	2902.2	2897.8
Reheat steam specific entropy	°C	$s_b = f(p_b, T_b)$	6.9542	6.8931	6.8377
Low pressure cylinder exhaust pressure	MPa	Given	0.004	0.004	0.004
Low pressure cylinder exhaust specific entropy	kJ/(kg·K)	$s_2 = s_b$	6.9542	6.8931	6.8377
Low pressure cylinder exhaust steam dryness		$x_2 = \frac{s_2 - s_2^{'}}{s_2^{''} - s_2^{'}}$	0.811	0.80	0.797

(continued)

Table 4.6 (continued)

Parameters	Unit	Calculation formula or source	Computational values		
Low pressure cylinder exhaust specific enthalpy	kJ/kg	$h_2 = h_2' + x_2\left(h_2'' - h_2'\right)$	2094.65	2067.88	2606.58
Heat absorption during the Cycle	kJ/kg	$q_1 = (h_1 - h_4) + (h_b - h_a)$	3075.74	3051.33	3029.79
Cycle Output Power	kJ/kg	$w_t = (h_1 - h_a) + (h_b - h_2)$	1106.52	1108.88	1094.64
Pump power	kJ/kg	$w_p = h_4 - h_3$	4.01	4.01	4.01
Thermal cycle efficiency		$\eta_t = \dfrac{w_1 - w_p}{q_1}$	0.358	0.362	0.36
Thermal cycle steam consumption rate	kg/(kW·h)	$d_0 = \dfrac{1}{w_t - w_p}$	3.265	3.258	3.300

From the above calculation results, it can be seen that compared with the ideal Rankine cycle, the thermal efficiency does not significantly improve after using the reheat cycle, but the exhaust steam dryness increases, and the cycle steam consumption rate decreases.

4.5 Regenerative Cycle with Reheating

Modern large-scale PWR NPPs generally use regenerative cycles with reheating, which can reduce the humidity of steam at the expansion endpoint of the steam turbine while improving the cycle thermal efficiency. The thermal efficiency of the ideal regenerative cycle with reheating is

$$\eta_t = 1 - \frac{h_2 - h_3}{\left(h_1 - h_4\right) + \Delta q_{\mathrm{rh}} \prod_{i=1}^{z} \frac{1}{1 + \frac{\tau_i}{q_{hi}}}} \tag{4.64}$$

The corresponding steam consumption rate is

$$d_0 = \frac{3600}{\left(h_1 - h_2\right) + \Delta q_{\mathrm{rh}}} \frac{1}{\left(1 - \sum_{i=1}^{z} \alpha_i Y_i\right)} \tag{4.65}$$

The thermal economy of the regenerative cycle with reheating involves many factors, namely: initial steam pressure and temperature, exhaust steam pressure, relative internal efficiency of the steam turbine before and after intermediate reheating, pressure, temperature and pressure loss after intermediate reheating, feed water temperature, regenerative distribution and losses caused by the irreversibility of the regenerative system, losses caused by the irreversibility of the feed water pump working process, etc. The regenerative distribution of the regenerative cycle with reheating has certain special features, mainly the following: (1) In order to reduce the regenerative steam extraction port of the unit, usually a part of the exhaust steam from the high-pressure cylinder is used as the first-stage regenerative steam extraction. The exhaust pressure of the high-pressure cylinder is determined according to the optimal reheat pressure and optimal regenerative distribution. If the end difference of the second-stage heater is selected, the feed water temperature of this stage is also determined accordingly. (2) Generally, the high-pressure cylinder of the reheat unit has a first-stage regenerative steam extraction port to ensure the need for optimal feed water temperature. Some units' high-pressure cylinders do not have steam extraction ports due to special internal cylinder structures. (3) The temperature, superheat, and enthalpy of each stage of extraction steam after reheating have all increased, which enlarges the temperature difference in the heat exchange process, increases the loss, and weakens the regeneration effect. The usual practice is to increase the enthalpy rise of the heater that uses the steam from the cold section as the regeneration extraction, and appropriately reduce the enthalpy rise of the regeneration extraction heater of each stage of the low-pressure cylinder after reheating, in order to reduce the irreversible loss of the entire regeneration process and improve the thermal economy of feedwater regeneration. (4) Although the reheating process has the effect of weakening regeneration, as long as the parameters of reheating and regeneration are properly matched, there is still sufficient thermal economic effect.

The thermal system diagram of a typical PWR NPP is shown in Fig. 4.18, and Table 4.7 shows the main features of the thermal systems of the Qinshan Phase I and Daya Bay NPPs.

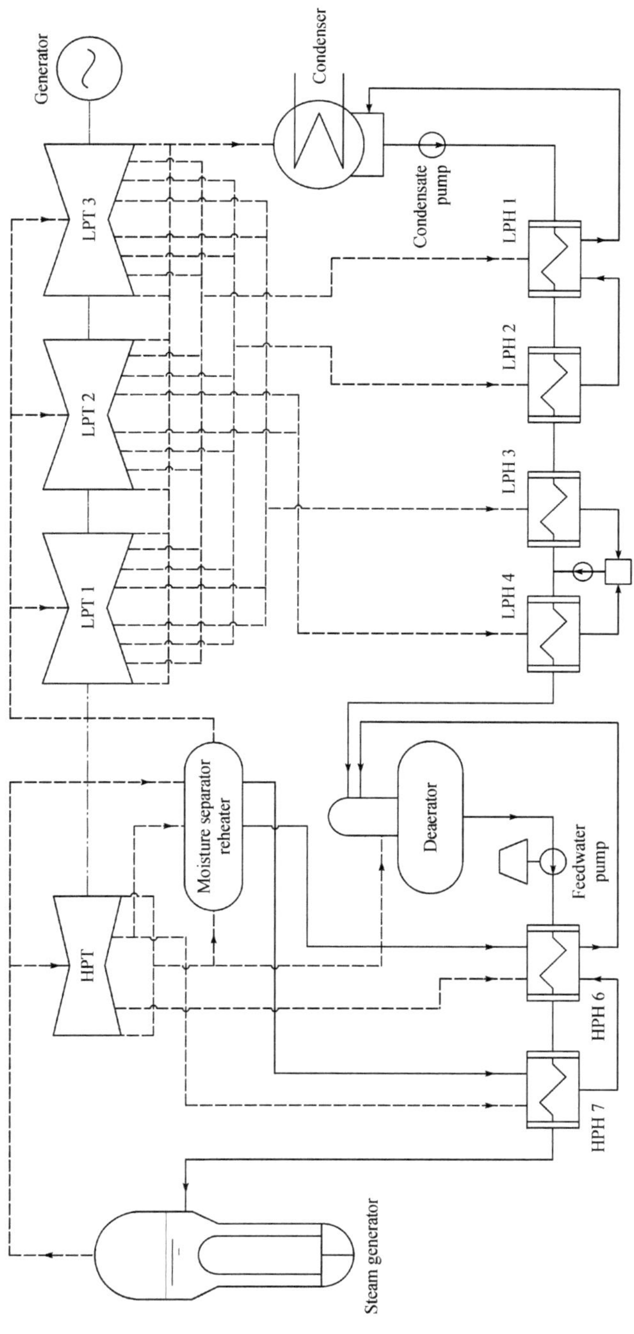

Notes: HPT High pressure steam turbine; LPT Low pressure steam turbine; HPH High pressure feedwater heater; LPH Low pressure feedwater heater

Fig. 4.18 Schematic process of the secondary loop thermodynamic system in a typical PWR NPP

Table 4.7 Main characteristics of the thermal systems in typical PWR NPPs

No.	Item		Qinshan Phase I	Daya Bay
1	Reactor thermal power/MW		966	2775
2	Net electrical power/MW		300	925
3	Total thermal power of nuclear steam supply system/MW		1035	2785
4	Number of loops in the primary coolant system		2	3
5	Steam turbo-generator Unit	High pressure turbine	Dual Flow Paths, 1 unit	Dual Flow Paths, 2 units
		Low pressure turbine	Dual Flow Paths, 2 unit	Dual Flow Paths, 3 unit
6	Number of extraction stages	High pressure cylinder	Three stages	Three stages
		Low pressure cylinder	Four stages	Four stages
7	Moisture separation/Reheat system		Intermediate moisture separation/Second-stage reheater	Intermediate moisture separation/second-stage reheater
8	Feedwater regenerative System	High pressure heater	3	2
		Deaerator	1	1
		Low pressure heater	3	4

4.6 Brayton Cycle

The Brayton cycle is a gas power cycle, in which the working fluid does not change phase during the cycle. Energy systems that use this thermodynamic cycle in engineering practice include gas turbine power units, high-temperature gas-cooled reactor NPPs, etc.

4.6.1 Simple Brayton Cycle

4.6.1.1 Ideal Brayton Cycle

The ideal simple Brayton cycle consists of four processes: adiabatic compression, isobaric heating, adiabatic expansion, and isobaric cooling. Figure 4.19 shows the principle flow of a gas turbine power unit, where (a) is an open cycle and (b) is a closed cycle.

In the open cycle, air drawn from the atmosphere is compressed by a centrifugal compressor and sent to the combustion chamber to mix and burn with the injected fuel, producing high-temperature and high-pressure gas. The gas enters the gas turbine to expand and do work, and the exhaust gas is discharged into the atmosphere. In the closed cycle, the exhaust gas discharged from the gas turbine enters the cooler, is cooled and lowered in temperature by the cooling medium, and then enters

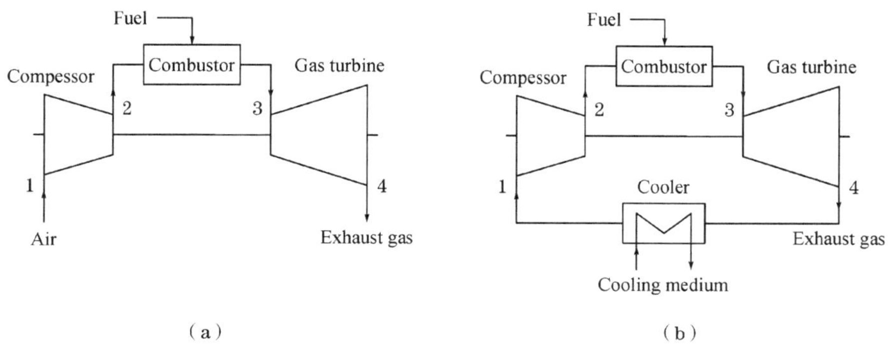

Fig. 4.19 Schematic process of gas turbine power plant: (**a**) Open cycle; (**b**) Closed cycle

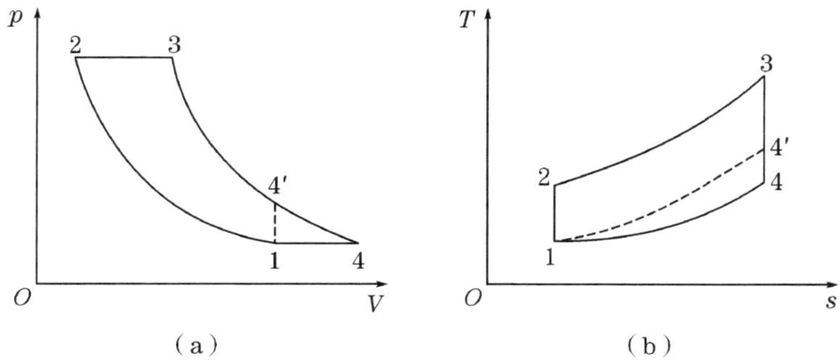

Fig. 4.20 *p-V* diagram and *T-s* diagram of Brayton cycle: (**a**) *p-V* diagram; (**b**) *T-s* diagram

the compressor for the next cycle. Figure 4.20 shows the *p-V* and *T-s* diagrams of the Brayton cycle.

Process 1–2 is a reversible adiabatic compression process, and the power consumption of the compressor is

$$w_c = h_2 - h_1 \tag{4.66}$$

Where h_1 and h_2 are the enthalpies of the air at the inlet and outlet of the compressor, respectively.

If the specific heat at constant pressure cp of the air is constant, for an ideal gas, we have

$$w_c = h_2 - h_1 = c_p\left(T_2 - T_1\right) \tag{4.67}$$

Where T_1 and T_2 are the temperatures of the air at the inlet and outlet of the compressor, respectively.

Process 2-3 is an isobaric combustion process, and the heat absorbed by the air is

$$q_1 = h_3 - h_2 = c_p\left(T_3 - T_2\right) \tag{4.68}$$

Where h_3 is the enthalpy of the gas at the outlet of the combustion chamber;

T_3 is the temperature of the gas at the outlet of the combustion chamber.

Process 3-4 is a reversible adiabatic expansion process, and the work done by the gas turbine to the outside is

$$w_{\mathrm{t}} = h_3 - h_4 = c_p\left(T_3 - T_4\right) \tag{4.69}$$

Where h_4 is the enthalpy of the exhaust gas at the outlet of the gas turbine;

T_4 is the temperature of the exhaust gas at the outlet of the gas turbine.

Process 4-1 is the isobaric cooling process of exhaust gas being discharged into the atmosphere, and the heat discharged into the atmosphere is

$$q_2 = h_4 - h_1 = c_p\left(T_4 - T_1\right) \tag{4.70}$$

The thermal efficiency of the ideal Brayton cycle is

$$\begin{aligned}\eta_{\mathrm{t}} &= \frac{w_{\mathrm{t}} - w_{\mathrm{e}}}{q_1} = \frac{c_p\left(T_3 - T_4\right) - c_p\left(T_2 - T_1\right)}{c_p\left(T_3 - T_2\right)} \\ &= 1 - \frac{T_4 - T_1}{T_3 - T_2} = 1 - \frac{T_1\left(\frac{T_4}{T_1} - 1\right)}{T_2\left(\frac{T_3}{T_2} - 1\right)}\end{aligned} \tag{4.71}$$

For the isentropic processes 1-2 and 3-4, the following relationship exists between the pressure and temperature of the ideal gas:

$$\frac{T_2}{T_1} = \left(\frac{p_2}{p_1}\right)^{\frac{\kappa-1}{\kappa}} \quad \frac{T_3}{T_4} = \left(\frac{p_3}{p_4}\right)^{\frac{\kappa-1}{\kappa}} \tag{4.72}$$

Since $p_1 = p_4$, $p_2 = p_3$, we get

$$\frac{T_2}{T_1} = \frac{T_3}{T_4} \quad \text{or} \quad \frac{T_3}{T_2} = \frac{T_4}{T_1} \tag{4.73}$$

Let the compression ratio $r_p = \frac{p_2}{p_1} = \frac{p_3}{p_4}$, then the thermal efficiency of the Brayton cycle can be expressed as

$$\eta_{\mathrm{t}} = 1 - \frac{T_1}{T_2} = 1 - \frac{1}{r_p^{\frac{\kappa-1}{\kappa}}} \tag{4.74}$$

The net work output of the Brayton cycle can be expressed as

$$w = c_p\left(T_3 - T_4\right) - c_p\left(T_2 - T_1\right) = c_p T_1\left[\frac{T_3}{T_1}\left(1 - \frac{1}{r_p^{\frac{\kappa-1}{\kappa}}}\right) - \left(r_p^{\frac{\kappa-1}{\kappa}} - 1\right)\right] \tag{4.75}$$

According to Eq. (4.74), the trend of the thermal efficiency of the Brayton cycle with the change of air compression ratio is shown in Fig. 4.21a ($\kappa = 1.4$); according to Eq. (4.75), the net output power of the cycle is a function of the ratio of the highest temperature to the lowest temperature and the compression ratio, its trend is shown in Fig. 4.21b, where the temperature ratio $r_T = \frac{T_3}{T_1}$, and the intake air temperature $T_1 = 300\text{K}$ is assumed.

For a determined r_T, by taking the derivative of Eq. (4.75) $\left(\frac{\partial w}{\partial r_p}\right) = 0$, the compression ratio corresponding to the maximum net output power can be obtained:

$$r_{p,\max} = \frac{p_2}{p_1} = \left(\frac{T_3}{T_1}\right)^{\frac{\kappa}{2(\kappa-1)}} \tag{4.76}$$

4.6.1.2 Actual Brayton Cycle

In the ideal Brayton cycle, both expansion and compression are isentropic processes, but in the actual process, there are energy losses due to irreversible factors.

The isentropic efficiency of the compressor η_C is defined as the ratio of the isentropic compression work w_C to the actual compression work $w_{C,a}$ that is

$$\eta_C = \frac{w_C}{w_{C,a}} = \frac{h_{2s} - h_1}{h_2 - h_1} = \frac{T_{2s} - T_1}{T_2 - T_1} \tag{4.77}$$

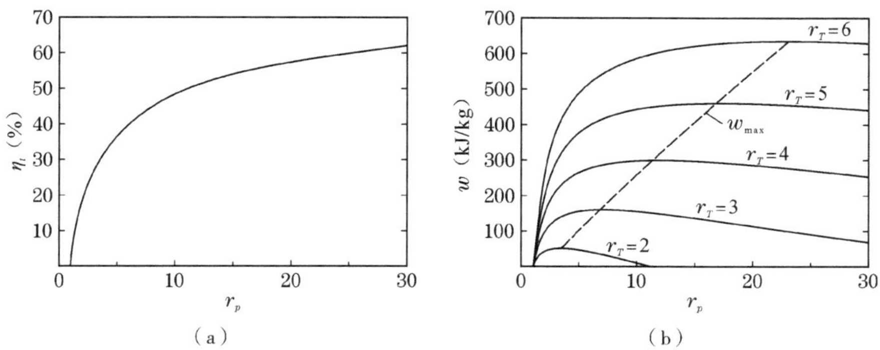

Fig. 4.21 Relationship between Brayton cycle efficiency, net output power and compression ratio: (**a**) Relationship between cycle efficiency and compression ratio; (**b**) Relationship between net output power and compression ratio, temperature ratio [28]

The isentropic efficiency of the gas turbine η_T is defined as the ratio of the actual work $w_{T,a}$ to the theoretical work w_T that is

$$\eta_T = \frac{w_{T,a}}{w_T} = \frac{h_3 - h_4}{h_3 - h_{4s}} = \frac{T_3 - T_4}{T_3 - T_{4s}} \tag{4.78}$$

Using the definition of isentropic compression work, the expression for the actual compression work of the compressor can be obtained:

$$w_{C,a} = \frac{c_p T_1}{\eta_C}\left(r_p^{\frac{\kappa-1}{\kappa}} - 1\right) \tag{4.79}$$

Similarly, the expression for the actual work of the gas turbine can be obtained:

$$w_{T,a} = \eta_T c_p T_3 \left(1 - \frac{1}{r_p^{\frac{\kappa-1}{\kappa}}}\right) \tag{4.80}$$

The actual net work is

$$w_a = w_{T,a} - w_{C,a} = c_p T_1 \left[\eta_T \frac{T_3}{T_1}\left(1 - \frac{1}{r_p^{\frac{\kappa-1}{\kappa}}}\right) - \frac{1}{\eta_C}\left(r_p^{\frac{\kappa-1}{\kappa}} - 1\right)\right] \tag{4.81}$$

The actual cycle thermal efficiency is

$$\eta_{t,a} = \frac{\eta_T \frac{T_3}{T_1}\left(1 - \frac{1}{r_p^{\frac{\kappa-1}{\kappa}}}\right) - \frac{1}{\eta_C}\left(r_p^{\frac{\kappa-1}{\kappa}} - 1\right)}{\frac{T_3}{T_1} - \frac{1}{\eta_C}\left(r_p^{\frac{\kappa-1}{\kappa}} - 1\right) - 1} \tag{4.82}$$

Figure 4.22 shows a comparison of the characteristics related to the ideal Brayton cycle and the actual Brayton cycle.

4.6.2 Improved Brayton Cycle

In order to improve the efficiency of the Brayton cycle, measures such as regenerative heating, reheat, and intercooling can be adopted [28].

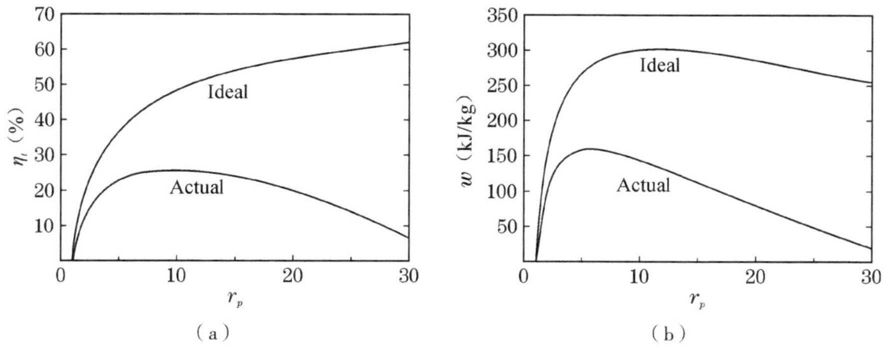

Fig. 4.22 Comparison of characteristics between ideal and actual cycles: (**a**) Relationship between cycle efficiency and compression ratio; (**b**) Relationship between net output power and compression ratio

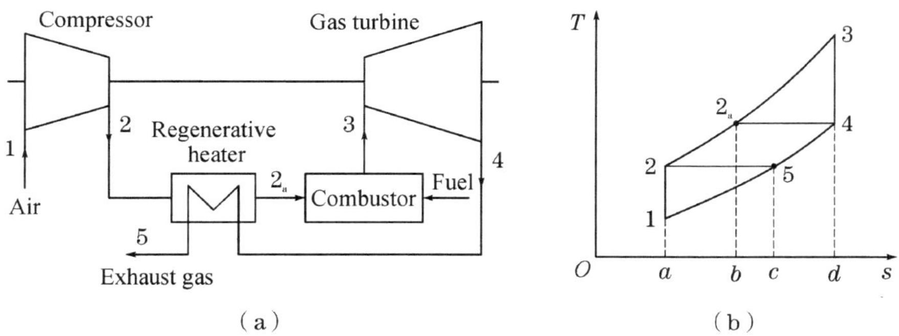

Fig. 4.23 Regenerative Brayton cycle: (**a**) Schematic diagram of thermal system; (**b**) *T*-*s* diagram

4.6.2.1 Regenerative Cycle

From the *T*-*s* diagram of the simple Brayton cycle, it can be seen that the exhaust temperature T_4 of the gas turbine is much higher than the outlet temperature T_2 of the compressor, the heat loss of the exhaust q_2 is large, and the cycle thermal efficiency is low.

In order to improve the thermal efficiency, the waste heat of the gas turbine exhaust can be used to heat the compressed air, which is a regenerative measure. The Brayton cycle with regenerative heating is shown in Fig. 4.23.

Process 1-2 is the air compression process, and the power consumed by the compressor is

$$w_e = h_2 - h_1$$

Process $2 - 2_a$ is the regenerative heating process, and the heat absorbed by the air is

$$q_{rg,2} = h_{2_a} - h_2$$

Process $2_a - 3$ is the heat absorption process of air in the combustion chamber, and the heat absorbed is

$$q_1 = h_3 - h_{2_a}$$

Process 3-4 is the process of gas expansion work in the gas turbine, and the work done is

$$w_t = h_3 - h_4$$

Process 4-5 is the heat release process of the exhaust gas in the regenerator, and the heat released is

$$q_{rg,1} = h_4 - h_5$$

Process 5-1 is a heat release process from exhaust gas to the environment, with the heat release being

$$q_2 = h_5 - h_1$$

According to the definition of cycle thermal efficiency, the thermal efficiency of the Brayton cycle with regeneration is

$$\eta_{t,rg} = 1 - \frac{q_2}{q_1} = 1 - \frac{h_5 - h_1}{h_3 - h_{2_a}} = 1 - \frac{(h_4 - h_1) - (h_4 - h_5)}{(h_3 - h_2) - (h_{2_a} - h_2)} \tag{4.83}$$

The limit of regeneration is when the air temperature T_{2a} at the exit of the heat exchanger equals the temperature T_4 of the exhaust gas at the exit of the gas turbine. Ignoring the heat loss during the heat transfer process in the heat exchanger, we have $h_{2_a} - h_2 = h_4 - h_5 = q_{rg}$, so

$$T_{2_a} = T_4 \ T_5 = T_2$$

$$h_{2_a} = h_4 \, h_5 = h_2$$

Combining the above temperature and heat balance relationships, Eq. (4.83) can be rewritten as

$$\eta_{t,rg} = 1 - \frac{(h_4 - h_1) - q_{rg}}{(h_3 - h_2) - q_{rg}} \geqslant 1 - \frac{h_4 - h_1}{h_3 - h_2} = \eta_t \tag{4.84}$$

As can be seen, the thermal efficiency of the Brayton cycle with regeneration is higher than that of the simple Brayton cycle.

4.6.2.2 Reheat Cycle

The output power of the Brayton cycle increases with the increase of the maximum temperature ratio, but the maximum temperature of the working fluid is limited by the heat resistance temperature of the structural material. One way to improve cycle efficiency under constant maximum temperature conditions is to use reheating, i.e., sending the exhaust gas from the gas turbine to the reheater for heating, and then sending it to the next stage gas turbine for expansion and work. One-stage or two-stage reheating can be used. Because of the reheating measures, the exhaust gas temperature increases, which also improves the heat exchange effect of the heat exchanger. Figure 4.24 shows the Brayton cycle with reheating.

Process 1-2 is the gas compression process, with the power consumed by the compressor being

$$w_c = h_2 - h_1$$

Process 2-3 is the heat absorption process of the gas in the combustion chamber, with the heat absorption being

$$q_1^{'} = h_3 - h_2$$

Process 3-a is the expansion process of the gas in the No.1 gas turbine, with the work done being

$$w_{t,1} = h_3 - h_a$$

Process a-b is the heat absorption process of the gas in the reheater, with the heat absorption being

$$q_1^{''} = h_b - h_a$$

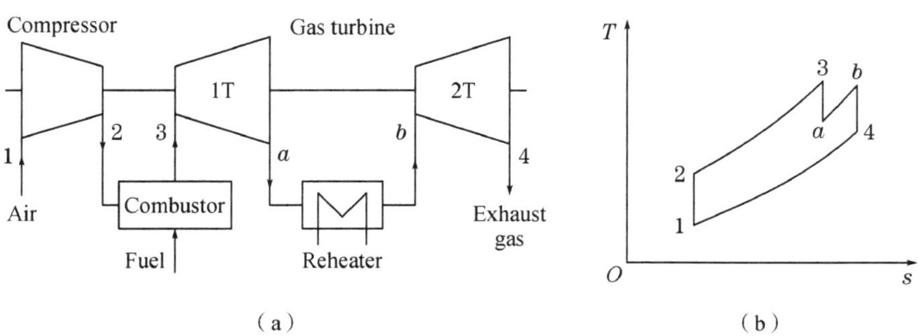

Fig. 4.24 Brayton cycle with reheating: (**a**) Schematic diagram of thermal system; (**b**) T-s diagram

Process b-4 is the expansion process of the gas in the No.2 gas turbine, with the work done being

$$w_{t,2} = h_b - h_4$$

Process 4-1 is a heat release process from exhaust gas to the environment, with the heat release being

$$q_2 = h_4 - h_1$$

According to the definition of cycle thermal efficiency, the thermal efficiency of the Brayton cycle with reheating is

$$\begin{aligned}\eta_{t,rh} &= 1 - \frac{q_2}{q_1} = 1 - \frac{q_2}{\left(q_1^{'} + q_1^{''}\right)} \\ &= 1 - \frac{h_4 - h_1}{(h_3 - h_2) + (h_b - h_a)} \geqslant 1 - \frac{h_4 - h_1}{h_3 - h_2} = \eta_t\end{aligned} \tag{4.85}$$

As can be seen, the thermal efficiency of the Brayton cycle with reheating is higher than that of the simple Brayton cycle.

4.6.2.3 Brayton Cycle with Intercooling, Reheating, and Regeneration

Unlike the steam power cycle, the compressor of the gas turbine cycle consumes a large amount of power. To reduce the power consumption of the compressor, the intermediate gas is cooled. Figure 4.25 shows the Brayton cycle with intercooling, reheating, and regeneration [29, 30].

Process $1 - 1_a$ is the compression process of air in the first stage compressor (1C), with the power consumed by the compressor being

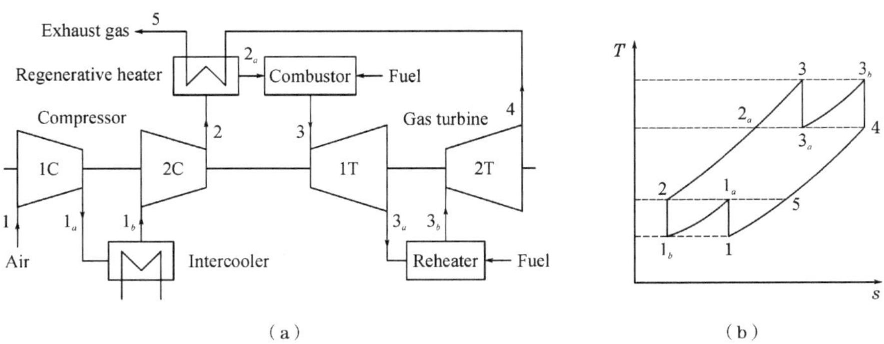

Fig. 4.25 Brayton cycle with intercooling, reheat and regeneration: (**a**) Schematic diagram of thermal system; (**b**) *T*-*s* diagram [30]

$$w_{c,1} = h_{1_a} - h_1$$

Process $1_a - 1_b$ is the heat release process of the air compressed by 1C in the intercooler, with the heat release being

$$q_{2,1} = h_{1_a} - h_{1_b}$$

Process $1_b - 2$ is the compression process of air in the second stage compressor (2C), with the power consumed by the compressor being

$$w_{c,2} = h_2 - h_{1_b}$$

Process $2 - 2_a$ is the heat absorption process of the air compressed by 2C in the heat exchanger, with the heat absorption being

$$q_{rg,1} = h_{2_a} - h_2$$

Process $2_a - 3$ is the heat absorption process of air in the fuel chamber, with the heat absorption being

$$q_{1,1} = h_3 - h_{2_a}$$

Process $3 - 3_a$ is the work done by the gas expansion in the first stage gas turbine (1 T), the work done is

$$w_{t,2} = h_{3_b} - h_5$$

Process $3_a - 3_b$ is the heat absorption process of 1 T exhaust entering the reheater, the heat absorption is

$$q_{1,2} = h_{3_b} - h_{3_a}$$

Process $3_b - 4$ is the work done by the reheated gas expansion in the second stage gas turbine (2 T), the work done is

$$w_{t,2} = h_{3_b} - h_4$$

Process 4–5 is the heat release process of 2 T exhaust entering the heat exchanger, the heat release is

$$q_{rg,2} = h_4 - h_5$$

Process 5–1 is the heat release process of the exhausted gas to the environment, the heat release is

$$q_{2,2} = h_5 - h_1$$

According to the definition of cycle thermal efficiency, we have

$$\eta_t = \frac{w_t - w_c}{q_1} = \frac{(w_{t,1} + w_{t,2}) - (w_{c,1} + w_{c,2})}{q_{1,1} + q_{1,2}}$$
$$= \frac{(h_3 - h_{3_a}) + (h_{3_b} - h_4) - (h_{1_a} - h_1) - (h_2 - h_{1_b})}{(h_3 - h_{2_a}) + (h_{3_b} - h_{3_a})} \tag{4.86}$$

For two-stage compression and two-stage expansion, when both the two-stage compressor and the two-stage gas turbine have the same pressure ratio, the compressor consumes the least work, and the gas turbine outputs the most work, that is

$$\frac{p_{1_a}}{p_1} = \frac{p_2}{p_{1_b}} \quad \text{and} \quad \frac{p_3}{p_{3_a}} = \frac{p_{3_b}}{p_4}$$

Assuming that the air temperature entering the first and second stage compressors is the same, and each stage compressor has the same efficiency, then the air temperature at the outlet of each stage compressor is also the same. For the gas turbine, there are similar assumptions, then

$$T_1 = T_{1_b} \quad T_2 = T_{1_a} = T_5 \quad T_{2_a} = T_{3_a} = T_4 \quad T_3 = T_{3_b}$$
$$h_1 = h_{1_b} \quad h_2 = h_{1_a} = h_5 \quad h_{2_a} = h_{3_a} = h_4 \quad h_3 = h_{3_b}$$

According to the above temperature and enthalpy relationship, Eq. (4.86) can be rewritten as

$$\eta_{t.B} = \frac{(h_3 - h_2) - (h_4 - h_1)}{h_3 - h_4} > 1 - \frac{h_4 - h_1}{h_3 - h_2} = \eta_t \tag{4.87}$$

It can be seen that the thermal efficiency of the Brayton cycle with intermediate cooling, reheating, and regeneration is higher than that of the simple Brayton cycle.

The combination of high-temperature gas-cooled reactor and Brayton closed cycle mainly adopts thermal measures of regeneration and intercooling to greatly improve cycle efficiency and specific work. Therefore, this system has high initial temperature, high efficiency, high specific work, good variable working condition characteristics, and the greatest potential to improve system performance.

4.6.3 Brayton-Rankine Combined Cycle

The combination of the Brayton closed helium cycle with high initial cycle temperature and the Rankine closed steam cycle with low exhaust heat temperature, using the steam cycle to absorb the high-temperature exhaust heat of the helium

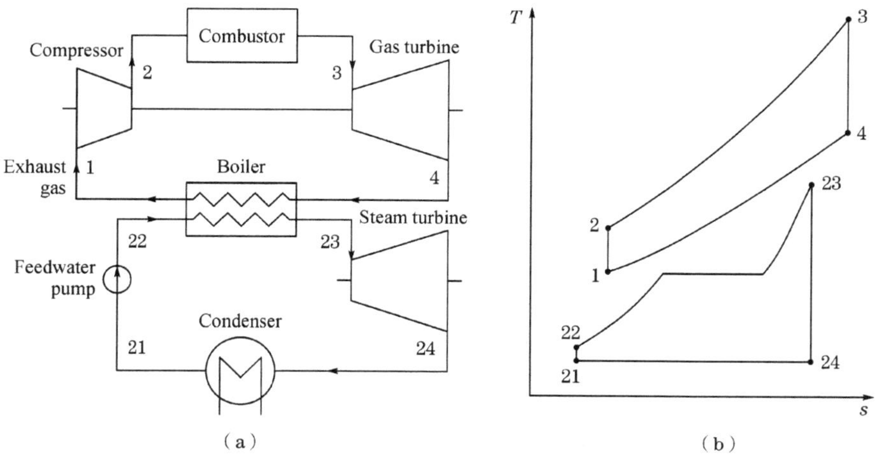

Fig. 4.26 Combined Brayton-Rankine cycle: (**a**) Schematic diagram of thermal system; (**b**) *T*-*s* diagram

cycle, to achieve the cascade utilization of energy, and thus improve the level of energy utilization. Figure 4.26 shows the ideal Brayton-Rankine combined cycle [31, 32].

The heat absorption of the Brayton cycle is

$$Q_{1,g} = G_g (h_3 - h_2) \tag{4.88}$$

Where G_g is the mass flow rate of the working fluid in the Brayton cycle.

The heat discharged in the Brayton cycle's constant pressure heat release process 4-1 is all absorbed by the Rankine cycle's constant pressure heat absorption process 22–23, that is

$$Q_{2,g} = Q_{1,v} \tag{4.89}$$

$$G_g (h_4 - h_1) = G_v (h_{23} - h_{22}) \tag{4.90}$$

Where G_v is the mass flow rate of the working fluid in the Rankine cycle.

The heat release of the Rankine cycle is

$$Q_{2,g} = G_v (h_{24} - h_{21}) \tag{4.91}$$

The net work done by the combined cycle is

$$W = Q_{1,g} - Q_{2,v} = G_g (h_3 - h_2) - G_v (h_{24} - h_{21}) \tag{4.92}$$

Then the thermal efficiency of the combined cycle is

$$\eta_t = 1 - \frac{Q_{2,g}}{Q_{1,g}} = 1 - \frac{G_v (h_{24} - h_{21})}{G_g (h_3 - h_2)} = 1 - \frac{(h_4 - h_1)(h_{24} - h_{21})}{(h_{23} - h_{22})(h_3 - h_2)} \tag{4.93}$$

The relationship between the thermal efficiency of the combined cycle and the efficiency of the Brayton cycle ηt, g and the efficiency of the Rankine cycle η_t, v is as follows:

$$\begin{aligned}\eta_t &= 1-\frac{Q_{2,v}}{Q_{1,g}} = 1-\frac{Q_{2,v}}{Q_{2,g}}\frac{Q_{1,v}}{Q_{1,g}} = 1-\left(1-\eta_{t,v}\right)\left(1-\eta_{t,g}\right) \\ &= \eta_{t,v}+\eta_{t,g}\left(1-\eta_{t,v}\right) = \eta_{t,g}+\eta_{t,v}\left(1-\eta_{t,g}\right)\end{aligned} \quad (4.94)$$

Since $\eta_{t,\,g}(1 - \eta_{t,\,v})$ and $\eta_{t,\,v}(1 - \eta_{t,\,g})$ are both greater than zero, we have $\eta_t > \eta_{t,\,g}$ and$\eta_t > \eta_{t,\,v}$, that is, the efficiency of the combined cycle is greater than the efficiency of the Brayton cycle and the efficiency of the Rankine cycle.

4.6.3.1 Direct Combined Cycle

High-temperature gas-cooled reactors use helium as a coolant, and the outlet temperature of the helium in the reactor can reach 850–1300 °C, thus enabling the use of various thermodynamic cycles for energy conversion. The helium turbine cycle is combined with the steam turbine cycle, and Fig. 4.27 shows the thermodynamic system process using a direct combined cycle.

The high-temperature helium (6.9 MPa, 900 °C) from the reactor outlet first flows through the helium turbine that drives the compressor, then enters the main power generation helium turbine to expand and do work, driving the generator to produce electricity. The helium discharged from the main helium turbine still has a high pressure and temperature (2.6 MPa, 556 °C), enters the direct steam generator to heat the water on the secondary side, is cooled to about 150 °C, then cooled to about 30 °C by the pre-cooler, pressurized to 7.0 MPa by the compressor, and sent to the reactor for the next round of helium cycle.

The superheated steam (4 MPa, 526 °C) produced by the direct steam generator is sent to the steam turbine generator set. The exhausted steam discharged after expanding and doing work in the steam turbine enters the condenser. The condensate formed after cooling is heated by the feedwater heater, pressurized by

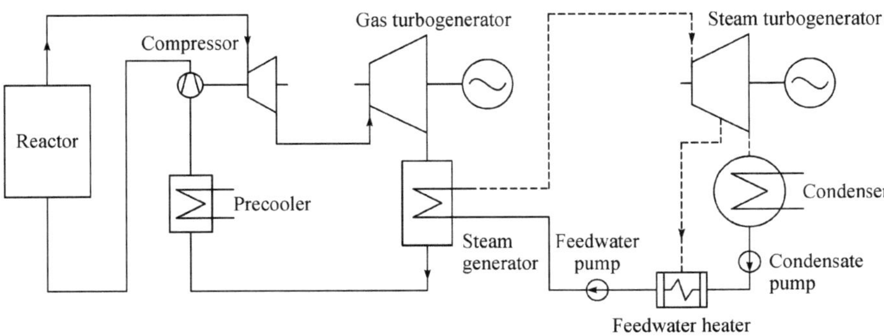

Fig. 4.27 Schematic process of thermal system in NPP using direct combined cycle

the feedwater pump, and sent to the steam generator for the next round of steam-water cycle.

The thermal efficiency of the direct combined cycle can reach 48%, but it increases the investment cost of the system and cannot rule out the possibility of a core water ingress accident.

4.6.3.2 Indirect Combined Cycle

In order to avoid the water from the steam-water circuit entering the core under accident conditions, an indirect combined cycle composed of three circuits is also used in engineering. The thermodynamic system process is shown in Fig. 4.28. The working fluids of the first and second circuits are helium, and the working fluid of the third circuit is water and steam.

In the first circuit, helium circulates in the circuit under the drive of the compressor, flows through the reactor to absorb the heat released by the fission of the core nuclear fuel. The high-temperature helium (about 900 °C) from the reactor outlet is cooled to 300 °C after heating the nitrogen in the second circuit through the intermediate heat exchanger, and then sent back to the reactor by the compressor.

In the second circuit, the helium in the second circuit that has absorbed heat from the intermediate heat exchanger enters the helium turbine to expand and do work, driving the generator to produce electricity. The exhausted gas discharged from the helium turbine enters the steam generator, heats the feedwater of the third circuit and generates steam. The exhausted gas after heat release is sent to the intermediate heat exchanger by the compressor to start the next cycle.

In the third circuit, the steam generated by the steam generator expands and does work in the turbine, driving the generator to produce electricity. The exhausted steam discharged from the turbine enters the condenser to condense and release heat. The condensate formed is pumped by the condensate pump to the feedwater regenerative system for heating, then pressurized by the feedwater pump and sent to the steam generator.

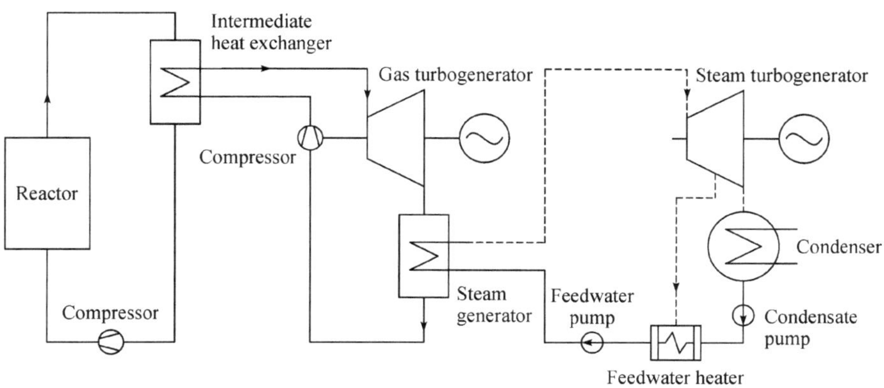

Fig. 4.28 Schematic process of thermal system in NPP using indirect combined cycle

Questions and Exercises

1. What are the differences and connections between the Rankine cycle and the Carnot cycle? Why doesn't the actual power cycle use the Carnot cycle?
2. What impact does the change in initial and final parameters of steam have on the thermal economy of a NPP?
3. The outlet saturated steam pressure of the steam generator is $p_s = 6.5$ MPa, and the steam dryness is $x_s = 0.9975$; the inlet steam pressure of the turbine is $p_1 = 6.3$ MPa. The steam pressure is $p_1 = 0.7$ MPa, with a dryness fraction of $x_1 = 0.97$; the exhaust steam pressure at the turbine outlet is $p_2 = 0.005$ MPa; the relative internal efficiency of the turbine is $\eta_{oi} = 0.85$. Ignoring the work of the water pump, try to calculate:

    1. The internal work of the turbine;
    2. The thermal efficiency of the cycle;
    3. The pipeline loss coefficient.

4. What are the benefits of using feedwater regeneration to improve cycle thermal efficiency?
5. What is the purpose of using a reheat cycle? What is its impact on cycle thermal efficiency?
6. What are the main factors affecting the thermal efficiency of the Brayton cycle?
7. In a combined cycle composed of the Brayton cycle and the Rankine cycle, what factors influence its thermal efficiency?

Chapter 5
Thermal Balance Analysis of NPP Thermal Systems

The thermal balance calculation of the thermal system of a NPP plays an important role in the design, operation, and technical transformation of the NPP. The online operation monitoring and diagnostic system of the NPP needs to determine the changes in economic indicators caused by certain operating parameters and system changes through thermal balance calculations.

According to the calculation results of thermal balance analysis, we can point out the quantity and reasons for energy loss in the process of energy conversion, transmission, and distribution, evaluate the perfection of energy utilization in thermodynamics, and also provide a basis for the exergy analysis of the thermal system.

5.1 Traditional Thermal Balance Analysis Methods

The methods for quantitatively calculating the thermal efficiency of the power plant thermal system mainly include the efficiency method, equivalent enthalpy drop method, cycle function method, matrix method, etc. This chapter focuses on the thermal efficiency calculation method, and also briefly introduces the equivalent enthalpy drop method and the cycle function method.

1. Thermal balance method [33]

The thermal balance method can usually be divided into series and parallel methods.

Based on the first law of thermodynamics, a thermal balance equation is listed for each stage of the heater in the extraction heating system, and the extraction amount of each stage of the heater is calculated based on this.

In the series algorithm, the calculation order starts from the heater with the highest extraction pressure, and the extraction amount of each stage of the heater is calculated in turn. In the calculation of the extraction amount of each stage of the

M. Peng et al., *Thermal Analysis of Nuclear Power Plants*, Nuclear Science and Technology, https://doi.org/10.1007/978-981-96-8983-5_5

heater, there is generally only one unknown, and manual calculation is also very convenient.

In the parallel algorithm, the thermal balance equations of all heaters are solved together to obtain all extraction amounts at once.

Since the thermal balance equation of the feedwater heater is a linear algebraic equation about the extraction amount, it can always be represented by a matrix equation, which is very suitable for computer solution. In practical applications, the series method and the parallel method are often used interchangeably. The mathematical expression of the parallel method is simpler, and the series method is more advantageous in algorithm implementation.

2. Cycle function method [34, 35]

The cycle function method is a method that divides the feedwater regenerative system into several heating units and seeks the unit inlet coefficient as the basic feature. The entire thermal system is divided into the main cycle and the auxiliary cycle, and the main cycle and the auxiliary cycle can be superimposed.

When calculating performance indicators, it can be done in steps, and the impact of the main loop and auxiliary loop on performance indicators also has additivity. When there are local changes in the thermal system, it is equivalent to adding an auxiliary loop to the main loop. Only need to calculate the changed auxiliary loop to find out the change in performance indicators, which brings great convenience to the calculation of variable conditions and comparison of different schemes. Especially when comparing and selecting different schemes for power plants, it is simpler than the thermal balance method.

3. Equivalent enthalpy drop method [36]

The equivalent enthalpy drop method focuses on the local quantitative analysis of the thermal system, aiming to avoid the need to calculate the entire thermal coefficient when the structure or parameters of the thermal system change locally.

The equivalent enthalpy drop method is based on the premise that the fresh steam flow rate remains unchanged, and is divided into the constant heat equivalent enthalpy drop method and the variable heat equivalent enthalpy drop method.

In the equivalent enthalpy drop method, the fresh steam parameters, reheating parameters, exhaust steam parameters, and heater parameters should all be known and remain unchanged. This condition is difficult to guarantee in actual unit performance analysis, which will cause certain errors in the calculation results of the equivalent enthalpy drop method. When the unit operating conditions change greatly, the error will be larger and not suitable for online performance diagnosis. In addition, the steam-water parameters required in the equivalent enthalpy drop method still need to be obtained by the thermal balance method, so the equivalent enthalpy drop method is not complete as a method of analyzing the thermal system of a power plant.

4. Matrix method [37, 38]

The matrix method is a matrix expression of the thermal balance method, using matrix operations to calculate the extraction volume, and then calculate the economic indicators of the thermal system, which is more suitable for computer processing.

5.2 Constant Power Calculation Method for Thermal System of NPPs

The percentage of energy effectively utilized by the energy system or equipment to the energy provided by the outside world is called the efficiency of the energy system or equipment. The efficiency method is based on the first law of thermodynamics, using thermal efficiency to quantitatively express the effective degree of energy utilization of the energy system or equipment. The efficiency method for thermal balance analysis of the thermal system of a power plant is divided into constant power method and constant flow method.

The constant power calculation method takes the electric power of the unit as a constant, and calculates the required steam volume through calculation. It is widely used in design and operation departments. The constant flow calculation method takes the steam volume entering the turbine as a constant, calculates how much electric power can be generated, and is mostly used by turbine manufacturers.

The thermal balance calculation method commonly used in engineering is the constant power method, that is, under the conditions of known turbine type, capacity, initial and final parameters, the connection method of the unit's regenerative system and the steam-water parameters of each stage of extraction, the relative internal efficiency of the high and low pressure turbines, cooling water temperature, etc., calculate the unit's steam consumption and extraction volume at rated conditions, and then determine the unit's thermal economic indicators.

5.2.1 Initial Conditions Required for Calculation

Modern PWR NPPs all adopt regenerative cycles with steam intermediate reheating. The thermal balance analysis of their thermal systems is mainly to perform thermal calculations on turbine generator sets with regenerative-reheating cycles.

To carry out thermal calculations, it is necessary to know the composition form of the thermal system of the NPP, the mechanism of energy transmission and conversion in the main systems, the thermodynamic connection between the main equipment, and the capacity of the power plant, the main thermal parameters under the calculation conditions, etc. The specific content includes the following aspects:

5.2.1.1 Type and Overall Parameters of the NPPs

The main reactor type, NPP system composition, main parameters of the reactor and primary loop system, rated electrical power, etc.

The majority of currently operating NPPs are PWR types, most of which use natural circulation steam generators, with a small number of NPPs using direct steam generators. According to the capacity of the NPP, the reactor and primary

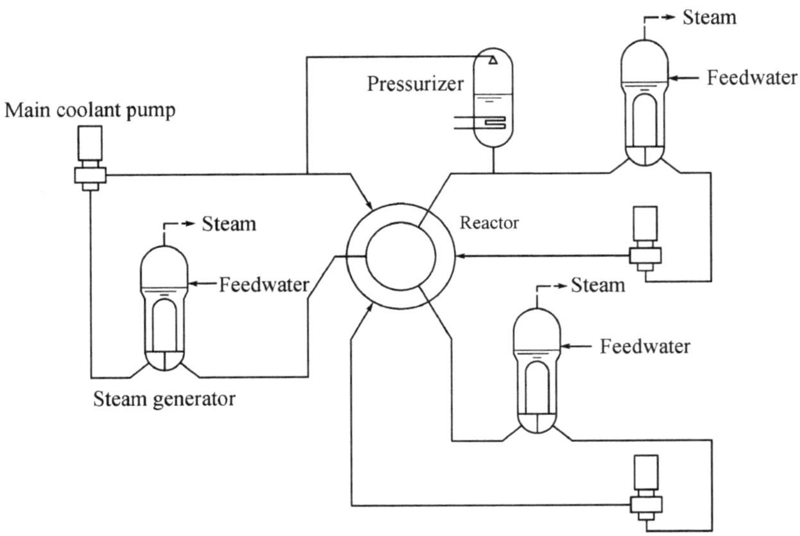

Fig. 5.1 Reactor coolant system of 1000 MW PWR NPP

loop system are generally divided into two or three loops, each coolant loop sharing an electrical power of 300 MW.

The working pressure of the main loop system of a general PWR NPP is about 15 MPa, the coolant temperature at the reactor inlet is 280–300 °C, the outlet temperature is 310–330 °C, and the coolant flow rate is 15,000–21,000 t/h. Figure 5.1 shows a schematic diagram of the reactor coolant system for a 1000 MW PWR NPP.

5.2.1.2 Type of Turbine and Initial and Final Steam Parameters

PWR NPPs usually use condensing saturated steam turbines, consisting of a double-circuit high-pressure turbine and two or three double-circuit low-pressure turbines.

The initial steam parameters refer to the parameters of the fresh steam at the outlet of the steam generator. The saturated steam pressure produced by the natural circulation steam generator is 5–7 MPa, with a humidity of 0.4%–0.5%; the slightly superheated steam pressure produced by the direct steam generator is 5–7 MPa, with a superheat of 20–30 °C.

The final steam parameters refer to the back pressure of the turbine or the working pressure of the condenser, generally 4.4–5.4 kPa.

The turbines of PWR NPPs usually use saturated steam with a pressure of 5–7 MPa and a humidity of 0.4%–0.5%, or slightly superheated steam with a superheat of 20–30 °C. Its available enthalpy is only about 60% of that of conventional high-temperature and high-pressure thermal power units (fresh steam pressure 16–17 MPa, temperature 500–550 °C), so the fresh steam mass flow rate of the nuclear turbine is about 170%–190% of the same power conventional thermal power unit, and the volume flow rate is about 165%–175%. Due to the large exhaust volume flow rate, in order to improve the efficiency of the turbine, it is necessary to increase the number of low-pressure cylinders and increase the height of the final stage blades.

5.2.1.3 Principle Thermodynamic Diagram of NPPs

Calculate the principle thermodynamic diagram of the thermal system of the NPP under operating conditions, including the connection method of the unit regenerative system and the steam-water parameters of each stage of extraction, and the connection method of the steam reheating system and the corresponding reheating steam parameters.

The thermodynamic diagram of a NPP is a line diagram that connects the main and auxiliary thermal equipment with line segments representing the relationship of fluid flow according to a given thermodynamic cycle form, using specified symbols. Depending on the purpose of use and the method of preparation, it is divided into principle thermodynamic diagrams and comprehensive thermodynamic diagrams.

The principle thermodynamic diagram shows the connection relationship between various thermal equipment that the working substance must flow through when completing a given thermodynamic cycle, and has the following characteristics:

1. The same type and same parameter equipment is only represented once on the thermodynamic diagram.
2. Only the main connections between equipment are represented, spare equipment and pipelines, accessories are not represented.
3. Except for the necessary accessories for the calculation condition, general accessories are not represented.

The principle thermodynamic diagram essentially shows the process of energy conversion and heat utilization of the working substance, reflecting the technical perfection and thermal economy of the power plant's energy conversion process.

5.2.1.4 Efficiency of the Steam Turbine Generator Set

The internal efficiency, mechanical efficiency, and generator efficiency of the turbine can be referred to Table 5.1 for values.

Modern PWR NPPs have a large rated power, such as the Daya Bay NPP, where each unit has a rated electrical power of 900 MW. The first experimental NPP in China, the Qinshan Phase I NPP, has a relatively small rated electrical power of 300 MW. Therefore, the mechanical efficiency of the turbine $\eta_m = 0.99 - 0.995$, and the generator efficiency $\eta_{ge} = 0.98 - 0.99$.

Table 5.1 Efficiency of steam turbine generator set

Rated power/MW	Internal efficiency η_i	Mechanical efficiency η_m	Generator efficiency η_{ge}
0.75–6	0.76–0.82	0.965–0.985	0.93–0.96
12–25	0.82–0.85	0.985–0.99	0.965–0.975
50–100	0.85–0.87	~0.99	0.98–0.985
>125	>0.87	>0.99	>0.985

5.2.1.5 Cycle Cooling Water Temperature

This is determined according to the hydrological data of the NPP's operating region. For example, the cycle cooling water temperature of the Guangdong Daya Bay NPP is set at 24 °C, while that of the Zhejiang Qinshan Phase I NPP is set at 18 °C.

5.2.1.6 Other Related Parameters

1. Pressure loss of fresh steam, reheated steam, and extraction steam at each stage of regenerative

The pressure loss of fresh steam is generally taken as 3%–7% of the initial steam pressure; the steam passing through the reheater and return pipe generates approximately 8%–12% pressure drop.

The heating steam is extracted from a certain stage of the turbine and flows into the heater. Due to the existence of flow resistance, the steam side pressure in the heater is lower than the extraction port pressure, and even lower than the stage pressure. The pressure drop from the stage to the extraction port is called the internal pressure drop, and the pressure drop from the extraction port to the heater is called the external pressure drop. Usually, the sum of the internal and external pressure drops does not exceed 10% of the absolute extraction pressure. In thermal calculations, the pressure loss of extraction steam at each stage of regenerative is taken as 4%–8% of the extraction pressure at that stage.

2. Heater type, terminal difference, and heat dissipation rate

Heaters are divided into surface type and mixed type. Surface type heat exchangers are mostly used in engineering, where the heating medium and the heat-absorbing medium do not mix. Figure 5.2 shows the working principle and temperature curve of the surface heater and the terminal difference.

The upper terminal difference of each heater is θ_u (the difference between the saturated water temperature under the pressure of each heater and the temperature of the heated water), and the lower terminal difference is θ_d (the difference between the water temperature leaving the drain cooler and the temperature of the heated water inlet). The size of the heater terminal difference is usually determined by technical and economic calculations, generally 3–8 °C. For small power units, the terminal difference is 6–8 °C; for high-parameter large power units, the terminal difference is 0–3 °C.

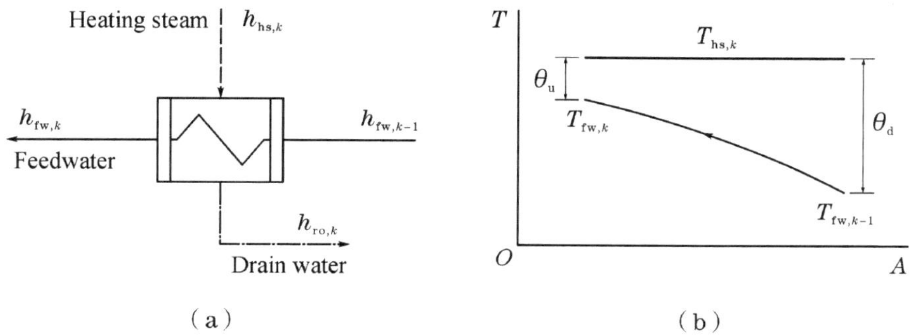

Fig. 5.2 Working principle of surface heater: (**a**) working flow; (**b**) temperature variation curve and terminal difference

In the mixed type heater, the heating steam and feed water directly mix. The steam condenses into water after contacting the lower temperature feed water, releasing heat to raise the feed water temperature. The mixed type heater can heat the feed water to the saturated temperature corresponding to the heating steam pressure, with a heating terminal difference of zero. Deaerators generally use mixed type heaters, high-pressure heaters and low-pressure heaters generally use surface type heaters, where high-pressure heaters are usually counterflow drain type with drain coolers, and low-pressure heaters are usually counterflow drain type.

The thermal economy of the unit improves as the terminal difference decreases. If the feedwater temperature is constant and other conditions remain unchanged, due to the terminal difference, the extraction pressure and its enthalpy value both increase, thereby reducing the thermal economy of the unit. Reducing the heater terminal difference requires increasing the heat transfer area, which increases investment, so the terminal difference should be determined after a comprehensive technical and economic comparison.

3. The heat loss of each level of heater: when considering the efficiency of the heater, it is taken as 0.97–0.99, and when considering the utilization coefficient of the enthalpy of each level of heating steam, it is taken as 0.985.
4. Plant load

The plant load of a NPP refers to the electrical load required by the NPP itself during normal startup, operation, shutdown, maintenance, and accident shutdown. It is characterized by large demand and high reliability requirements. During normal operation of the NPP, the plant load mainly comes from the electrical energy output of the generator set. Figure 5.3 shows a schematic diagram of the power transmission system of a NPP.

The plant load of a NPP is mostly pump loads, followed by electric heating elements and fans for supply and exhaust. The main electrical equipment and systems include: main coolant pump, stabilizer electric heating element, top-up pump, equipment cooling water pump, high-pressure safety injection pump, shutdown cooling pump, containment spray pump, containment hydrogen removal system

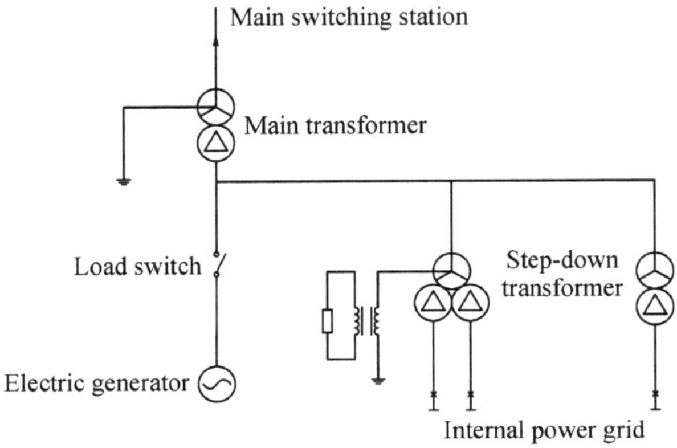

Fig. 5.3 Schematic diagram of NPP transmission system

suction fan, electric main feedwater pump, condensate pump, power plant supporting facilities (i.e., BOP system), etc. Generally speaking, the plant load is about 5%–10% of the total electric power of the NPP.

5.2.2 *Main Calculation Steps*

To perform thermal balance calculations on the thermal system of a PWR NPP, three types of basic equations need to be listed during the calculation process: mass balance equations, thermal balance equations for each heater, and power equations for the turbine unit.

Thermal balance calculations generally use a simple iterative method, with the basic steps as follows:

1. The net electrical power of the NPP is known as N_{net}. Assuming the effective efficiency of the power station is $\eta_{\text{e, npp}}$, then the thermal power of the reactor is

$$Q_{\text{R}} = \frac{N_{\text{net}}}{\eta_{\text{e,npp}}} \tag{5.1}$$

2. Determine the heating power Q_{mcp} of all main pumps to the coolant, ignore the impact of heat loss, and calculate the thermal power of the nuclear steam supply system:

$$Q_{\text{s}} = Q_{\text{R}} + Q_{\text{mcp}} \tag{5.2}$$

3. Based on Q_{s}, perform thermal balance calculations for the steam generator and determine the total feedwater flow rate G_{fw}:

$$G_{\mathrm{fw}} = \frac{Q_{\mathrm{s}}\eta_{\mathrm{sg}}\left(1+\xi_{\mathrm{P}}\right)}{\left(h_{\mathrm{fh}}-h_{\mathrm{s}}^{'}\right)+\left(1+\xi_{\mathrm{P}}\right)\left(h_{\mathrm{s}}^{'}-h_{\mathrm{fw}}\right)} \tag{5.3}$$

where ξ_{p} is the steam generator blowdown rate, generally taken as about 1% of the fresh steam output; h_{fh} is the enthalpy of fresh steam at the outlet of the steam generator, kJ/kg; $h_{\mathrm{s}}^{'}$ is the saturated water enthalpy under the operating pressure of the steam generator, kJ/kg; h_{fw} is the feedwater enthalpy of the steam generator, kJ/kg.

4. Perform steam consumption calculations for each equipment in the secondary loop system and determine the total steam consumption:

$$D_{\mathrm{s}} = \sum_{i=1}^{n} G_{\mathrm{s},i} \tag{5.4}$$

(a) Based on the total feedwater amount G_{fw} calculated earlier from the steam generator, perform thermal balance calculations for the feedwater regenerative system, and determine the extraction amount $G_{\mathrm{se},\,k}$at each extraction point of the steam turbine and the condensate amount G_{cd} at the outlet of the main condenser.

The heat balance equation for each stage of the heater is listed according to the absolute amount of soda water flow, and a heat balance equation must be established for each stage of the heater. In engineering calculations, the relative amount of steam consumption of the steam turbine is usually used to represent the share of steam extraction and condensation at each stage, and then the steam consumption of the steam turbine is obtained according to the power equation, and the absolute values of the steam extraction and condensation flow at each stage are calculated. Sometimes it is first roughly estimated, and the absolute value of the steam extraction at each stage is obtained according to the heat balance equation of the heater at each stage, and then the electric power, efficiency and other thermal economic indicators of the unit are calculated.

In the Z level regenerative heating system, there are Z steam extraction coefficients and one condensation coefficient, a total of $(Z + 1)$ unknowns, which need to be solved by $(Z + 1)$ first-order equations, that is, using the heat balance equations of Z heaters and a power equation of a steam turbine, a total of $(Z + 1)$ equations are used to solve $(Z + 1)$ unknowns. The calculation procedure depends on the connection method of the regenerative system. Generally, the steam extraction coefficient is calculated from the high-pressure heater, and then the lower-pressure heater is calculated.

(b) Calculate the steam consumption of the steam turbine generator set, and determine the steam consumption of the high and low pressure cylinders and the steam-water separator/reheater during rated full power operation;

(c) Calculate the feedwater pump, determine the steam consumption of the feedwater pump steam turbine or the input power of the feedwater pump motor;
(d) Calculate the heat balance of the main condenser, and determine the condensate amount G'_{cd} in the main condenser;
(e) Compare G_{cd} with G'_{cd}, if

$$\left|\frac{G_{\mathrm{cd}} - G'_{\mathrm{cd}}}{G_{\mathrm{cd}}}\right| > 0.01$$

Then correct G_{fw}, return to step (1) for iterative calculation until the accuracy is met.

5. Calculate the heat balance of the steam generator according to D_{s}, and determine the thermal power Q'_{s} of the nuclear steam supply system;

$$Q'_{\mathrm{s}} = \frac{D_{\mathrm{s}}\left[\left(h_{\mathrm{fh}} - h'_{\mathrm{s}}\right) + \left(1 + \xi_{\mathrm{P}}\right)\left(h'_{\mathrm{s}} - h_{\mathrm{fw}}\right)\right]}{\eta_{\mathrm{sg}}} \tag{5.5}$$

6. Set Q_{mcp} as a constant, calculate the thermal power Q'_{R} of the reactor, and further determine the effective efficiency of the power station:

$$\eta'_{\mathrm{e,npp}} = \frac{N_{\mathrm{net}}}{Q'_{\mathrm{R}}} = \frac{N_{\mathrm{net}}}{Q'_{\mathrm{s}} - Q_{\mathrm{mcp}}} \tag{5.6}$$

7. Compare the assumed efficiency with the calculated efficiency, if

$$\left|\eta_{\mathrm{e,npp}} - \eta'_{\mathrm{e,npp}}\right| < \varepsilon \quad (\text{usually}\ \varepsilon = 0.001)$$

holds, the heat balance calculation ends, otherwise, reset $\eta_{\mathrm{e,\,npp}}$, repeat the calculation of the above steps until the requirements are met.

5.3 Heat Balance Analysis Model of Main Equipment

PWR NPPs are complex energy systems composed of various thermal equipment according to thermodynamic principles and containing various thermal processes such as energy conversion, transfer and distribution. Although there are many thermal equipment, they can be roughly divided into reactors, pumps, heat exchange equipment, steam turbine generator sets and other equipment according to their working principles and energy usage. Classifying and establishing the heat balance analysis model of thermal equipment helps to simplify the thermal analysis of complex thermal systems.

5.3.1 Reactor

The reactor is the core equipment of the NPP, mainly realizing the conversion of nuclear fission energy to thermal energy and the transfer of heat from nuclear fuel to coolant. During the operation of the reactor, in order to prevent the leakage of radioactive materials in the core, there is usually good shielding, and the heat dissipation of the reactor body to the environment is very small. Since the heat dissipation loss of the reactor is not easy to calculate and the value is very small, it is often ignored in the heat balance analysis process, and the heat loss of the reactor is not considered.

5.3.2 Steam Generator

There are three types of steam generators widely used in PWR NPPs: vertical U-tube natural circulation steam generators, horizontal natural circulation steam generators, and vertical direct-flow steam generators. Among them, vertical U-tube natural circulation steam generators are widely used in European and American countries, and the steam generators used in the Daya Bay NPP and Qinshan NPP that have been put into operation in China are also of this type.

Figure 5.4a shows a schematic diagram of the structure of a vertical natural circulation steam generator, and Fig. 5.4b shows a schematic diagram of the structure of a direct flow steam generator.

5.3.2.1 Heat Transfer Calculation of Vertical U-tube Natural Circulation Steam Generator [39]

In modern PWR NPPs, the reactor and primary loop system include 2–3 coolant loops, each loop has a steam generator and a main coolant pump. Under normal operation, the coolant flow in each loop is 1/2 or 1/3 of the total coolant flow of the reactor.

If the heat dissipation of the main pipe and the heat added to the coolant during the operation of the main pump are ignored, the thermal power allocated to each steam generator by the nuclear steam supply system is

$$Q_{s1} = G_{c1} c_p \left(T_{si} - T_{so}\right) = \frac{G_c}{N_c} c_p \left(T_{co} - T_{ci}\right) \tag{5.7}$$

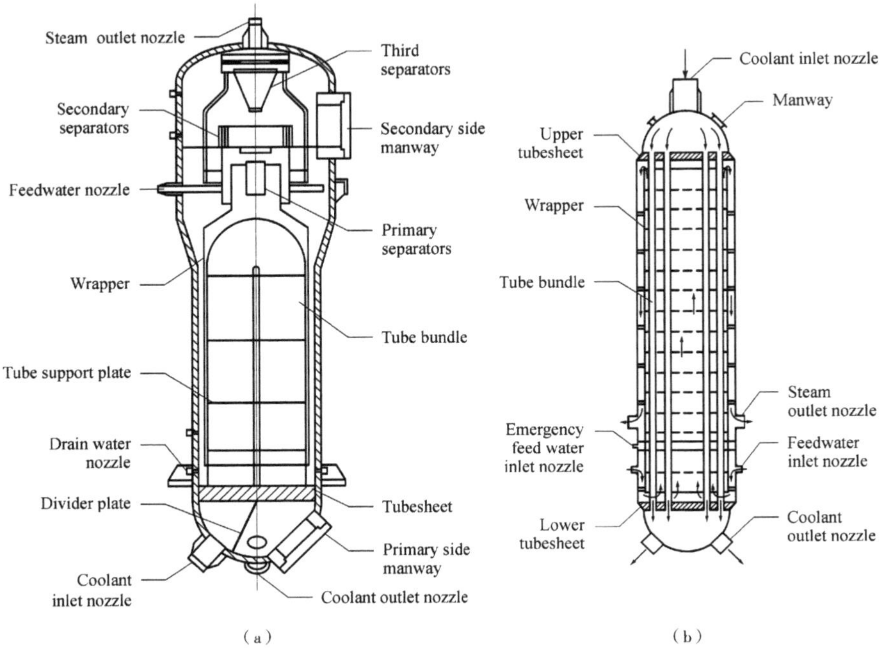

Fig. 5.4 Vertical U-tube natural circulation steam generator and once through steam generator: (**a**) U-tube natural circulation steam generator; (**b**) Once through steam generator

where G_c, G_{c1} are the coolant flow through the reactor and a single steam generator, respectively, kg/s; c_p is the average specific heat at constant pressure of the reactor coolant, kJ/(kg · K); T_{si}, T_{so} are the inlet and outlet coolant temperatures of the steam generator, respectively, °C; T_{ci}, T_{co} are the inlet and outlet coolant temperatures of the reactor, respectively, °C; N_c is the number of coolant loops.

Considering the heat loss of the steam generator, the thermal power absorbed by the working fluid on the secondary side is

$$Q_{sg} = Q_{s1} \cdot \eta_{sg} \tag{5.8}$$

where η_{sg} the thermal efficiency of the steam generator, 0.98–0.99.

The heat transfer equation of the steam generator is

$$Q_{sg} = 10^{-3} F \cdot K \cdot \Delta T_m \tag{5.9}$$

where F is the effective heat transfer area of the steam generator, m^2; K is the total heat transfer coefficient of the steam generator, W/(m^2 · K); ΔT_m is the average heat transfer temperature difference between the primary and secondary sides of the steam generator, °C.

The heat transfer process in the steam generator consists of several parts, such as the forced convective heat transfer from the primary loop coolant to the heat transfer tube wall, the heat conduction of the heat transfer tube wall and the fouling layer, and the boiling heat release of the outer wall of the heat transfer tube to the secondary loop working fluid. If the heat transfer tube is regarded as a thin-walled cylinder, the total heat transfer coefficient is

$$K = \frac{1}{\frac{1}{\alpha_1} \cdot \frac{d_{ca}}{d_{si}} + R_w + R_s + \frac{1}{\alpha_2} \cdot \frac{d_{ca}}{d_{so}}} \tag{5.10}$$

where α_1 is the convective heat transfer coefficient on the primary side, W/(m^2 · K); R_w, $R_s$are the thermal resistance of the heat transfer tube wall and the fouling resistance, respectively, (m^2 · K)/W; α_2 is the boiling heat transfer coefficient on the secondary side, W/(m^2 · K); d_{si}, d_{so} are the inner and outer diameters of the heat transfer tube, respectively, m; d_{ca} is the calculated diameter of the heat transfer tube, usually the outer diameter of the heat transfer tube is taken as the calculated diameter, that is, $d_{ca} = d_{so}$.

When the working fluid works in a co-current or counter-current manner, the heat transfer temperature difference is determined by the following formula:

$$\Delta T_m = \frac{\Delta T_{max} - \Delta T_{min}}{\ln \frac{\Delta T_{max}}{\Delta T_{min}}} \tag{5.11}$$

where ΔT_{max}, ΔT_{min} are the maximum and minimum temperature differences at both ends of the calculated section, respectively.

If $\frac{\Delta T_{max}}{\Delta T_{min}} \leq 1.7$, then

$$\Delta T_m = \frac{\Delta T_{max} + \Delta T_{min}}{2} \tag{5.12}$$

Due to limitations in heat transfer mechanisms, the parameters of fresh steam in a vertical U-tube natural circulation steam generator will vary with changes in power load. Figure 5.5 shows the static characteristics of the vertical U-tube natural circulation steam generator used in the M310 unit.

At the calculated power level, assuming that the working pressure of the secondary side of the steam generator is p_s, calculate the total heat transfer coefficient K based on this condition, then according to the heat load Q_{sg} allocated to a single steam generator by the nuclear steam supply system, the average heat transfer temperature difference of the primary and secondary sides of the steam generator can be determined

$$\Delta T_m = 10^3 \frac{Q_{sg}}{F \cdot K} \tag{5.13}$$

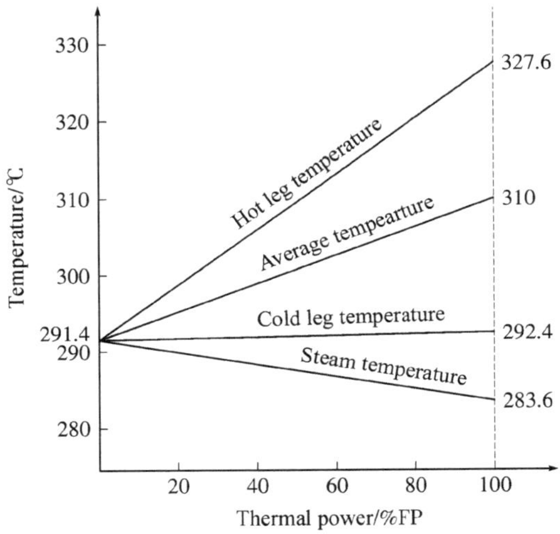

Fig. 5.5 Static characteristics of a vertical U-tube natural circulation steam generator

The temperature difference of the coolant at the inlet and outlet of the steam generator is

$$\Delta T_c = T_{si} - T_{so} = \frac{Q_{sg1}}{\eta_{sg} G_{c1} c_p} \tag{5.14}$$

Knowing that the average temperature of the coolant in the reactor and the primary circuit system is T_m, the temperatures of the coolant at the inlet and outlet of the steam generator are respectively

$$T_{si} = T_m + \frac{\Delta T_c}{2}, \qquad T_{so} = T_m - \frac{\Delta T_c}{2} \tag{5.15}$$

According to the calculation formula (5.11) of the logarithmic average temperature difference on the primary and secondary sides of the steam generator, the evaporation temperature of the secondary side working fluid can be determined

$$T_s = \frac{T_{si} - T_{so} e^A}{1 - e^A} \tag{5.16}$$

where A is a coefficient, $A = \dfrac{\Delta T_c}{\Delta T_m}$.

The saturation pressure $p_s^{'}$ corresponding to T_s is the fresh steam pressure on the secondary side of the steam generator. If it satisfies

$$\frac{\left| p_s - p_s^{'} \right|}{p_s} < 0.01$$

Then the heat balance calculation of the steam generator is completed, otherwise, reset p_s and repeat the above calculation process until the requirements are met.

The steam output of the steam generator is related to the heat load and the fresh steam parameters. After determining the fresh steam parameters under the given load through the above iterative calculation, the steam output of a single steam generator can be determined by the following formula:

$$G_{s1} = \frac{Q_{sg}}{h_{fh} - h_s' + (1+\xi_P)(h_s' - h_{fw})} \tag{5.17}$$

where h_{fh} is the enthalpy value of fresh steam at the outlet of the steam generator, kJ/kg; h_s' is the enthalpy value of saturated water under evaporation pressure p_s, kJ/kg; ξ_p is the discharge coefficient, usually taken as 1% of the fresh steam flow; h_{fw} is the feed water enthalpy value, kJ/kg.

The fresh steam output of the entire nuclear steam supply system is

$$D_s = G_{s1} \cdot N_c \tag{5.18}$$

When calculating the total heat transfer coefficient K of the steam generator, it is necessary to first determine the heat transfer coefficients of the primary and secondary sides and the thermal resistance of the tube wall material and dirt. The specific calculation process is introduced as follows.

5.3.2.2 Primary Side Heat Transfer Coefficient

The heat exchange inside the heat transfer tube belongs to the turbulent heat release process of forced convection inside the tube. The heat release coefficient α_1 is generally calculated by the Dittus-Bolter equation:

$$\alpha_1 = 0.023 \frac{\lambda_c}{d_{si}} \mathrm{Re}^{0.8} \mathrm{Pr}^{0.4} \tag{5.19}$$

where λ_c is the thermal conductivity of the coolant, W/(m · K); Pr is the Prandtl number of the coolant; Re is the Reynolds number of the coolant, calculated by the following formula:

$$\mathrm{Re} = \frac{w_c \cdot d_{si}}{v_c} \tag{5.20}$$

where v_c is the kinematic viscosity coefficient of the coolant, m²/s; w_c is the flow rate of the coolant inside the heat transfer tube, m/s, the calculation formula is

$$w_c = \frac{4}{\pi d_{si}^2 N_s} \frac{G_{c1}}{\rho_c} \tag{5.21}$$

where N_s is the number of heat transfer tubes; ρ_c is the density of the coolant, kg/m³.

5.3.2.3 Tube Wall and Fouling Thermal Resistance

In a steam generator, the thermal resistance of the tube wall generally accounts for 40%–50% of the total thermal resistance. If the outer diameter of the heat transfer tube is used as a reference, the calculation formula for the thermal resistance of the tube wall is

$$R_w = \frac{d_{so}}{2\lambda_w} \ln \frac{d_{so}}{d_{si}} \tag{5.22}$$

where λ_w is the thermal conductivity of the heat transfer tube material, W/(m · K).

In engineering calculations, the following formula is usually used for approximate calculation:

$$R_w = \frac{d_{so} - d_{si}}{2\lambda_w} \tag{5.23}$$

For commonly used heat transfer tube materials in steam generators such as Inconel-600 and Incoloy-800, their thermal conductivity can be found from Table 5.2 based on the temperature of the heat transfer tube.

The thermal conductivity of Inconel-600 can also be calculated using the formula recommended by CE company:

$$\lambda_w = 14.244 + 1.555 \times 10^{-2} t_w \tag{5.24}$$

where t_w is the average temperature of the heat transfer tube wall, °C.

The thermal conductivity of Incoloy-800 can also be calculated using the formula recommended by KWU company:

$$\lambda_w = 11.628 + 1.57 \times 10^{-2} t_w \tag{5.25}$$

Fouling thermal resistance refers to the thermal resistance caused by fouling on the tube wall, which is related to factors such as the material of the heat transfer tube and the quality of the operating water. In engineering calculations, there are three methods to consider fouling thermal resistance:

Table 5.2 Thermal conductivity of commonly used materials for heat transfer tubes in steam generators

Temperature (°C)	21	93	204	316	427	538	649	760	871
Inconel-600	14.65	15.91	17.58	19.26	20.93	22.61	24.70	26.80	28.89
Incoloy-800	17.72	12.98	14.65	16.75	18.42	20.0	21.77	23.86	25.96

1. Reduce the convective heat transfer coefficient to consider the impact of fouling on the corresponding side.
2. When calculating the total heat transfer coefficient, do not consider the fouling thermal resistance, but introduce a reserve factor when determining the heat transfer area. This factor is usually considered in combination with factors such as tube blockage allowance and thermal calculation errors, and can generally be taken as about 10%.
3. Use empirical data.

   Early steam generator designs usually took the fouling thermal resistance value as 5.25×10^{-5} m$^2 \cdot{}^\circ$ C/W. Later, due to the improvement of heat transfer tube materials and strict control measures for secondary circuit water quality, the fouling on the secondary side was significantly reduced. The fouling thermal resistance value recommended by Westinghouse in the Reference Safety Analysis Report (RESAR) is 8.77×10^{-6} m$^2 \cdot{}^\circ$ C/W.

Usually in the calculation process, according to the recommendation of relevant data, the value of R_s is taken as

$$R_s = (5.16 \quad 6.88) \times 10^{-5}\ \mathrm{m^2K/W} \tag{5.26}$$

5.3.2.4 Secondary Side Heat Transfer Coefficient

In a vertical U-tube natural circulation steam generator, the secondary side working fluid has to go through the preheating area and boiling area with different heat transfer characteristics in the flow process. The preheating area is single-phase convective heat transfer and underheated boiling heat transfer, and the boiling area is inter-tube boiling heat transfer. When performing detailed thermal design and heat transfer calculations, the heat transfer coefficient of the secondary side of the steam generator should be calculated in sections.

1. Convective heat transfer coefficient of the preheating area

   When the secondary side fluid longitudinally impinges on the heat transfer tube bundle from the outside, the heat transfer coefficient can be calculated according to the Dittus-Bolter formula, and the inner diameter of the tube in the formula should be replaced with the equivalent diameter of the flow channel; when the secondary side fluid flows outside the tube, and the preheating section is also equipped with a support plate used to fix the tube bundle and increase the heat dissipation intensity, the heat transfer coefficient can be calculated according to the heat transfer coefficient formula of the shell side of the shell-and-tube heat exchanger with a support plate; for the situation where the tube bundle enters the area and is equipped with a support plate, the heat transfer coefficient formula can be calculated when the tube bundle is impinged transversely, and the impact angle should be considered.

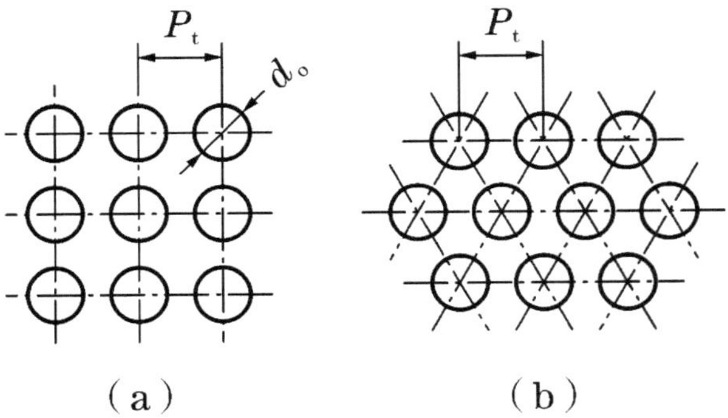

Fig. 5.6 Arrangement of heat transfer tubes: (**a**) square arrangement; (**b**) triangular arrangement

Table 5.3 The value of coefficients in the convection heat transfer equation

	Axial direction			Radial direction		
Formula coefficient	C_1	C_2	C_3	C_1	C_2	C_3
EPRI78	0.03	0.8	0.333	0.29	0.6	0.333
RCPE81	$0.013 + 0.033\beta_v$	0.8	0.4	1.5	0.563	0.333
Square arrangement	$0.042\frac{P_t}{d_{so}} - 0.024$	0.8	0.4	0.547	0.563	0.333
Triangular arrangement	$0.026\frac{P_t}{d_{so}} - 0.006$	0.8	0.4	0.547	0.563	0.333

Figure 5.6 shows the common arrangement of the heat transfer tubes of the steam generator, and the calculation of the heat transfer coefficient is slightly different under different arrangements. The ATHOS program developed by the Electric Power Research Institute (EPRI) for steady-state and transient analysis of steam generators recommends a set of calculation formulas with the same form when performing single-phase convective heat transfer calculations, considering both longitudinal impingement and transverse impingement of the tube bundle.

$$\alpha_{2,a} = C_1 \,\mathrm{Re}^{C_2}\,\mathrm{Pr}^{C_3}\,\frac{\lambda_s}{d_e} \tag{5.27}$$

where C_1, C_2, C_3 are coefficients, listed in Table 5.3; λ_s is the thermal conductivity of the feedwater in the preheating area on the secondary side of the steam generator, W/(m · °C); d_e is the equivalent diameter of the flow channel between the tube bundles on the secondary side, m.

For U-tube natural circulation steam generators, the actual flow between the tube bundles is neither longitudinal nor strictly transverse impingement. After the ATHOS program calculates the heat dissipation coefficients of longitudinal and transverse impingement separately, it uses their root mean square value as the effective heat transfer coefficient.

2. Subcooled boiling heat transfer coefficient in the preheating area

The calculation of the subcooled boiling heat transfer coefficient often uses the following relatively simple formulas.

(a) Jens-Lottes formula

$$\alpha_{2,b} = 1.264 q_s^{0.75} e^{p_s/6.2} \tag{5.28}$$

where p_s is the secondary side pressure, MPa; q_s is the heat flux density, W/m^2.

The experimental conditions of the formula are: tube inner diameter 3.63–5.74 mm, tube length is 21–168 times the tube inner diameter, system pressure 0.7–17.2 MPa, water temperature 115–340 °C, mass flow rate range 11–1.05 × 10^4 kg/(m^2 · s), heat flux density up to 12.5 × 10^6 W/m^2.

(b) Thom formula

$$\alpha_{2,b} = 44.4 q_s^{0.5} e^{p_s/8.7} \tag{5.29}$$

The experimental conditions of the formula are: system pressure 5.17–13.8 MPa, heat flux density is (2.8–6.0) × 10^5 W/m^2.

3. Inter-tube boiling heat transfer coefficient in the boiling area

The heat transfer in the boiling area on the secondary side of the natural circulation steam generator belongs to inter-tube boiling heat transfer. At present, most of the engineering uses the large space bubble nucleate boiling heat transfer calculation formula. In addition to the Jens-Lottes formula and the Thom formula, the recommended calculation formulas in engineering design also include the following commonly used formulas.

(a) Rohsenow formula

$$\alpha_{2,c} = \left(\frac{c_p'}{C_1 r Pr'}\right)\left(\mu' r \sqrt{\frac{g(\rho' - \rho'')}{\sigma}}\right)^{0.33} q_s^{0.67} \tag{5.30}$$

where c_p' is the specific heat capacity of saturated water at constant pressure, J/(kg · K); C_1 is a constant depending on the heating surface-liquid combination, for water-nickel stainless steel, it can be taken as 0.013; r is the latent heat of vaporization, J/kg; Pr' is the Prandtl number of saturated water; μ' is the dynamic viscosity of saturated water, kg/(m · s); ρ', ρ'' are the densities of saturated water and saturated steam respectively, kg/m^3; σ is the surface tension of the water-steam interface, N/m.

(b) Kutateradze simplified formula

$$\alpha_{2,c} = 5 p_s^{0.2} q_s^{0.7} \tag{5.31}$$

This formula is applicable for calculating the heat transfer coefficient of water boiling in a large space within a pressure range of 0.02–10 MPa.

Since the mechanism of boiling heat transfer is complex and related to the condition of the heat transfer surface, the boiling heat transfer coefficients calculated by different formulas vary greatly. In general, to simplify the calculation, the influence of the preheating section is often ignored, and the Kutateradze simplified formula is directly used to calculate the average heat transfer coefficient of the secondary side of the steam generator. For thermal analysis, the calculation accuracy is still acceptable.

5.3.3 *Feedwater Regenerative Heating System*

The regenerative cycle can improve the efficiency of the thermal cycle, so PWR NPPs generally use the regenerative cycle. The feedwater regenerative heating system, together with the steam turbine extraction system, constitutes the regenerative cycle, which is an important part of the secondary thermal system of the NPP. It consists of multi-stage feedwater heaters, drain coolers, drain tanks, drain pumps and corresponding piping systems, and plays an important role in improving the thermal economy of the NPP.

Feedwater heaters are divided into low-pressure heaters, deaerators, and high-pressure heaters. The low-pressure heater is located between the condensate pump and the deaerator, and the high-pressure heater is located between the feedwater pump and the steam generator. The condensate from the condenser passes through the low-pressure heater, deaerator, and high-pressure heater in sequence, is heated to a certain temperature, and then sent to the steam generator. The heating steam mainly comes from the extraction of various stages of the high and low-pressure cylinders of the steam turbine.

Depending on the different ways in which the heating steam and feedwater come into contact, feedwater heaters can be divided into two types: mixed and surface.

5.3.3.1 Surface Feedwater Heater

Both the low-pressure and high-pressure feedwater heaters in PWR NPPs are surface heaters, which achieve heat transfer through the metal heating surface. Depending on the different ways of heat transfer between steam and water in the feedwater heater, feedwater heaters are generally divided into two types: serial flow drain heaters and parallel flow drain heaters.

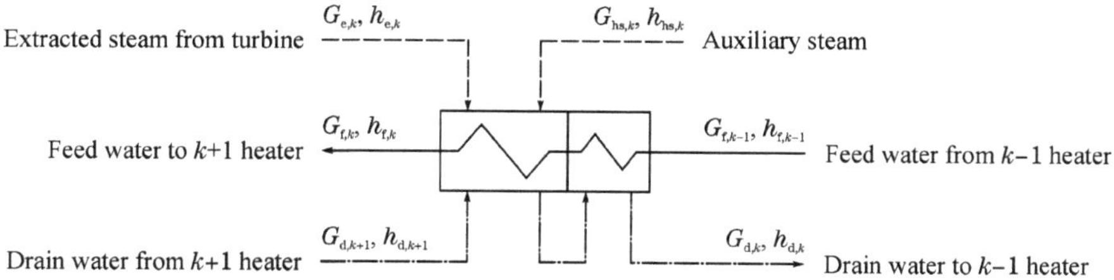

Fig. 5.7 Serial flow drain heater with drain cooler

Serial Flow Drain Heater

Figure 5.7 shows a series flow drain heater with a drain cooler. This type of heater has a drain cooler connected in series at the front end. The steam extracted from the turbine, auxiliary steam, and the drain from the downstream heater heat the feedwater in the heater. The resulting steam-water mixture is discharged into the drain cooler, which preheats the feedwater entering the heater, and then is drained as the drain of this stage heater.

Assuming the feedwater heater is the kth stage, the heat released by the heating medium per unit mass flow rate in it is respectively

Steam turbine extraction	$q_{h,k} = h_{e,k} - h_{d,k}$
Auxiliary steam	$q'_{h,k} = h_{hs,k} - h_{d,k}$
Hydrophobic	$\gamma_k = h_{d,k+1} - h_{d,k}$

The enthalpy rise of the feedwater per unit mass flow rate in the heater is

$$\tau_k = h_{f,k} - h_{f,k-1}$$

According to the working process of the feedwater heater, its mass balance equation and heat conservation equation are

$$G_{d,k} = G_{e,k} + G_{hs,k} + G_{d,k+1} \tag{5.32}$$

$$\left(G_{e,k} q_{h,k} + G_{hs,k} q'_{h,k} + G_{d,k+1} \gamma_k\right) \eta_k = G_{f,k} \tau_k \tag{5.33}$$

where η_k is the thermal efficiency of the k th stage feedwater heater.

The flow rate and enthalpy value of the auxiliary heating steam are usually given, then the amount of extraction steam consumed by the feedwater heater is

$$G_{e,k} = G_{f,k} \frac{\tau_k}{\eta_k q_{h,k}} - G_{hs,k} \frac{q'_{h,k}}{q_{h,k}} - G_{d,k+1} \frac{\gamma_k}{q_{h,k}} \tag{5.34}$$

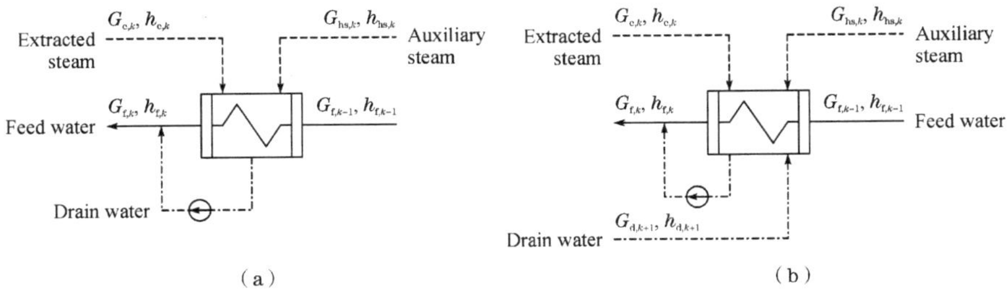

Fig. 5.8 Drainage confluence type feedwater heater

For feedwater heaters without a drain cooler, the calculation method is exactly the same.

Drainage Confluence Feedwater Heater

Figure 5.8 shows a drainage confluence feedwater heater, where the drain of the turbine extraction steam and auxiliary steam is collected with the feedwater at the heater outlet and sent to the downstream feedwater heater.

Taking the feedwater heater shown in Fig. 5.8b as an example, the heat released by the heating medium per unit mass flow rate in the heater is respectively

Steam turbine extraction	$q_{h,k} = h_{e,k} - h_{f,k-1}$
Auxiliary steam	$q'_{h,k} = h_{hs,k} - h_{f,k-1}$
Hydrophobic	$\gamma_k = h_{d,k+1} - h_{f,k-1}$

The enthalpy rise of the feedwater per unit mass flow rate in the heater is

$$\tau_k = h_{f,k} - h_{f,k-1}$$

According to the working principle of the drainage confluence feedwater heater, its mass balance equation and heat balance equation are

$$G_{f,k} = G_{f,k-1} + G_{e,k} + G_{hs,k} + G_{d,k+1} \tag{5.35}$$

$$\left(G_{e,k} q_{h,k} + G_{hs,k} q'_{h,k} + G_{d,k+1} \gamma_k\right) \eta_k = G_{f,k} \tau_k \tag{5.36}$$

Then the amount of extraction steam consumed by the feedwater heater is

$$G_{e,k} = G_{f,k} \frac{\tau_k}{\eta_k q_{h,k}} - G_{hs,k} \frac{q'_{h,k}}{q_{h,k}} - G_{d,k+1} \frac{\gamma_k}{q_{h,k}} \tag{5.37}$$

The feedwater flow rate at the inlet of the feedwater heater is

$$G_{\mathrm{f},k-1} = G_{\mathrm{f},k} - G_{\mathrm{hs},k} - G_{\mathrm{d},k+1} - \left(G_{\mathrm{f},k} \frac{\tau_k}{\eta_k q_{\mathrm{h},k}} - G_{\mathrm{hs},k} \frac{q'_{\mathrm{h},k}}{q_{\mathrm{h},k}} - G_{\mathrm{d},k+1} \frac{\gamma_k}{q_{\mathrm{h},k}} \right) \tag{5.38}$$

5.3.3.2 Mixed Feedwater Heater

The mixed heater does not have a heat transfer surface made of metal materials. The heating steam and feedwater are directly mixed and heat transferred, which is a type of drainage confluence heater.

In the feedwater regenerative system of a PWR NPP, the thermal deaerator is a mixed heater, which heats the feedwater to the saturation state under the operating pressure of the deaerator to remove dissolved oxygen and other gases in the water.

According to the size of the operating pressure, deaerators can be divided into vacuum, atmospheric pressure, and high pressure types. The vacuum deaerator is only an auxiliary device because condensate water may still leak air when passing through some equipment and pipes under vacuum. In medium-parameter units, atmospheric pressure deaerators of 0.12 MPa or 0.25 MPa are generally used, in high-parameter units, high-pressure deaerators of 0.6 MPa are generally used, and in ultra-high-pressure units, to ensure good feedwater quality, sometimes the condenser is used as a vacuum deaerator, as a supplement to the high-pressure deaerator.

The thermal deaerator used by the generator set usually consists of a deaerator tower (deaerator head) and a feedwater tank. The deoxygenation of feedwater mainly takes place in the deaerator tower. According to the structure classification, deaerators can be divided into water spray tray type, spray packing type, etc.; according to the heat transfer mechanism, they can be divided into atomization, bubble boiling, and rotary film types. Figure 5.9 shows a water spray tray type deaerator, which belongs to the atmospheric pressure deaerator.

The basic working principle of the shower tray deaerator is: The water that needs to be deoxygenated enters the condensate water distribution box from the pipe at the

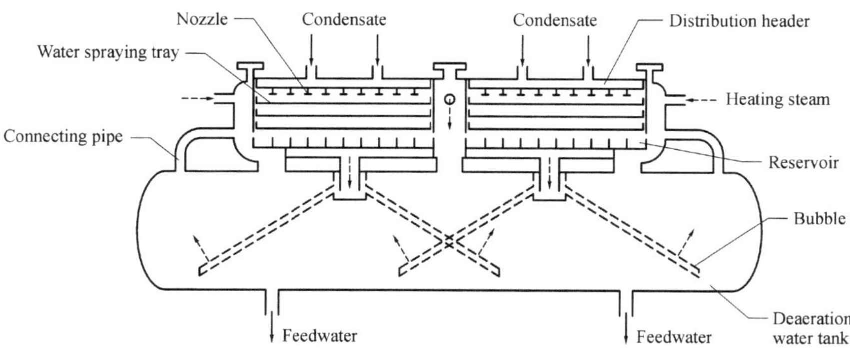

Fig. 5.9 Shower tray deaerator

top of the deaerator, is sprayed out in the form of droplets by the nozzle, and is heated to saturation by the counter-flowing steam during the falling process. 80%–90% of the oxygen and other gases dissolved in the water are separated. The water droplets fall onto the shower tray, are dispersed into fine water flows through the small holes on the shower tray, and fall into the deoxygenation water tank after passing through each level of the shower tray. During the falling process, they continue to be heated by the upward flowing steam, and most of the remaining gases in the water are removed. The separated gases are continuously discharged from the exhaust port at the top of the deaerator.

The heating steam enters through the central pipe and the inlet pipe, a part of it rises from the steam hole at the bottom layer of the shower tray and heats the fine water flow, and another part escapes from the steam pipe at the bottom of the deoxygenation water tank, bubbling and heating the deoxygenated water to prevent secondary oxygen dissolution due to heat dissipation and cooling.

The feed water entering the deaerator usually includes feed water delivered from the low-pressure feed water heater, moisture separator reheater, and the drainage of the high-pressure feed water heater. The heating steam comes from the extraction steam of the high-pressure cylinder of the steam turbine. The water in the deoxygenation water tank is drawn out by the feed water pump, flows through each level of the high-pressure feed water heater in stages, is heated to the final feed water temperature, and then enters the steam generator. Figure 5.10 shows the thermal balance relationship of the deaerator.

If the deaerator is set as the kth stage heater in the feed water regenerative system, the heat released by the unit mass flow rate of the steam turbine extraction steam in the deaerator is

$$q_{\mathrm{h},k} = h_{\mathrm{e},k} - h_{\mathrm{f},k}$$

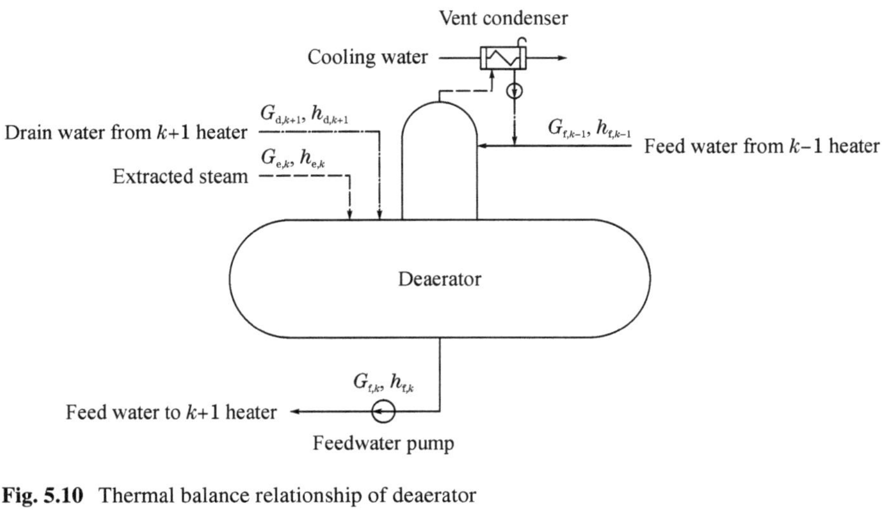

Fig. 5.10 Thermal balance relationship of deaerator

The heat released by the unit mass flow rate of the drainage in the deaerator is

$$\gamma_k = h_{\mathrm{d},k+1} - h_{\mathrm{f},k}$$

The enthalpy rise of the unit mass flow rate of the feed drainage in the deaerator is

$$\tau_k = h_{\mathrm{fw},k} - h_{\mathrm{fw},k-1}$$

According to the working principle of the deaerator, ignoring the steam loss caused by the deaerator exhaust and the heat loss of the deaerator body, the mass balance equation and heat balance equation of the deaerator are

$$G_{\mathrm{f},k-1} + G_{\mathrm{d},k+1} + G_{\mathrm{e},k} = G_{\mathrm{f},k} \tag{5.39}$$

$$G_{\mathrm{e},k} q_{\mathrm{h},k} + G_{\mathrm{d},k+1}\gamma_k = G_{\mathrm{f},k}\tau_k \tag{5.40}$$

Combining the above two formulas, we get

$$G_{\mathrm{e},k} = G_{\mathrm{f},k}\frac{\tau_k}{q_{\mathrm{h},k}} - G_{\mathrm{d},k+1}\frac{\gamma_k}{q_{\mathrm{h},k}} \tag{5.41}$$

$$G_{\mathrm{f},k-1} = G_{\mathrm{f},k} - G_{\mathrm{d},k+1} - G_{\mathrm{e},k} \tag{5.42}$$

From the above heat balance equation, it can be seen that there are differences in the definitions of the two physical quantities, heat q released per unit mass of heated steam and heat γ released per unit mass of hydrophobic water, in surface heat exchangers and hybrid heat exchangers.

5.3.4 Moisture Separator Reheater

The steam discharged from the high-pressure cylinder of the steam turbine has a high humidity, about 12%. To prevent the water carried in the steam from eroding the low-pressure turbine blades, ensure the normal working life of the steam turbine, and improve the relative internal efficiency of the low-pressure turbine, a moisture separator reheater is usually used to increase the dryness and temperature of the steam entering the low-pressure cylinder. The reheating of steam can not only improve the dryness of the steam at the end stage of the low-pressure steam turbine, reduce the corrosion of the blades of the large low-pressure steam turbine, but also increase the power output of the low-pressure steam turbine. Compared with the non-reheating cycle, the two-stage reheating method can increase the cycle thermal efficiency by 1.8%–2.5%.

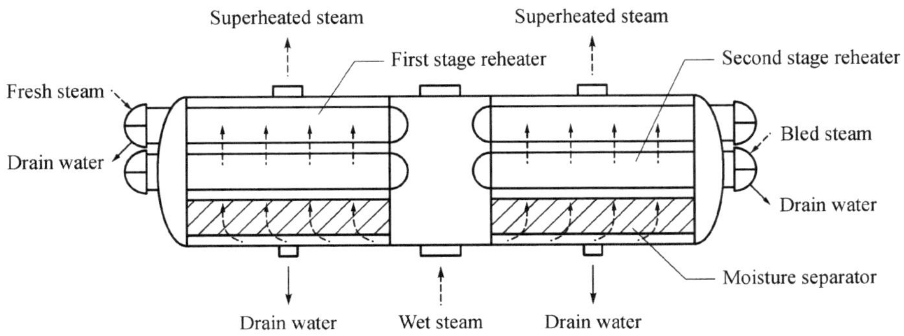

Fig. 5.11 Schematic diagram of horizontal moisture separator reheater

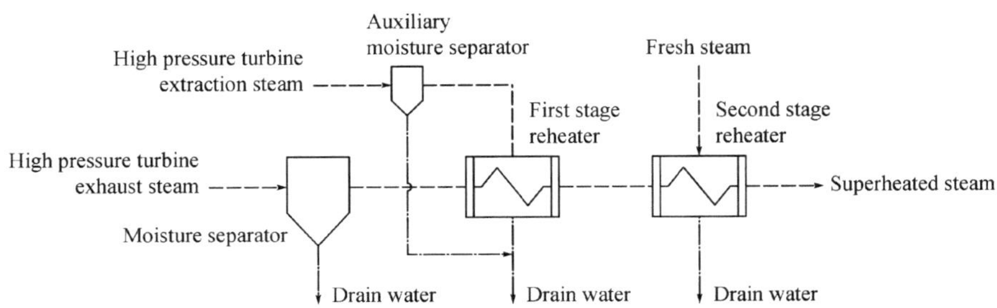

Fig. 5.12 Steam reheater with auxiliary moisture water separator

Figure 5.11 is a schematic diagram of a horizontal moisture separator reheater commonly used in NPPs. The low-pressure saturated steam discharged from the high-pressure cylinder of the steam turbine enters the moisture separator reheater from the lower pipe, passes through the moisture separation component, and about 98% of the water is removed, significantly improving the dryness of the steam. It then sequentially enters the first and second stage steam reheaters, is heated to a slightly superheated state, and then sent to the low-pressure cylinder of the steam turbine. The heating steam of the first stage steam reheater is usually the first stage extraction steam from the high-pressure cylinder of the steam turbine, and the heating steam of the second stage steam reheater is the fresh steam from the main steam pipe. The steam-water mixture formed after the heating steam releases heat is collected in a specially designed drain tank and finally sent to the feedwater heater.

Some steam reheaters are also equipped with auxiliary moisture separators and venting systems to separate the steam and water from the extraction steam and discharge the non-condensable gases from the reheater, As shown in Fig. 5.12, the heating steam of the second stage steam reheater is fresh steam, so no auxiliary moisture separator is needed, and the rest of the structure is similar to the first stage reheater.

Assuming that the steam parameters discharged from the high-pressure cylinder of the steam turbine are (p_{zh}, x_{zh}), the flow rate of the wet steam entering the moisture separator is G_{zh}, and the dryness of the steam after moisture separation is x_{spo}. In the moisture separator, the water in the wet steam is mainly separated on the surface of the metal wire mesh or corrugated plate bundle, so there must be flow resistance during the separation process. If the steam pressure drop in the moisture separator is Δp_{sp}, then the steam pressure at the separator outlet is

$$p_{spo} = p_{zh} - \Delta p_{sp}$$

Assuming that the separator drain is saturated water corresponding to the outlet pressure p_{spo}, the steam enthalpy h_{spo} and drain enthalpy h_{spw} at the separator outlet can be found by looking up the steam table. If heat loss is ignored, the energy and mass balance equations in the moisture separator are

$$G_{zh} = G_{spw} + G_{spo} \tag{5.43}$$

$$G_{zh} h_{zh} = G_{spw} h_{spw} + G_{spo} h_{spo} \tag{5.44}$$

where h_{zh} is the enthalpy value of the steam discharged from the high-pressure steam turbine,kJ/kg; G_{spw}, G_{spo} are the mass flow rates of the drain and outlet steam from the moisture separator respectively, kg/s.

Solving the above equation, we can get:

$$G_{spo} = G_{zh} \frac{h_{zh} - h_{spw}}{h_{spo} - h_{spw}} \tag{5.45}$$

Further, G_{spw} can be calculated.

The heat balance calculation of the steam reheater is similar to that of general heat exchange equipment, which will not be repeated here.

5.3.5 Steam Turbine Generator

The steam turbine generator is an important equipment in the PWR NPP to convert thermal energy into electrical energy. In the working process, the steam turbine converts the thermal energy of the steam generated by the steam generator into the kinetic energy of the steam flow, and then into mechanical energy to drive the generator to generate electricity.

During the operation of the steam turbine, there are various internal energy losses. In addition to nozzle losses, moving blade losses, and residual speed losses, there are also friction fan losses, stage internal steam leakage losses, wet steam losses, and throttling losses. The internal loss of the steam turbine causes the steam's ability to work inside the steam turbine to decrease, and part of the steam's kinetic energy is dissipated as heat and absorbed by the steam flow, increasing the enthalpy of the steam. If the steam turbine's shaft seal steam leakage loss and heat dissipation loss are ignored, then there is no heat loss in the expansion process of the steam in the flow part of the steam turbine.

The internal power of the steam turbine is:

$$N_{i} = N_{ih} + N_{il} \tag{5.46}$$

where N_{ih}, N_{il} are the internal powers of the high and low-pressure steam turbines, their values are respectively

$$\begin{cases} N_{ih} = G_{0h}h_{0h} - \sum_{k=1}^{m} G_{seh,k}h_{seh,k} - G_{zh}h_{zh} \\ N_{il} = G_{01}h_{01} - \sum_{k=1}^{n} G_{sel,k}h_{sel,k} - G_{zl}h_{zl} \end{cases} \tag{5.47}$$

The mechanical energy generated by the turbine is dissipated into heat energy due to reasons such as bearing friction during the transmission process, which is absorbed and carried away by the lubricating oil. This energy loss can be represented by the mechanical efficiency of the turbine:

$$\eta_{mt} = \frac{N_{t}}{N_{i}} \tag{5.48}$$

where N_{t} is the effective output power of the turbine, kW.

The motor generates electrical energy under the drive of the turbine. Due to mechanical losses and electrical losses, part of the energy is converted into heat energy, which is carried away by the cooling hydrogen. Its energy loss is represented by the effective efficiency of the generator:

$$\eta_{me} = \frac{N_{e}}{N_{t}} \tag{5.49}$$

where N_{e} is the effective output electrical power of the generator, kW.

Therefore, the energy loss in the turbine generator set is mainly the loss in the power transmission process of the turbine and the generator, which can be represented as

$$\Delta N_{te} = N_{i} - N_{e} = \left(\frac{1}{\eta_{mt}\eta_{me}} - 1 \right) N_{e} \tag{5.50}$$

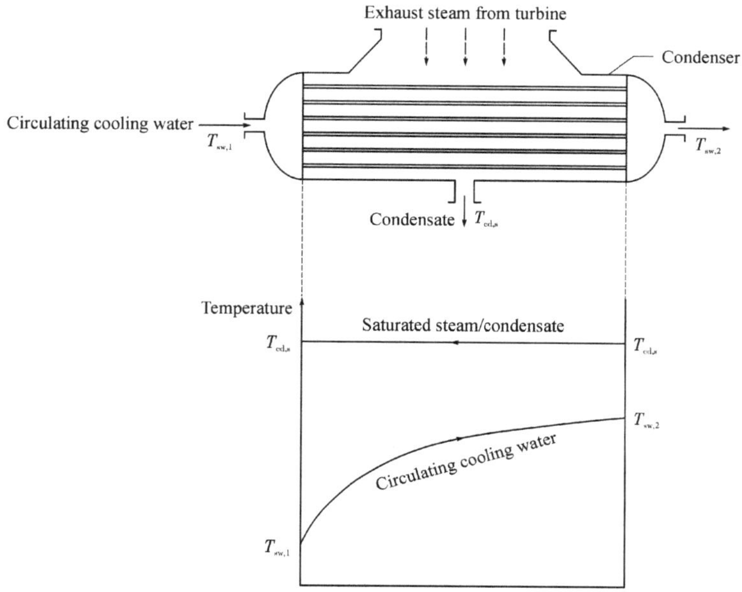

Fig. 5.13 Working principle diagram of condenser

5.3.6 *Condenser*

The condenser is an important component of the turbine generator set. Its function is to accept steam from the low-pressure turbine and bypass system, turbine feedwater pump, and blowdown box, recover the drainage during startup and the drainage of the feedwater heating system, condense it into water, and establish the required back pressure. Figure 5.13 shows the working principle diagram of the condenser [39].

Assuming that the condenser pressure under the calculated operating condition is p_{cd}, if the condensation water supercooling is δT_{cd}, then the temperature and enthalpy of the condensation water are respectively

$$T_{cd} = T_{cd,s} - \delta T_{cd}, \quad h_{cd} = f\left(p_{cd}, T_{cd}\right)$$

where $T_{cd,s}$ is the saturation temperature corresponding to the working pressure of the condenser, °C; h_{cd} is the enthalpy of the condensate, which is found from the steam table based on p_{cd} and δT, kJ/kg.

The heat load of the condenser is

$$Q_{cd} = \sum_{j=1}^{n} G_{cd,j}\left(h_{cd,j} - h_{cd}\right) \tag{5.51}$$

where $G_{\mathrm{cd},j}$ and $h_{\mathrm{cd},j}$ are the mass flow rate and enthalpy of the jth working substance entering the condenser, respectively.

Ignoring the heat dissipation loss, the heat balance equation of the condenser is

$$Q_{\mathrm{cd}} = G_{\mathrm{sw}} c_p \left(T_{\mathrm{sw},2} - T_{\mathrm{sw},1}\right) \tag{5.52}$$

where G_{sw} and c_p are the circulating cooling water flow rate and its specific heat at constant pressure; $T_{\mathrm{sw},1}$ and $T_{\mathrm{sw},2}$ are the inlet and outlet temperatures of the circulating cooling water.

The heat released by the drainage and steam discharged into the condenser is completely absorbed by the circulating cooling water and taken out of the thermal cycle, dissipated in the environment, therefore, the heat loss in the condenser is

$$\Delta Q_{\mathrm{cd}} = Q_{\mathrm{cd}}$$

According to the heat transfer equation of the condenser, the average heat transfer temperature difference of the condenser can be obtained

$$\Delta T_{\mathrm{m}} = 10^{-3} \frac{Q_{\mathrm{cd}}}{F_{\mathrm{c}} K} \tag{5.53}$$

where F_{c} is the effective heat transfer area of the condenser, m^2; K is the overall heat transfer coefficient of the condenser, W/(m^2 ·° C).

Considering

$$\Delta T_{\mathrm{m}} = \frac{\Delta T_{\mathrm{sw}}}{\ln\left(1 + \frac{\Delta T_{\mathrm{sw}}}{\delta T}\right)} \tag{5.54}$$

where ΔT_{sw} is the temperature rise of the circulating cooling water, $\Delta T_{\mathrm{sw}} = T_{\mathrm{sw},2} - T_{\mathrm{sw},1}$; δT is the terminal difference of the condenser, $\delta T = T_{\mathrm{cd}} - T_{\mathrm{sw},2}$.

Let $C_1 = \frac{\Delta T_{\mathrm{sw}}}{\Delta T_{\mathrm{m}}}$, then

$$T_{\mathrm{cd,s}} = T_{\mathrm{cd}} + \delta T_{\mathrm{cd}} = T_{\mathrm{sw},2} + \frac{\Delta T_{\mathrm{sw}}}{e^{C_1} - 1} + \delta T_{\mathrm{cd}} \tag{5.55}$$

According to the temperature determined by the above formula, the steam table can be consulted to obtain the working pressure of the condenser $p'_{\mathrm{cd}} = f\left(T_{\mathrm{cd,s}}\right)$. This is compared with the previously assumed p_{cd}. If the error meets the accuracy requirements, the calculation can be ended. Otherwise, p'_{cd} replaces p_{cd} for the next round of calculations.

In engineering design, a simplified calculation method for the overall heat transfer coefficient of the condenser is commonly used, assuming that the entire cooling area has an average heat transfer coefficient. In fact, in different sections of the cooling area in the condenser, the steam parameters, cooling water parameters, air concentration, steam flow rate, cooling water flow rate, and the arrangement of local condensing tubes are all different, so the heat transfer coefficients in each section are also different. However, the overall heat transfer coefficient calculation formula is based on a large amount of simulation experiments and physical test data, so it still has a certain accuracy in use.

5.3.6.1 Heat Exchange Institute (HEI) Formula

The HEI formula is a widely used formula for calculating the overall heat transfer coefficient of power station condensers in Western countries. It is simple and clear, easy to use, and has a complete set of correction factors for various types of cooling tube materials, specifications, and cooling water temperatures. It does not require any pre-assumed parameters and can directly calculate the results. The disadvantage is that it considers fewer influencing factors.

The expression of the HEI formula is

$$K = C\zeta_{\mathrm{c}}\beta_T\beta_{\mathrm{m}}\sqrt{v_{\mathrm{w}}} \tag{5.56}$$

where β_T is the cooling water temperature correction factor, selected according to the curve shown in Fig. 5.14; ζ_{c} is the cleanliness factor, selected according to the quality of cooling water, steam purity and its impact on cooling tube materials, see Table 5.4; β_{m} is the correction factor for cooling tube material and wall thickness, see Table 5.5; v_{w} is the flow rate of cooling water inside the cooling tube, m/s; C is the calculation coefficient determined by the outer diameter of the cooling tube, see Table 5.6.

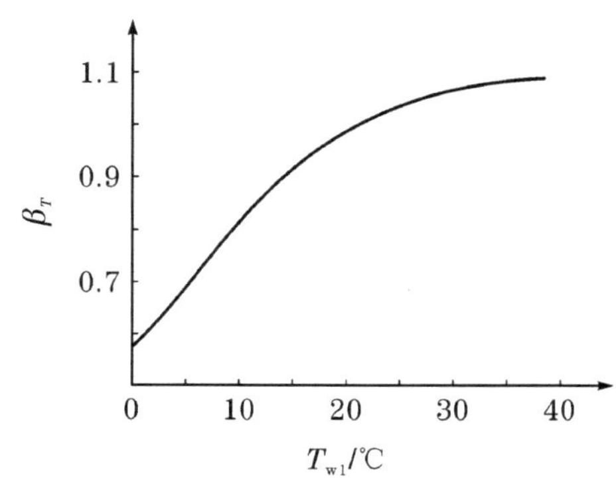

Fig. 5.14 Cooling water temperature correction coefficient β_T

Table 5.4 Cleanliness coefficient ζ_c

Item	ζ_c
Once through water supply and clean water	0.80–0.85
Recycle water supply and chemically treated water	0.75–0.80
Dirty sewage and water that forms mineral deposits	065–0.75
New tube	0.80–0.85
Condenser with continuous cleaning	0.85
Titanium condenser tube	0.90

Table 5.5 Correction coefficient for cooling pipe material and wall thickness β_m

Wall thickness (mm) Materials	Navy brass	Aluminium brass	B10	B30	Stainless steel	Titanium
0.56	1.06	1.03	0.99	0.93	0.83	0.85
0.71	1.04	1.02	0.97	0.90	0.79	0.81
0.89	1.02	1.00	0.94	0.87	0.75	0.77
1.24	1.00	0.97	0.90	0.82	0.69	0.71
1.65	0.96	0.94	0.85	0.77	0.63	–

Table 5.6 Value of calculation coefficient *C*

Outer diameter of cooling tube (mm)	15.9–19.1	22.2–25.4	28.6–31.8	34.9–38.1
Calculation coefficient	2747	2706	2665	2623

5.3.6.2 Bierman Formula

The Bierman formula is a widely used formula for calculating the overall heat transfer coefficient of power station condensers in the former Soviet Union and some countries including China. It considers the impact of various operating factors and design parameters on the heat transfer coefficient of the condenser and their interrelationships. It is applicable to the calculation of the overall heat transfer coefficient of fixed turbine condensers with cooling water temperature $T_{w1} \leq 45\ ^\circ\mathrm{C}$, tube flow rate $v_w = 1 - 2.5$ m/s, and normal vacuum system tightness.

The disadvantage of the Bierman formula is that it requires a large amount of calculation. When calculating the heat transfer coefficient of the condenser under design conditions, it is necessary to pre-assume the steam load, which makes the calculation more troublesome.

The expression of the Bierman formula is

$$K = 4070\zeta\beta_{\mathrm{w}}\beta_T\beta_Z\beta_{\mathrm{d}} \tag{5.57}$$

where ζ is the correction factor considering the cleanliness of the inner surface of the cooling tube, material and wall thickness; β_w is the correction factor considering the flow rate inside the cooling pipe; β_T is the correction factor considering the temperature of the cooling water; β_Z is the correction factor considering the number of cooling water flows; β_d is the correction factor considering the change in steam load of the condenser.

Table 5.7 The value of ζ_m when the wall thickness of the cooling tube is 1 mm

Cooling tube material	Orichalcum	B5	B30	Stainless steel
Value	1.0	0.95	0.92	0.85

The correction factor ζ is determined by the following formula:

$$\zeta = \zeta_s \zeta_m \tag{5.58}$$

where ζ_s is a coefficient which depends on the cooling water supply mode, its value is: when the direct water supply and the salt content in the water is low, $\zeta_s = 0.85 - 0.90$; When the water is supplied in a cycle, $\zeta_s = 0.75 - 0.85$.

ζ_m is a coefficient which depends on the cooling pipe material and wall thickness, see Table 5.7 for values.

The correction factor β_w is determined by the following formula:

$$\beta_w = \left(\frac{1.1 v_w}{\sqrt[4]{d_i}} \right)^{\chi} \tag{5.59}$$

where d_i is the inner diameter of the cooling pipe, mm; χ is the calculation index, the calculation formula is

$$\chi = \begin{cases} 0.12\zeta\left(1 + 0.15T_{w1}\right) & \left(\dfrac{\chi}{\zeta} \le 0.6\right) \\ 0.6\zeta & \left(\dfrac{\chi}{\zeta} > 0.6\right) \end{cases} \tag{5.60}$$

The correction factor β_T mainly depends on the cooling water temperature T_{w1}, and is also related to the correction factor ζ and the specific steam load g_s of the condenser, the calculation formula is

$$\beta_T = \begin{cases} 1 - \dfrac{\left(0.52 - 0.0072 g_s\right)\sqrt{\zeta}}{10^3}\left(35 - T_{w1}\right)^2 & \left(T_{w1} \le 35^\circ\mathrm{C}\right) \\ 1 + 0.002\left(T_{w1} - 35\right) & \left(35^\circ\mathrm{C} < T_{w1} < 45^\circ\mathrm{C}\right) \end{cases} \tag{5.61}$$

where g_s is the specific steam load of the condenser, g/(m^2 · s).

The correction factor β_Z mainly depends on the number of cooling water flows Z, and is also related to the cooling water temperature T_{w1}, the calculation formula is

$$\beta_Z = 1 + \frac{Z-2}{10}\left(1 - \frac{T_{w1}}{45}\right) \tag{5.62}$$

From the above formula, when $Z = 2$, $\beta_Z = 1$; when $Z = 1$ and $T_{w1} < 45\ °C$, $\beta_Z < 1$.

The coefficient β_d is used to consider the correction of the steam load when the condenser is calculated under variable conditions. When the condenser operates within the variable condition range from the rated value g_s to $g_s^{'}$, $\beta_d = 1$, where

$$g_s^{'} = (0.8 - 0.01 T_{w1}) g_s \tag{5.63}$$

When the condenser operates under the condition that the specific steam load is reduced to $g_s^{''} < g_s^{'}$, then

$$\beta_d = \frac{g_s^{''}}{g_s^{'}}\left(2 - \frac{g_s^{''}}{g_s^{'}}\right) \tag{5.64}$$

5.3.6.3 Partial Calculation Relationship

The sectional calculation formula was introduced by engineers from the former Soviet Union. It is an empirical formula derived from a large number of heat transfer experiments, industrial tests of condensers, and operational experience. It directly uses the theoretical calculation formula of heat release during steam condensation in heat transfer, and organizes the heat exchange coefficients of the cooling water side and the steam side into expressions suitable for the entire condenser, taking into account all influencing factors. The physical meaning of the heat transfer process of the condenser expressed is very clear.

The shortcoming is that it is necessary to pre-assume the steam load ratio and the geometric characteristics of the cooling tube bundle during calculation, so the calculation process is relatively complicated and the calculation workload is large.

The basic idea of the sectional calculation formula is to calculate the convective heat transfer coefficients of the cooling water side α_w and the steam side applicable to the entire condenserα_s, and then directly use the following formula to calculate the total heat transfer coefficient:

$$K = \frac{1}{\frac{1}{\alpha_w}\frac{d_o}{d_i} + \frac{1}{\alpha_s} + \frac{d_o}{2\lambda}\ln\frac{d_o}{d_i}} \tag{5.65}$$

where d_o is the outer diameter of the cooling tube, mm; λ is the thermal conductivity of the cooling tube wall material, W/(m·°C).

The heat exchange between the cooling water and the inner wall of the cooling tube is a forced convection heat exchange inside the tube. The heat release coefficient is calculated using the Dittus-Bolter equation commonly used in engineering:

$$\alpha_w = 0.023\,\mathrm{Re}^{0.8}\,\mathrm{Pr}^{0.4}\frac{\lambda_w}{d_i} \tag{5.66}$$

where λ_w is the thermal conductivity of the cooling water, W/(m ·° C).

In Eq. (5.66), the physical parameters of the cooling water *Re*, *Pr* and λ_w are qualitatively determined by the average temperature of the cooling water.

The convective heat transfer coefficient between the steam on the steam side and the outer wall of the cooling tube is calculated as

$$\alpha_s = 12.9\prod^{0.1} Nu^{-0.5}\left(1+\frac{Z}{2}\right)^{1/3}\overline{S}^{0.15}\alpha_n\varepsilon_0^{-0.04} \tag{5.67}$$

where $\prod$ is a coefficient used to consider the dynamic effect of steam flow on the condensate film layer on the cooling tube; *Nu* is the Nusselt number when pure static steam condenses on the outer wall of a single horizontal round tube; $\overline{S}$ is a dimensionless coefficient; α_n is the convective heat transfer coefficient when pure static steam undergoes film condensation on the outer wall of a single horizontal round tube; ε_0 is the relative air content when steam just begins to condense in the condenser tube bundle.

The calculation formula for the coefficient $\prod$ is

$$\prod = \frac{\rho_s v_s^2}{\rho_c g d_o} = \frac{\rho_s\left(\frac{G_{cd}}{\rho_s f_s}\right)^2}{\rho_c g d_o} \tag{5.68}$$

where ρ_s is the steam density determined by the condenser pressure p_{cd}, kg/m^3; ρ_c is the condensate density, kg/m^3; v_s is the average steam velocity at the narrowest passage between the cooling tubes at the steam inlet, m/s; f_s is the sum of the smallest passage area between each cooling tube at the steam inlet of the tube bundle, m^2.

The definition of the dimensionless coefficient $\overline{S}$ is

$$\overline{S} = \frac{WL}{\pi d_o NL} = \frac{W}{\pi d_o N} \tag{5.69}$$

where *W* is the sum of the narrowest channel widths between each cooling tube at the steam inlet of the tube bundle, m; *L* is the length of the cooling tube, m; *N* is the number of cooling tubes.

The convective heat transfer coefficient α_n is calculated according to the Nusselt formula:

$$\alpha_n = 0.728\sqrt[4]{\frac{\rho_c r \lambda_c^3 g}{v_c \Delta T_s d_o}} \tag{5.70}$$

where λ_c is the thermal conductivity of condensate, W/(m ·° C); ν_c is the kinematic viscosity of condensate, m^2/s; *r*is the latent heat of vaporization, J/kg.

The Nusselt number Nu is calculated as follows

$$Nu = \frac{\alpha_n d_o}{\lambda_c} \tag{5.71}$$

The calculation formula for the coefficient ε_0 is

$$\varepsilon_0 = \frac{G_{air}}{G_{cd}} \tag{5.72}$$

where G_{air} is the amount of air leaking into the condenser, kg/s.

5.3.7 Pump

The main pumps in the thermal system of a PWR NPP include the main coolant pump, feedwater pump, condensate pump, circulating water pump, etc. Depending on the operating conditions, the pump drive type uses either an electric motor drive or a steam turbine drive. It is generally believed that the following losses occur when the pump is working:

1. Hydraulic losses caused by friction and local resistance of the fluid in the flow path;
2. Flow losses due to poor sealing and liquid backflow;
3. Mechanical losses caused by component friction.

These losses are actually the dissipation of some mechanical energy into heat due to irreversible factors during the operation of the pump. Since the working parts of the pump are in direct contact with the fluid, the dissipated energy is basically absorbed by the fluid, causing the temperature of the fluid to rise. According to the principle of energy balance, it can be approximately considered that the energy of the pump is basically not lost in quantity.

The pump efficiency is usually used to evaluate the perfection of the pump's energy utilization:

$$\eta_p = \frac{N_{pe}}{N_p} \tag{5.73}$$

where N_p is the input shaft power of the pump, kW; N_{pe} is the effective output power of the pump, which can be calculated as follows:

$$N_{pe} = \frac{G_f H_f}{102} \tag{5.74}$$

where G_f is the mass flow rate of the fluid, kg/s; H is the pump head, mH_2O.

5.3.7.1 Electric Pump

For electric pumps, such as main pumps and condensate pumps, when the drive motor is working, energy loss is caused by copper loss, iron loss, and mechanical loss in the armature winding. The lost energy is converted into heat energy or taken away by the equipment cooling water or dissipated into the surrounding air. The efficiency of the motor is

$$\eta_{\mathrm{m}} = \frac{N_{\mathrm{p}}}{N_{\mathrm{m}}} \tag{5.75}$$

where N_{m} is the input power of the motor, kW.

Therefore, the energy loss of the electric pump during operation is mainly the loss of the motor:

$$\Delta N_{\mathrm{mp}} = N_{\mathrm{m}} - N_{\mathrm{p}} = \left(\frac{1}{\eta_{\mathrm{p}}\eta_{\mathrm{m}}} - \frac{1}{\eta_{\mathrm{p}}} \right) \frac{G_{\mathrm{f}} H_{\mathrm{f}}}{102} \tag{5.76}$$

5.3.7.2 Steam Pump

The steam pump is driven by a small power steam turbine. In large-capacity PWR NPPs, feedwater pumps generally use steam. Figure 5.15 shows a schematic diagram of the steam feedwater pump device used in the Daya Bay Nuclear Power Station.

During the operation of the steam turbine, due to the irreversibility of the adiabatic expansion process of the steam, part of the mechanical energy is dissipated as heat energy and absorbed by the steam. If leakage and heat dissipation losses are ignored, there is no heat loss in the expansion process of the steam in the steam turbine; mechanical power is dissipated as heat energy due to the friction of bearings and gear reducers during the transmission process and is taken away by lubricating oil. This part of the energy loss can be represented by mechanical efficiency:

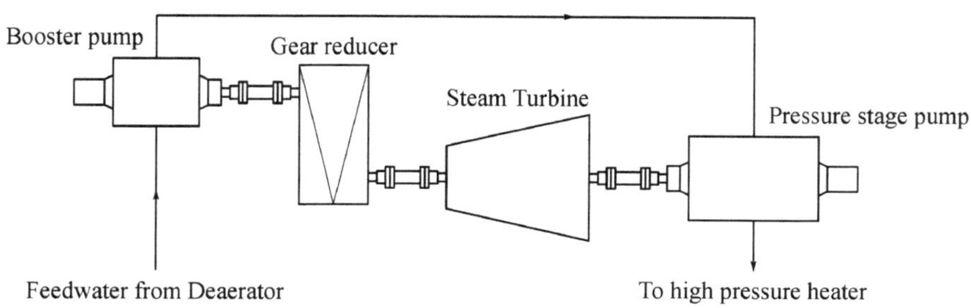

Fig. 5.15 Schematic diagram of steam driven feedwater pump device

$$\eta_{\mathrm{mt}} = \frac{N_{\mathrm{p}}}{N_{\mathrm{i}}} = \frac{N_{\mathrm{p}}}{G_{\mathrm{s}} H_{\mathrm{i}}} \tag{5.77}$$

where N_i is the internal power of the steam turbine, kW; G_s is the steam consumption of the steam turbine, kg/s; H_i is the internal enthalpy drop of steam in the steam turbine, kJ/kg.

It can be seen that the energy loss of the steam pump during operation is:

$$\Delta N_{\mathrm{tp}} = N_{\mathrm{i}} - N_{\mathrm{p}} = \left(\frac{1}{\eta_{\mathrm{p}} \eta_{\mathrm{mt}}} - \frac{1}{\eta_{\mathrm{p}}} \right) \frac{G_{\mathrm{f}} H_{\mathrm{f}}}{102} \tag{5.78}$$

In a primary coolant system, because the main pump works on the coolant, the thermal power of the nuclear steam supply system is generally larger than the thermal power of the PWR. The heating power of the main pump to the coolant is equal to the input electric power of the main pump motor minus the cooling and loss power of the motor, which is equivalent to the input power of the pump. In this way, the total heating power of the main pump to the coolant in the primary loop system can be calculated by the following formula:

$$Q_{\mathrm{mcp}} = \frac{G_{\mathrm{c}} \cdot H_{\mathrm{mcp}}}{102 \eta_{\mathrm{mcp}}} \tag{5.79}$$

where G_c is the total coolant flow through the core, kg/s; H_{mcp} is the total head of the main pump, m water column (mH_2O); η_{mcp} is the effective efficiency of the main pump.

5.4 Example of Thermal Balance Analysis

In order to illustrate the basic process of thermal balance calculation of nuclear power devices, here is an example of local thermal balance calculation of a nuclear power device. Figure 5.16 shows the thermal line diagram of a nuclear power device using three-stage steam extraction regenerative. Most of the calculation work of the thermal line diagram has been completed, and the results are marked in the thermal line diagram.

In addition to the parameters given in the thermal line diagram, other given parameters are shown in Table 5.8.

The heat balance calculation is required to be carried out using the constant power method to determine the following parameters: efficiency of the nuclear power unit, reactor power, total steam output of the steam generator, steam consumption of the main steam turbine, total steam consumption of the steam turbine generator, steam consumption of the feedwater pump steam turbine, feedwater

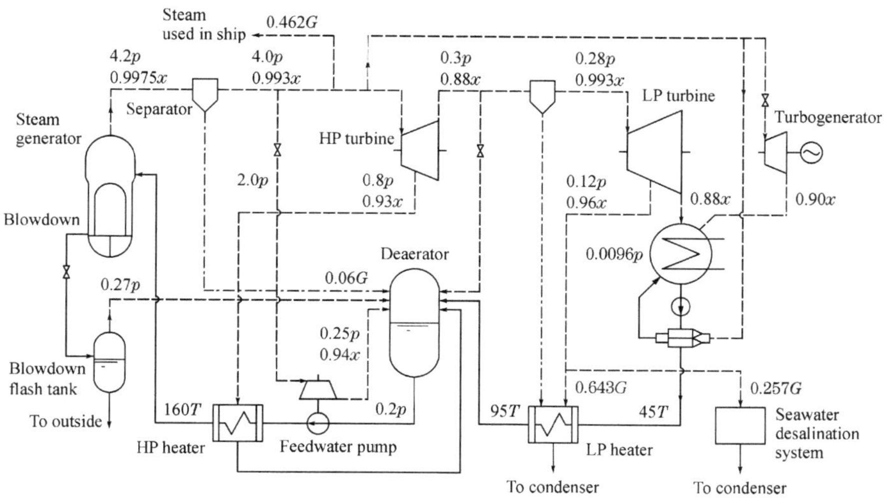

Fig. 5.16 Thermal line diagram of NPP with three stage extraction and reheat

Table 5.8 Partial known parameters

No.	Item	Symbol	Unit	Value
1	Plant effective power	N_e	kW	6615
2	Ratio of steam generator thermal power to reactor thermal power	η_{sg}	–	0.9969
3	Steam generator blowdown rate	ξ_{pw}	–	0.01164
4	Blowdown tank pressure	p_{pw}	MPa	0.27
5	Main condenser pressure	p_c	kPa	4.9
6	Auxiliary condenser pressure	p_{ac}	kPa	6.45
7	Pressure drop from exhaust to condenser of low pressure turbine	Δp_c	kPa	0.27
8	Mechanical efficiency of high pressure turbine	$\eta_{m, ht}$	–	0.9802
9	Mechanical efficiency of low pressure turbine	$\eta_{m, lt}$	–	0.9802
10	Gear reducer efficiency of high pressure turbine	$\eta_{g, ht}$	–	0.9751
11	Gear reducer efficiency of low pressure turbine	$\eta_{g, lt}$	–	0.9751
12	Efficiency of shafting	η_w	–	0.9479
13	Turbo-generator power	$N_{e, g}$	kW	2×640
14	Generator efficiency	η_g	–	0.932
15	Effective efficiency of power generation turbine	$\eta_{e, gt}$	–	0.5427
16	Pressure drop from power generation turbine exhaust to auxiliary condenser	Δp_{ac}	kPa	0.27
17	Pressure drop of feed pump outlet to steam generator inlet at feed water flow rate equal to 18 kg/s	$\Delta p_{fw, v0}$	MPa	0.5734
18	Static pressure drop from feed pump outlet to steam generator inlet	$\Delta p_{fw, c}$	MPa	0.2942
19	Feed pump efficiency	η_{fwp}	–	0.568
20	Mechanical efficiency of feed pump turbine	$\eta_{m, fwpt}$	–	0.891

output of the feedwater pump, extraction steam of the steam turbine, steam consumption of the low-pressure steam turbine, and head of the feedwater pump.

I. Determining the thermodynamic parameters at each state point

1. Steam generator
   The operating pressure of the steam generator p_s = 4.2 MPa. The enthalpy of saturated water and saturated steam at this pressure are as follows:

$$h_s^{'} = 1101.6\,\text{kJ/kg} \qquad h_s^{''} = 2799.4\,\text{kJ/kg}$$

Then, the enthalpy of the fresh steam is:

$$h_{fh} = h_s^{'} + x_{fh}\left(h_s^{''} - h_s^{'}\right) = 1101.6 + 0.9975 \times (2799.4 - 1101.6) = 2795.2\,\text{kJ/kg}$$

Given that the feedwater temperature T_{fw} = 160 ° C, and the feedwater pressure is:

$$p_{fw} = p_s + 0.1 = 4.3\,\text{MPa}$$

By checking the steam table, the enthalpy of the feedwater, h_{fw} = 677.64 kJ/kg.

2. Waste discharge tank
   The waste water entering the waste discharge tank is the saturated water under the working pressure of the steam generator.
   The operating pressure of the waste discharge tank, p_{pw} = 0.27 MPa. The enthalpy of the flash steam and the water discharged overboard are equal to the enthalpy of the saturated steam and saturated water at this pressure:

$$h_{pw,s} = 2719.9\,\text{kJ/kg} \quad h_{pw,w} = 546.24\,\text{kJ/kg}$$

3. Main separator
The steam pressure at the outlet of the main separator, p_{msp} = 4.0 MPa. The enthalpy and entropy of saturated water and saturated steam at this pressure are as follows:

$$h_{msp}^{'} = 1087.4\,\text{kJ/kg} \quad s_{msp}^{'} = 2.7965\,\text{kJ/(kg K)}$$
$$h_{msp}^{''} = 2800.3\,\text{kJ/kg} \quad s_{msp}^{''} = 6.0685\,\text{kJ/(kg K)}$$

Then, the enthalpy and entropy of the steam at the outlet of the main separator are:

$$h_{msp,s} = 1087.4 + 0.993 \times (2800.3 - 1087.4) = 2788.3\,\text{kJ/kg}$$
$$s_{msp,s} = 2.7965 + 0.993 \times (6.0685 - 2.7965) = 6.0456\,\text{kJ/(kg K)}$$

4. Steam turbine generator
The exhaust steam pressure$p_{g,to} = p_{ac} + \Delta p_{ac}$ = 6.45 + 0.27 = 6.72 kPa

5. High pressure cylinder of the main steam turbine
   The extraction steam enthalpy $h_{\mathrm{ht,\,e}} = 720.94 + 0.93 \times (2767.5 - 720.94) = 2624.24$ kJ/kg
   The exhaust steam enthalpy $h_{\mathrm{ht,\,o}} = 561.43 + 0.88 \times 2163.2 = 2465.05$ kJ/kg
6. Low pressure cylinder of the main steam turbine
   The inlet steam enthalpy $h_{\mathrm{lt,\,i}} = 551.44 + 0.993 \times 2170.1 = 2706.35$ kJ/kg
   The extraction steam enthalpy $h_{\mathrm{lt,\,e}} = 439.36 + 0.96 \times 2244.1 = 2593.7$ kJ/kg
   The exhaust steam pressure $p_{\mathrm{lt,\,o}} = p_{\mathrm{c}} + \Delta p_{\mathrm{c}} = 4.9 + 0.27 = 5.17$ kPa
   Exhaust enthalpy $h_{\mathrm{lt,\,o}} = 137.77 + 0.869 \times 2423.8 = 2244.05$ kJ/kg
7. Intermediate separator
   The enthalpy of the drain is approximately the saturated water under the inlet steam pressure of the low-pressure cylinder, with an enthalpy value of $h_{\mathrm{spw}} = 551.44$ kJ/kg
8. Steam turbine feed pump
   The pressure and temperature of the feed water at the feed pump inlet are $p_{\mathrm{fwp,\,i}} = 0.31$ MPa and $T_{\mathrm{fwp,\,i}} = 120$ °C, respectively, so the corresponding enthalpy value is $h_{\mathrm{fwp,\,i}} = 503.80$ kJ/kg; the exhaust pressure and steam dryness of the feed pump turbine are $p_{\mathrm{fwpt,\,o}} = 0.24$ MPa and $x_{\mathrm{fwpt,\,o}} = 0.94$, respectively, and the exhaust enthalpy value is $h_{\mathrm{fwpt,\,o}} = 2583.44$ kJ/kg.

II. Conducting Heat Balance Calculation

1. If the effective efficiency of the device is $\eta_{\mathrm{e,\,npp}} = 0.18$, then the thermal power of the reactor is

$$Q_{\mathrm{R}} = \frac{N_{\mathrm{net}}}{\eta_{\mathrm{e,npp}}} = \frac{6615}{0.18} = 36{,}750\,\mathrm{kW}$$

2. The heat load of the two steam generators is

$$Q_{\mathrm{sg}} = Q_{\mathrm{R}} \cdot \eta_{\mathrm{sg}} = 36{,}750 \times 0.9969 = 36{,}636\,\mathrm{kW}$$

The total steam output of the steam generator is

$$G_{\mathrm{s}} = \frac{Q_{\mathrm{sg}}}{\left(h_{\mathrm{fh}} - h_{\mathrm{s}}'\right) + \left(1 + \xi_{\mathrm{pw}}\right)\left(h_{\mathrm{s}}' - h_{\mathrm{fw}}\right)}$$

$$= \frac{36{,}636}{\left(2795.2 - 1101.6\right) + \left(1 + 0.01164\right)\left(1101.6 - 677.64\right)} = 17.26\,\mathrm{kg/s}$$

The total feed water amount is

$$G_{\text{fw}} = \left(1+\xi_{\text{pw}}\right)G_{\text{s}} = \left(1+0.01164\right)\times 17.26 = 17.46\,\text{kg/s}$$

3. Calculation of feed pump head and steam consumption of feed pump turbine
The resistance coefficient of the feed water pipe from the outlet of the steam turbine feed pump to the inlet of the steam generator is

$$\xi = \frac{\Delta p_{\text{fw,v}}}{G_{\text{fw}}^2} = \frac{\Delta p_{\text{fw,v}_0}}{G_{\text{fw}0}^2} = \frac{0.5734}{18^2} = 0.00177$$

The flow resistance of the feed water pipe is

$$\Delta p_{\text{fw,v}} = \xi G_{\text{fw}}^2 = 0.00177\times 17.46^2 = 0.5395\,\text{MPa}$$

The head of the feed pump is

$$H_{\text{fwp}} = \frac{p_{\text{fw}} + \Delta p_{\text{fw,v}} + \Delta p_{\text{fw,c}} - p_{\text{fwp,i}}}{\rho g}$$
$$= \frac{\left(4.3+0.5395+0.2942-0.31\right)\times 10^6}{927.17\times 9.807} = 530.5\,\text{mH}_2\text{O}$$

The effective power output of the feed pump is

$$N_{\text{e,fwp}} = G_{\text{fw}} H_{\text{fwp}} g = 17.46\times 530.5\times 9.807\times 10^{-3} = 90.84\,\text{kW}$$

The internal enthalpy drop of the feed pump turbine is

$$H_{\text{i,fwpt}} = h_{\text{msp,s}} - h_{\text{fwpt,o}} = 2788.3 - 2583.44 = 204.86\,\text{kJ/kg}$$

Therefore, the steam consumption of the feed pump turbine is

$$G_{\text{s,fwpt}} = \frac{N_{\text{e,fwp}}}{H_{\text{i,fwpt}}\eta_{\text{fwp}}\eta_{\text{m,fwpt}}} = \frac{90.84}{204.86\times 0.568\times 0.891} = 0.876\,\text{kg/s}$$

4. Calculation of steam consumption of high pressure feed water heater
The enthalpy of the drain of the high-pressure feed water heater h_{hw} = 570.9 kJ/kg
The steam consumption is

$$G_{\text{s,hte}} = \frac{h_{\text{fw}} - h_{\text{fwp,i}}}{h_{\text{ht,e}} - h_{\text{hw}}} G_{\text{fw}} = \frac{677.64 - 503.80}{2{,}624.24 - 570.9}\times 17.46 = 1.478\,\text{kg/s}$$

5. Calculation of steam consumption of power generation turbine

The adiabatic enthalpy drop of the turbine $H_{a,gt} = 917.98$ kJ/kg can be obtained from the enthalpy entropy diagram, the total steam consumption is

$$G_{s,gt} = \frac{N_{e,g}}{H_{a,gt}\eta_g\eta_{e,gt}} = \frac{2\times 640}{917.98\times 0.932\times 0.5427} = 2.757\,\text{kg/s}$$

6. Calculation of steam consumption of main turbine unit

The main turbine consists of a high-pressure cylinder and a low-pressure cylinder, and the power equation of the main turbine is

$$N_e = \left(N_{i,ht}\eta_{m,ht}\eta_{g,ht} + N_{i,lt}\eta_{m,lt}\eta_{g,lt}\right)\eta_w$$

$$N_{i,ht} = G_{s,ht}\left(h_{ht,i} - h_{ht,o}\right) - G_{s,hte}\left(h_{ht,e} - h_{ht,o}\right)$$

$$N_{i,lt} = G_{s,lt}\left(h_{lt,i} - h_{lt,o}\right) - G_{s,lte}\left(h_{lt,e} - h_{lt,o}\right)$$

Substituting the known parameters into the above equation, we get

$$N_{i,ht} + N_{i,lt} = \frac{6615}{0.9802\times 0.9751\times 0.9479} = 7301.32$$

$$N_{i,ht} = 323.25G_{s,ht} - 235.283$$

$$N_{i,lt} = 462.3G_{s,lt} - 314.685$$

Hence

$$323.25G_{s,ht} + 462.3G_{s,lt} = 7851.288$$

The mass conservation equation and energy balance equation of the intermediate separator are as follows

$$G_{s,hto} = G_{s,lt} + G_{sp,w} + 0.26$$

$$\left(G_{s,hto} - 0.26\right)h_{ht,o} = G_{s,lt}h_{lt,i} + G_{sp,w}h_{sp,w}$$

Eliminate $G_{sp,w}$, get

$$G_{s,hto} = \frac{h_{lt,i} - h_{sp,w}}{h_{ht,o} - h_{sp,w}}G_{s,lt} + 0.26 = 1.126G_{s,lt} + 0.26$$

Substitute into the mass conservation equation of the high-pressure cylinder, get

$$G_{\mathrm{s,ht}} = G_{\mathrm{s,hte}} + G_{\mathrm{s,hto}} = 1.478 + 1.126G_{\mathrm{s,lt}} + 0.26 = 1.738 + 1.126G_{\mathrm{s,lt}}$$

Solve equations (a) and (b) simultaneously, get $G_{\mathrm{s,ht}} = 11.67$ kg/s, $G_{\mathrm{s,lt}} = 8.82$ kg/s.

7. Total steam consumption of the secondary loop system

$$\begin{aligned} D_{\mathrm{s}} &= G_{ship} + G_{\mathrm{msp,w}} + G_{\mathrm{s,fwpt}} + G_{\mathrm{s,gt}} + G_{\mathrm{Jet}} + G_{\mathrm{s,ht}} \\ &= 0.462 + 0.06 + 0.876 + 2.757 + 0.132 + 11.67 = 15.957\,\mathrm{kg/s} \end{aligned}$$

8. Calculate the effective efficiency of the device

Heat load of the steam generator

$$Q_{\mathrm{sg}}' = D_{\mathrm{s}}\left[\left(h_{\mathrm{fh}} - h_{\mathrm{s}}'\right) + \left(1 + \xi_{\mathrm{pw}}\right)\left(h_{\mathrm{s}}' - h_{\mathrm{fw}}\right)\right] = 15.957 \times 2122.495 = 33{,}868.6\,\mathrm{kW}$$

Reactor thermal power

$$Q_{\mathrm{R}}' = \frac{Q_{\mathrm{sg}}'}{\eta_{\mathrm{sg}}} = \frac{33{,}868.6}{0.9969} = 33{,}973.9\,\mathrm{kW}$$

Effective efficiency of the device

$$\eta_{\mathrm{e,npp}}' = \frac{N_{\mathrm{e}}}{Q_{\mathrm{R}}'} = \frac{6615}{33{,}973.9} = 0.1947$$

9. Compare the assumed efficiency $\eta_{\mathrm{e,npp}}$ with the calculated efficiency $\eta_{\mathrm{e,npp}}'$

$$\varepsilon = \left|\eta_{\mathrm{e,npp}} - \eta_{\mathrm{e,npp}}'\right| = |0.18 - 0.1947| = 0.0147$$

Since $\varepsilon > 0.001$, it does not meet the convergence accuracy, so assume $\eta_{\mathrm{e,npp}} = 0.1947$, and perform the next round of iterative calculation.

After four iterations, the device efficiency meets the convergence accuracy requirements, the calculation is ended, and the calculation results are listed in Table 5.9.

Table 5.9 Calculation results of some undetermined parameters

No.	Item	Unit	1	2	3	4
1	NPP efficiency		0.18	0.1947	0.1965	0.1966
2	Reactor thermal power	kW	36,750	33,975.35	33,664.12	33,647
3	Total steam output of steam generator	kg/s	17.26	15.96	15.81	15.80
4	Steam consumption of main turbine	kg/s	11.67	11.60	11.60	11.60
5	Total steam consumption of turbo-generator	kg/s	2.757	2.757	2.757	2.757
6	Steam consumption of feed pump turbine	kg/s	0.876	0.797	0.788	0.788
7	Feed pump capacity	kg/s	17.46	16.14	15.99	15.99
8	Intermediate extraction flow rate of high pressure turbine	kg/s	1.478	1.366	1.356	1.356
9	Steam consumption of low pressure turbine	kg/s	8.82	8.86	8.87	8.87
10	Feed pump head	mH_2O	530.5	521.86	520.99	520.99

5.5 Analysis of the Thermal Balance Calculation Results of the Thermal System of the NPPs

According to the data provided by relevant materials, the thermal balance calculation of the thermal system of the Zhejiang Qinshan 300 MW and Guangdong Daya Bay 900 MW PWR NPPs is carried out, and the calculation results are shown in Tables 5.10 and 5.11.

From the Tables 5.10 and 5.11, it can be seen that in the thermal system of the PWR NPP, the heat loss of the heat exchange equipment accounts for about 66% of the reactor thermal power, of which about 65% of the heat is discharged from the condenser in the thermal cycle; and the energy loss of other thermal equipment (including reactors, steam turbines, pumps and other equipment) is only about 0.296%–1.585%. In the total heat loss of the thermal system, the heat loss of the heat exchange equipment accounts for more than 97%, becoming the largest class of equipment for heat loss in the thermal system of the PWR NPP, so the fundamental way to improve the thermal economy of the thermal system is to reduce the heat loss of the heat exchange equipment.

There are two main reasons for the heat loss of the heat exchange equipment. One is that a large amount of heat is discharged from the thermal cycle during the operation of the thermal system and lost to the environment; the other is the heat loss during the heat exchange process. From the results of the thermal balance analysis, the former is the fundamental reason for the system heat loss, so the main focus should be on trying to reduce the heat loss in the thermal cycle of the power device. At present, the thermal system of the PWR NPP generally adopts a steam thermal cycle with regenerative and reheat. To reduce the heat loss in the condenser, it is necessary to further improve the regenerative and reheat methods of the thermal system, or break through the limitations of the existing thermal cycle form and adopt a completely new concept of thermal cycle.

Table 5.10 Analysis results of thermal equilibrium in the thermal system of PWR NPPs

No.	Equipment name	Energy loss coefficient (%) Qinshan No. 1 NPP	Energy loss coefficient (%) Daya Bay NPP	Main cause of energy loss
1	PWR	0	0	–
2	Steam generator	0.638	0.201	Heat dissipation
3	Steam reheater	0.201	0.004	
4	Feed water heating system	0.978	0.314	
5	Condenser	64.944	65.539	Heat dissipation and circulation cooling water carry away heat
6	Steam turbo-generator	1.086	0.253	Heat dissipation, mechanical friction and electrical loss
7	Main coolant pump	0.022	0.020	Copper loss, iron loss and friction loss of the motor or heat dissipation and mechanical loss of the auxiliary steam turbine
8	Condensate pump	0.013	0.012	
9	Feed water pump	0.063	0.010	
10	Main air extractor	0	0	Heat dissipation
11	Moisture separator	0	0	Flow friction, heat dissipation and leakage
12	Main steam line	0.401	0	
13	Exhaust steam line	0	0	
Total		68.346	66.354	
Plant net efficiency %		31.654	33.646	

Table 5.11 The ratio of energy losses for several main types of equipment in thermal systems

No.	Equipment type		Energy loss coefficient (%)		Energy loss ratio (%)	
			Qinshan No. 1 NPP	Daya Bay NPP	Qinshan No. 1 NPP	Daya Bay NPP
1	Reactor		0	0	0	0
2	Heat exchange equipment	Steam generator	66.761	66.058	97.681	99.554
		Steam reheater				
		Feed water heater				
		Condenser				
3	Steam turbo-generator		1.086	0.253	1.588	0.382
4	Pump	Main coolant pump	0.098	0.043	0.144	0.064
		Feed water pump				
		Condensate pump				
5	Other	Main steam line	0.401	0	0.587	0
		Moisture separator				
		Main air extractor				
		Exhaust steam line				

During the operation of a PWR NPP, there exists a series of energy conversion processes such as nuclear fission energy → thermal energy → mechanical energy → electrical energy, as well as the distribution and transmission of energy. The results of thermal balance analysis reflect the amount of energy loss in these processes, but do not reflect the differences in energy in different forms or even the same form in various parts and links of the thermal system. It also cannot accurately indicate the direction and possibility of energy conversion. Therefore, the information on the degree of energy utilization in the thermal system of the PWR NPP reflected by the thermal balance analysis is one-sided and incomplete, making it difficult to provide a sufficient theoretical basis for accurately pointing out effective ways to improve the thermal economy of the system.

Based on the above analysis, the following conclusions can be drawn:

1. The condenser is the equipment with the largest heat loss in the thermal system of the PWR NPP. The main reason for the loss is that about 65% of the heat from the reactor thermal power is taken out of the thermal cycle by circulating cooling seawater in the condenser and dissipated into the environment.
2. The main reason for the energy loss in the thermal system of the PWR NPP is the heat loss of the thermal equipment, especially the heat exchange equipment in the system. Therefore, the fundamental way to improve the thermal efficiency of the entire thermal system is to reduce the heat dissipation loss of each equipment in the power cycle of the thermal system and reduce the heat discharged into the environment.
3. The thermal balance analysis method reflects the balance of energy in quantity, but cannot evaluate the differences and changes in the quality of energy, so it cannot fully reflect the rationality of energy use in the energy system.

5.6 Equivalent Enthalpy Drop Method

Conventional thermal calculations are constant power calculations, which require the simultaneous solution of $(Z + 1)$ equations, which is cumbersome; and whenever there is a change in any moisture flow or parameters, the entire system must be recalculated quantitatively, which not only increases the calculation workload but also easily causes errors in the calculation process.

The equivalent enthalpy drop method is a constant flow calculation method proposed by Soviet scholar A.M. Kuznetsov in the 1960s. It assumes that the fresh steam flow and fuel heating value of the unit are constant, and the impact of any local changes in the thermal system on thermal economy is attributed only to changes in the power within the unit, i.e.,

$$\delta\eta_i = \frac{\eta_i' - \eta_i}{\eta_i} = \frac{\frac{H'}{Q} - \frac{H}{Q}}{\frac{H}{Q}} = \frac{\Delta H}{H} \tag{5.80}$$

where η_i, η_i' are the efficiency of the unit before and after the change in the thermal system, respectively; H, H' are the actual enthalpy drop of the unit before and after the change in the thermal system, respectively.

Thus, when there are local changes in the thermal system, it will not cause all levels of extraction to change, but only cause changes in the extraction and heat of a few levels with local changes. Only the change in the work done (internal work, i.e., actual enthalpy drop) within this local range needs to be analyzed and calculated to obtain the thermal economic effect after the change in the entire thermal system. It can be seen that the equivalent enthalpy drop method replaces the cumbersome calculation of the entire thermal system with local quantitative calculations, thereby simplifying the calculation of the thermal system.

The equivalent enthalpy drop method is a new method for analyzing and calculating the thermodynamic system of a power plant. The premise is that the fresh steam flow and the heat supplied by the fuel remain unchanged, the initial and final parameters of the cycle and the steam line remain unchanged, and the changes in internal power are used to analyze and calculate its economy. For a determined thermodynamic system, all its steam-water parameters are known, so the quantitative results of H_j, η_j and thermal economy changes can be calculated. This simplifies the calculation of the thermodynamic system and makes local quantitative calculation possible.

The equivalent enthalpy drop method can be used for the calculation of the whole machine, is more suitable for the quantitative calculation when the local changes of the thermodynamic system, and can also be used for the analysis and derivation of the optimal parameters (such as the optimal heat recovery distribution).

5.6.1 Concept of Equivalent Enthalpy Drop

Taking the three-stage extraction regenerative system shown in Fig. 5.17 as an example, the concept of equivalent enthalpy drop is explained.

In the figure, Nos. 1 and 3 are surface heaters, No. 2 is a mixed heater, and the meanings of various variable symbols are as follows:

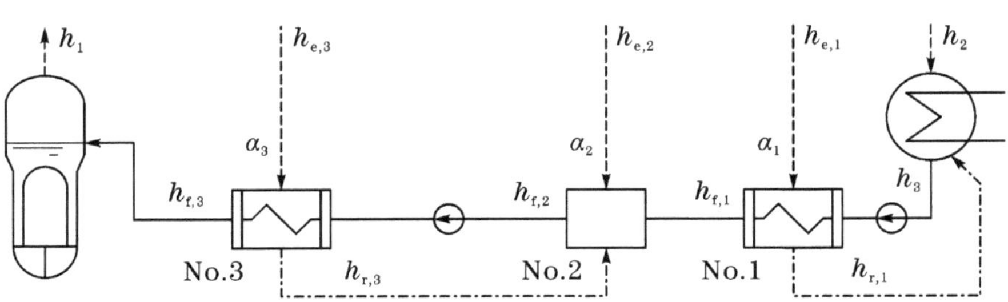

Fig. 5.17 Example of equivalent enthalpy drop during extraction

Table 5.12 The enthalpy rise of feedwater, heat release from steam extraction, and heat release from drainage in the regenerative system

Heater number	Enthalpy rise of unit mass feedwater	Heat release of unit mass extraction steam	Heat release of unit mass drainage
1	$\tau_1 = h_{f,1} - h_3$	$q_{h,1} = h_{e,1} - h_{r,1}$	$\gamma_1 = 0$
2	$\tau_2 = h_{f,2} - h_{f,1}$	$q_{h,2} = h_{e,2} - h_{f,1}$	$\gamma_2 = h_{r,3} - h_{f,1}$
3	$\tau_3 = h_{f,3} - h_{f,2}$	$q_{h,3} = h_{e,3} - h_{r,3}$	$\gamma_3 = 0$

h_1: fresh steam enthalpy at the outlet of the steam generator.
h_2: Exhaust steam enthalpy of the turbine into the condenser.
h_3: Condensate enthalpy at the outlet of the condenser.
$h_{e,1}$, $h_{e,2}$, $h_{e,3}$: Extraction enthalpy of each stage of the feedwater heater.
$h_{f,1}$, $h_{f,2}$, $h_{f,3}$: Outlet feedwater enthalpy of each stage of the feedwater heater.
α_1, α_2, α_3: Extraction coefficient of each stage of the feedwater heater.

The feedwater enthalpy rise, extraction heat release, and drain heat release of each stage of the feedwater heater are shown in Table 5.12.

The actual enthalpy drop of 1 kg of fresh steam from the steam generator in the reheat extraction steam turbine is

$$\begin{aligned} H &= (h_1 - h_{e,3}) + (1-\alpha_3)(h_{e,3} - h_{e,2}) \\ &\quad + (1-\alpha_3-\alpha_2)(h_{e,2} - h_{e,1}) + (1-\alpha_3-\alpha_2-\alpha_1)(h_{e,1} - h_2) \\ &= (h_1 - h_2) - \alpha_3(h_{e,3} - h_2) - \alpha_2(h_{e,2} - h_2) - \alpha_1(h_{e,1} - h_2) \\ &= (h_1 - h_2)\left(1 - \sum_{j=1}^{3}\alpha_j Y_j\right) \end{aligned} \tag{5.81}$$

where $Y_j = \dfrac{h_{e,j} - h_2}{h_1 - h_2}$, defined as the reheat extraction work deficiency coefficient.

The above formula is equivalent to $\left(1 - \sum_{j=1}^{3}\alpha_j Y_j\right)$ kg of fresh steam in a non-extraction pure condensing steam turbine with the same initial and final parameters. The actual enthalpy drop is called the equivalent enthalpy drop of the fresh steam.

Assume that the No. 1 heater obtains heat q_1, from the outside, reducing the extraction volume of this stage heater by 1 kg, and the reduced 1 kg extraction is called the extruded extraction. The internal work done by 1 kg of extruded extraction in the steam turbine from the extraction point of the No.1 heater to the exhaust port is called the equivalent enthalpy drop of the extruded extraction, that is,

$$H_1 = h_{e,1} - h_2 \tag{5.82}$$

Similarly, if heater No. 2 obtains heat q_2 from the outside, causing this stage of the heater to expel 1 kg of extraction steam, then the equivalent enthalpy drop of this part of the extraction steam is

$$H_2 = \left(h_{e,2} - h_2\right) - \alpha_{1,2}\left(h_{e,1} - h_2\right) = \left(h_{e,2} - h_2\right) - \alpha_{1,2} H_1 \tag{5.83}$$

where $\alpha_{1,2}$ is the share of the extraction steam expelled by heater No. 2 allocated to heater No. 1.

Since the feedwater flow in heater No. 1 has increased by 1 kg, the corresponding increase in extraction steam heating is

$$\alpha_{1,2} = \frac{h_{f,1} - h_3}{h_{e,1} - h_{r,1}} = \frac{\tau_1}{q_{h,1}} \tag{5.84}$$

Substituting into formula (5.83), we get

$$H_2 = \left(h_{e,2} - h_2\right) - \frac{\tau_1}{q_{h,1}} H_1 \tag{5.85}$$

If heater No. 3 obtains heat q_3, reducing its consumption of extraction steam by 1 kg, then the equivalent enthalpy drop of the expelled extraction steam is

$$\begin{aligned} H_3 &= \left(h_{e,3} - h_2\right) - \alpha_{2,3}\left(h_{e,2} - h_2\right) - \alpha_{1,3}\left(h_{e,1} - h_2\right) \\ &= \left(h_{e,3} - h_2\right) - \alpha_{2,3}\left(h_{e,2} - h_2\right) - \alpha_{1,3} H_1 \end{aligned} \tag{5.86}$$

where $\alpha_{2,3}$ and $\alpha_{1,3}$ are the shares of the extraction steam expelled by heater No. 3 allocated to heaters Nos. 2 and 1.

For the No. 2 mixed feedwater heater, the drain from heater No. 3 has decreased by 1 kg, and the heat released by the drain has correspondingly decreased, so the amount of extraction steam that needs to be increased is

$$\alpha_{2,3} = \frac{h_{r,3} - h_{f,1}}{h_{e,2} - h_{f,1}} = \frac{\gamma_2}{q_{h,2}} \tag{5.87}$$

For heater No. 1, since the feedwater flow has increased by $(1 - \alpha_{2,3})$kg, the corresponding increase in extraction steam is

$$\alpha_{1,3} = \left(1 - \alpha_{2,3}\right)\frac{h_{f,1} - h_3}{h_{e,1} - h_{r,1}} = \left(1 - \alpha_{2,3}\right)\frac{\tau_1}{q_{h,1}} \tag{5.88}$$

Substituting the expressions of $\alpha_{1,3}$ and $\alpha_{2,3}$ into the expression of H_3, we get

$$\begin{aligned} H_3 &= (h_{e,3} - h_2) - \frac{\gamma_2}{q_{h,2}}(h_{e,2} - h_2) - \left(1 - \frac{\gamma_2}{q_{h,2}}\right)\frac{\tau_1}{q_{h,1}} H_1 \\ &= (h_{e,3} - h_2) - \frac{\gamma_2}{q_{h,2}}\left[(h_{e,2} - h_2) - \frac{\tau_1}{q_{h,1}} H_1\right] - \frac{\tau_1}{q_{h,1}} H_1 \\ &= (h_{e,3} - h_2) - \frac{\gamma_2}{q_{h,2}} H_2 - \frac{\tau_1}{q_{h,1}} H_1 \end{aligned} \tag{5.89}$$

As can be seen, the equivalent enthalpy drop has the following general formula

$$H_Z = (h_{e,Z} - h_2) - \sum_{j=1}^{Z-1} \frac{A_j}{q_{h,j}} H_j \tag{5.90}$$

where A_j is taken as γ_j or τ_j, depending on the type of heater.

Obviously, the equivalent enthalpy drop $\Delta h_{i,j}$ is the real work ability of 1 kg of extraction steam flow returning from the extraction port of the jth stage heater to the steam turbine, indicating the energy level of the steam at each extraction port of the steam turbine.

5.6.2 Matrix Representation of the Equivalent Enthalpy Drop Method

The equivalent enthalpy drop method can be used for the calculation of the whole machine, and is more suitable for quantitative calculation when the thermal system changes locally, and can also be used for the analysis and derivation of optimal parameters (such as optimal regenerative distribution).

Figure 5.18 shows a seven-stage extraction regenerative system, in which No.2 and No.5 are both drain confluence heaters, and the definitions of q_h and γ are different from those of other heaters.

According to the principle of mass conservation and energy conservation, the heat balance equation of each stage heater is

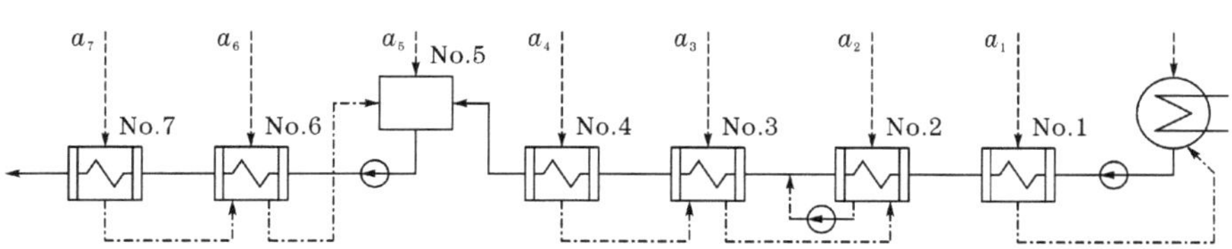

Fig. 5.18 Seven stage extraction and regenerative system

$$\left.\begin{aligned}\alpha_7 q_{h,7} &= \tau_7\\ \alpha_6 q_{h,6} + \alpha_7\gamma_6 &= \tau_6\\ \alpha_5 q_{h,5} + \alpha_6\gamma_5 + \alpha_7\gamma_5 &= \tau_5\\ \alpha_4 q_{h,4} + \alpha_5\tau_4 + \alpha_6\tau_4 + \alpha_7\tau_4 &= \tau_4\\ \alpha_3 q_{h,3} + \alpha_4\gamma_3 + \alpha_5\tau_3 + \alpha_6\tau_3 + \alpha_7\tau_3 &= \tau_3\\ \alpha_2 q_{h,2} + \alpha_3\gamma_2 + \alpha_4\gamma_2 + \alpha_5\tau_2 + \alpha_6\tau_2 + \alpha_7\tau_2 &= \tau_2\\ \alpha_1 q_{h,1} + \alpha_2\tau_1 + \alpha_3\tau_1 + \alpha_4\tau_1 + \alpha_5\tau_1 + \alpha_6\tau_1 + \alpha_7\tau_1 &= \tau_1\end{aligned}\right\} \tag{5.91}$$

It can be represented as

$$\mathbf{A}\alpha = \mathbf{\tau} \tag{5.92}$$

where $\boldsymbol{\tau} = [\tau_1, \tau_2, \ldots, \tau_m]^{\mathrm{T}}$; $\boldsymbol{\alpha} = [\alpha_1, \alpha_2, \ldots, \alpha_m]^{\mathrm{T}}$.

And the matrix **A** is represented as

$$A = \begin{pmatrix} q_{\mathrm{h},1} & \tau_1 & \tau_1 & \tau_1 & \tau_1 & \tau_1 & \tau_1\\ 0 & q_{\mathrm{h},2} & \gamma_2 & \gamma_2 & \tau_2 & \tau_2 & \tau_2\\ 0 & 0 & q_{\mathrm{h},3} & \gamma_3 & \tau_3 & \tau_3 & \tau_3\\ 0 & 0 & 0 & q_{\mathrm{h},4} & \tau_4 & \tau_4 & \tau_4\\ 0 & 0 & 0 & 0 & q_{\mathrm{h},5} & \gamma_5 & \gamma_5\\ 0 & 0 & 0 & 0 & 0 & q_{\mathrm{h},6} & \gamma_6\\ 0 & 0 & 0 & 0 & 0 & 0 & q_{\mathrm{h},7} \end{pmatrix} \tag{5.93}$$

It can be seen that the lower triangular elements of matrix **A** are generally zero, its diagonal elements are the corresponding q_h values, and the upper triangular elements are the corresponding γ or τ values.

For the work amount X_0 of unit fresh steam, we have

$$X_0 = (h_1 - h_n) - \sum_{j=1}^{m} \alpha_j (h_j - h_n) = u_0 - \sum_{j=1}^{m} \alpha_j u_j \tag{5.94}$$

where $u_0 = h_1 - h_n$; $u_j = h_j - h_n$.

Let $\mathbf{u} = [u_1, u_2, \ldots, u_m]^{\mathrm{T}}$, then

$$X_0 = u_0 - \boldsymbol{\alpha}^{\mathrm{T}}\mathbf{u}$$

Considering

$$\boldsymbol{\alpha} = \mathbf{A}^{-1}\boldsymbol{\tau}$$

We have

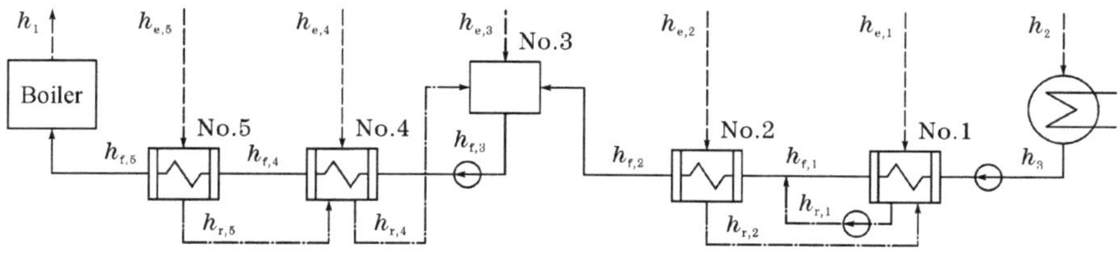

Fig. 5.19 Process flow of a certain extraction and regenerative thermal system

$$X_0 = u_0 - \left(\mathbf{A}^{-1}\boldsymbol{\tau}\right)^{\mathrm{T}} \mathbf{u} = u_0 - \boldsymbol{\tau}^{\mathrm{T}} \left(\mathbf{A}^{-1}\right)^{\mathrm{T}} \mathbf{u} = u_0 - \boldsymbol{\tau}^{\mathrm{T}}\mathbf{Y} \tag{5.95}$$

where $\mathbf{Y} = (\mathbf{A}^{-1})^{\mathrm{T}}\mathbf{u} = [Y_1, Y_2, \ldots, Y_m]$.

The meaning of each element Y_j in matrix $\mathbf{Y}$ is the increase in the work of the unit fresh steam of the steam engine after adding unit pure heat at the jth level, which is exactly the same as the equivalent extraction efficiency obtained by the conventional method.

5.6.3 *Application Example of Equivalent Enthalpy Drop Method*

Example 5.1 A steam extraction regenerative thermal system has 5 stages of steam extraction, which supply 2 high-pressure feedwater heaters, 1 deaerator, and 2 low-pressure feedwater heaters, as shown in Fig. 5.19.

The calculation of original data is as follows:

Generator power $N_g = 50,000$ kWh, mechanical efficiency $\eta_m = 0.9864$, generator efficiency $\eta_g = 0.984$, boiler efficiency $\eta_b = 0.9$, fresh steam enthalpy $h_1 = 3475$ kJ/kg, steam turbine exhaust steam enthalpy $h_2 = 2510.6$ kJ/kg, condenser outlet condensate enthalpy $h_3 = 301.4$ kJ/kg, feedwater pump power $N_{fwp} = 19.362$ kJ/kg, other data as shown in Table 5.13.

Use the equivalent enthalpy drop method to perform thermal calculations on this steam extraction regenerative system to determine the steam turbine inlet flow and unit device efficiency.

Solution: Use formula (5.90) to calculate the equivalent enthalpy drop, the calculation results are shown in Table 5.14.

Ignoring the influence of auxiliary components in the thermal system, the equivalent enthalpy drop of fresh steam is

$$H = h_1 - h_2 - \tau_1\eta_1 - \tau_2\eta_2 - \tau_3\eta_3 - \tau_4\eta_4 - \tau_5\eta_5 = 852.925\,\text{kJ/kg}$$

Unit expansion work

Table 5.13 Calculate basic data (Unit: kJ/kg)

No.	Enthalpy of extraction $h_{e,i}$	Feed-water enthalpy $h_{f,i}$	Drain enthalpy $h_{r,i}$	Feed-water enthalpy rise τ_i	Steam heat release q_i	Drain heat release γ_i
1	2731.0	488.3	500.0	186.9	2231.0	322.5
2	2826.5	571.5	623.9	83.2	2202.6	244.8
3	2921.9	667.2		95.7	2350.4	282.2
4	3079.0	817.6	816.3	150.4	2262.7	133.1
5	3211.4	948.2	949.4	130.6	2262.0	

Table 5.14 Calculation results of equivalent enthalpy drop

Equivalent enthalpy drop H_i/(kJ/kg)		Efficiency	Extraction fraction α_i	
Calculation formula	Value	$\eta_i = \frac{H_i}{q_i}$	Calculation formula	Value
$h_{e,1} - h_2$	220.4	0.091	$\frac{(1-\alpha_5-\alpha_4-\alpha_3)\tau_1-\gamma_1\alpha_2}{q_1}$	0.061
$h_{e,2} - h_2 - \gamma_1\eta_1$	286.649	0.130	$\frac{(1-\alpha_5-\alpha_4-\alpha_3)\tau_2}{q_2}$	0.032
$h_{e,3} - h_2 - \tau_2\eta_2 - \gamma_1\eta_1$	383.532	0.163	$\frac{\tau_3-\gamma_3(\alpha_5+\alpha_4)}{q_3}$	0.028
$h_{e,4} - h_2 - \gamma_3\eta_3 - \tau_2\eta_2 - \tau_1\eta_1$	500.730	0.221	$\frac{\tau_4-\alpha_5\gamma_4}{q_4}$	0.063
$h_{e,5} - h_2 - \gamma_4\eta_4 - \gamma_3\eta_3 - \tau_2\eta_2 - \tau_1\eta_1$	603.715	0.267	$\frac{\tau_5}{q_5}$	0.058

$$N_{T,i} = H + N_{fwp} = 852.925 + 19.362 = 872.287\,\text{kJ/kg}$$

Steam turbine inlet flow

$$G_{T,i} = \frac{N_g}{\eta_m \eta_g N_{T,i}} = \frac{50{,}000}{0.9864 \times 0.984 \times 872.287} = 59.056\,\text{kg/s}$$

Unit installation efficiency

$$\eta = \frac{H}{h_1 - h_{f,5}} \times 100\% = \frac{852.925}{3475 - 948.2} \times 100\% = 33.75\%$$

5.7 Cycle Function Method

The thermal efficiency of the steam extraction regenerative cycle can be represented as

$$\eta_{\mathrm{t}} = 1 - \frac{q_2}{q_1} = 1 - \frac{\alpha_k q_{\mathrm{c}}}{h_1 - h_{\mathrm{fw}}} \tag{5.96}$$

When the initial and final parameters of the steam, the reheat system, and its steam-water parameters are fixed, the values of q_{c}, h_1, and h_{fw} in the above formula are fixed, and only α_k is a variable. If α_k can be characterized by the corresponding q, τ, γ of the reheat system, it will have a single-valued cyclic function relationship with η_{t}, and the calculation of the thermodynamic system can be simplified. The cyclic function method is such a convenient new calculation method.

The cyclic function method divides any complex thermodynamic cycle and its system into a main cycle and several parallel auxiliary cycles. The main cycle refers to the reheat system that does not consider any additional components (shaft seals, valve stem steam leaks, ejector steam, etc.), and each additional component, as well as external heating, supplementary water, pressure reduction and temperature reduction devices, steam generator blowdown, working fluid leakage, etc., are treated as auxiliary cycles, and the thermal economic indicators of the main and auxiliary cycles are calculated separately, and finally integrated into the actual overall thermal economic indicators of the thermodynamic cycle.

5.7.1 *Basic Concepts*

5.7.1.1 Heating Unit

The feedwater regenerative system of NPPs and thermal power plant turbines can be divided into several heating units.

The heating unit is a basic component of the steam extraction regenerative cycle. Any heater in the feedwater regenerative system that collects condensate (mixed heater or surface heater with a condensate pump), together with all levels of surface heaters that discharge condensate to it, constitutes a heating unit. Apart from the condensate feedwater entering and leaving the heating unit, the heating unit is closed and does not discharge condensate to lower pressure heaters outside the unit. All the condensate collected by the heating unit is mixed with the condensate feedwater flowing into the unit and is pumped into a higher pressure heating unit.

The division of heating units follows several basic principles:

1. Heaters that discharge condensate, whether they are a single stage or several stages in series (discharging condensate stage by stage), cannot constitute a heating unit.

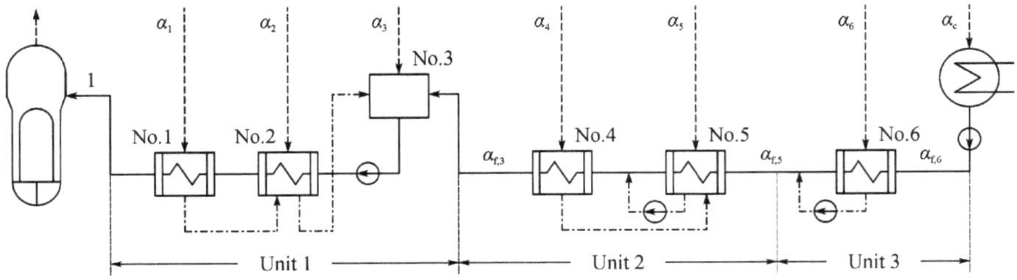

Fig. 5.20 Example 1 of heating unit division for steam extraction and regenerative system

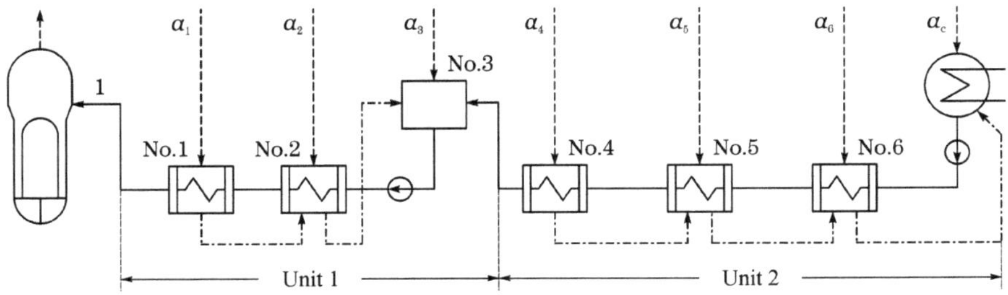

Fig. 5.21 Example 2 of heating unit division for steam extraction and regenerative system

2. Heaters that discharge condensate to the condenser, whether they are a single stage or several stages in series, together with the condenser, constitute a heating unit. The condensate discharged to the condenser is mixed with the condensate of the turbine exhaust and is pumped into a higher pressure heating unit by the condensate pump.
3. A mixed heater or a heater with a condensate pump, in the absence of other heaters discharging condensate to it, constitutes a heating unit on its own.
4. A heating unit composed of a single-stage heater is called a single-stage unit; a heating unit composed of two-stage heaters is called a two-stage unit. In this way, there can be three-stage units, four-stage units, and so on.

Figures 5.20 and 5.21 show two slightly different forms of steam extraction regenerative systems. According to the principle of heating unit division, the steam extraction regenerative system shown in Fig. 5.20 can be divided into three heating units. The first unit is a three-stage unit, the second unit is a two-stage unit, and the third unit is a single-stage unit. The steam extraction regenerative system shown in Fig. 5.21 can be divided into two heating units. The first unit is a three-stage unit, and the second unit is a four-stage unit.

5.7.1.2 Unit Coefficient

When the outlet water volume of a heating unit is 1 kg, the extraction volume of each stage of the heating unit is called the unit extraction coefficient $g_{e,\,i}$, the collected drainage volume is called the unit drainage coefficient $g_{d,\,i}$, and the water inflow volume is called the unit water inflow coefficient $g_{f,\,i}$.

Taking the feedwater regenerative system shown in Fig. 5.20 as an example, the mass balance equations and energy conservation equations of the three heating units are listed separately, and the extraction coefficient, drainage coefficient, and water inflow coefficient of each unit are determined based on this.

First Unit

The mass balance equation and heat balance equation of each stage of the heater in the first unit are as follows

$$\begin{cases} \alpha_1 + \alpha_2 + \alpha_3 + \alpha_{f,3} = 1 \\ \alpha_1 q_{h,1} = \tau_1 \\ \alpha_2 q_{h,2} + \alpha_1 \gamma_2 = \tau_2 \\ \alpha_3 q_{h,3} + (\alpha_1 + \alpha_2)\gamma_3 = \alpha_{f,3}\tau_3 \end{cases} \tag{5.97}$$

The unit extraction coefficients of each stage of the heater are as follows

$$\begin{cases} g_{e,1} = \alpha_1 = \dfrac{\tau_1}{q_{h,1}} \\ g_{e,2} = \alpha_2 = \dfrac{\tau_2}{q_{h,2}} - g_{e,1}\dfrac{\gamma_2}{q_{h,2}} \\ g_{e,3} = \alpha_3 = \dfrac{\tau_3}{q_{h,3}} - (g_{e,1} + g_{e,2})\dfrac{\gamma_3}{q_{h,3}} \end{cases} \tag{5.98}$$

The unit drainage coefficients of each stage of the heater are as follows

$$\begin{cases} g_{d,1} = 0 \\ g_{d,2} = \alpha_1 = g_{e,1} = \dfrac{\tau_1}{q_{h,1}} \\ g_{d,3} = \alpha_1 + \alpha_2 = g_{e,1} + g_{e,2} = \dfrac{\tau_2}{q_{h,2}} + g_{e,1}\left(1 - \dfrac{\gamma_2}{q_{h,2}}\right) \end{cases} \tag{5.99}$$

The unit water inflow coefficient of the first heating unit is

$$g_{f,3} = \alpha_{f,3} = \frac{q_{h,3} - g_{d,3}\left(q_{h,3} - \gamma_3\right)}{q_{h,3} + \tau_3} \tag{5.100}$$

Second Unit

The heat balance equation and mass balance equation of each stage of the heater in the second unit are as follows

$$\begin{cases} \alpha_4 q_{h,4} = \alpha_{f,3}\tau_4 \\ \alpha_5 q_{h,5} + \alpha_4\gamma_5 = \alpha_{f,3}\tau_5 \\ \alpha_{f,5} = \alpha_{f,3} - \left(\alpha_4 + \alpha_5\right) \end{cases} \tag{5.101}$$

The unit extraction coefficients of each stage of the heater are as follows

$$\begin{cases} g_{e,4} = \dfrac{\alpha_4}{\alpha_{f,3}} = \dfrac{\tau_4}{q_{h,4}} \\ g_{e,5} = \dfrac{\alpha_5}{\alpha_{f,3}} = \dfrac{\tau_5}{q_{h,5}} - g_{e,4}\dfrac{\gamma_5}{q_{h,5}} \end{cases} \tag{5.102}$$

The unit drainage coefficients of each stage of the heater are as follows

$$\begin{cases} g_{d,4} = 0 \\ g_{d,5} = g_{e,4} = \dfrac{\tau_4}{q_{h,4}} \end{cases} \tag{5.103}$$

The unit water inflow coefficient of the second heating unit is

$$g_{f,5} = \frac{q_{h,5} - g_{d,4}\left(q_{h,5} - \gamma_5\right)}{q_{h,5} + \tau_5} \tag{5.104}$$

Third Unit

The heat balance equation and mass balance equation of each stage of the heater in the third unit are as follows

$$\begin{cases} \alpha_{f,5}\tau_6 = \alpha_6 q_{h,6} \\ \alpha_{f,6} = \alpha_{f,5} - \alpha_6 \end{cases} \tag{5.105}$$

The unit extraction coefficient is

$$g_{e,6} = \frac{\alpha_6}{\alpha_{f,5}} = \frac{\tau_6}{q_{h,6}} \tag{5.106}$$

The unit water inflow coefficient of the third heating unit is

$$g_{f,6} = \frac{q_{h,6}}{q_{h,6} + \tau_6} = \frac{q_{h,6} - g_{d,6}\left(q_{h,6} - \gamma_6\right)}{q_{h,6} + \tau_6} \tag{5.107}$$

In summary, for each heating unit, the general formula for the unit water inflow coefficient is

$$g_{f,i} = \frac{q_{h,i} - g_{d,i}\left(q_{h,i} - \tau_i\right)}{q_{h,i} + \tau_i} \tag{5.108}$$

The general formulas for the unit extraction coefficient $g_{e,m}$ and the unit drainage coefficient $g_{d,m}$ of each stage of the heater in each heating unit are as follows

$$g_{e,m} = \frac{\tau_m}{q_{h,m}} - g_{d,m}\frac{\gamma_m}{q_{h,m}} = \frac{\tau_m}{q_{h,m}} - \frac{\gamma_m}{q_{h,m}}\sum_{i=1}^{m-1} g_{e,i} \tag{5.109}$$

$$g_{d,m} = \frac{\tau_{m-1}}{q_{h,m-1}} + \left(1 - \frac{\gamma_{m-1}}{q_{h,m-1}}\right) g_{d,m-1} = \frac{\tau_{m-1}}{q_{h,m-1}} + \left(1 - \frac{\gamma_{m-1}}{q_{h,m-1}}\right)\sum_{i=1}^{m-2} g_{e,i} \tag{5.110}$$

The relationship between the exhaust coefficient α_k of the extraction regenerative cycle and the water inflow coefficients of each heating unit is

$$\alpha_k = g_{f,1} g_{f,2} g_{f,3} \tag{5.111}$$

From the above formula, it can be seen that the circulating water inflow coefficient of the first heating unit is equal to its unit water inflow coefficient; the circulating water inflow coefficient of other units is equal to the product of its unit water inflow coefficient and the water inflow coefficient of the higher pressure unit; the circulating extraction coefficient of the second unit is equal to its unit extraction coefficient; the circulating extraction coefficient of other units is equal to its unit extraction coefficient multiplied by the water inflow coefficient of the higher pressure unit. Therefore, the steam-water flow of each part of the regenerative cycle can be obtained from the steam-water flow of each unit.

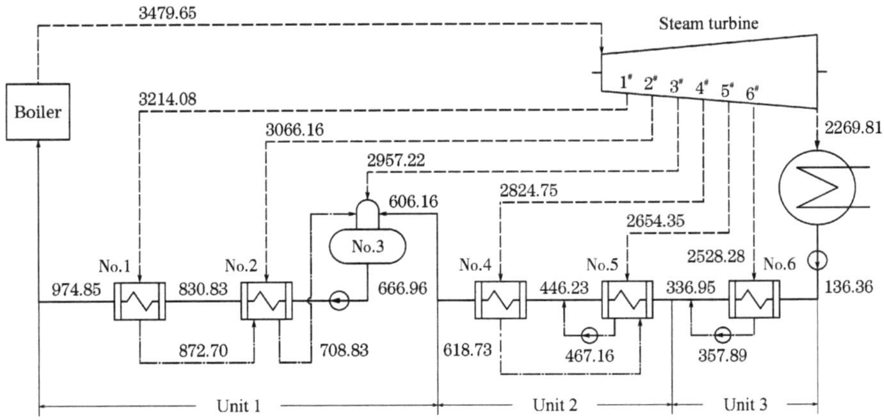

Fig. 5.22 Drawing of extraction and regenerative system

Table 5.15 Power performance parameter values of each heater

No.	Heat release of exhaust $q_{h,i}$/(kJ/kg)	Feed water enthalpy rise τ_i/(kJ/kg)	Heat release of drain γ_i/(kJ/kg)	Unit water inflow coefficient $g_{e,i}$	Unit extraction coefficient $g_{d,i}$	Unit drainage coefficient $g_{f,i}$
1	2431.38	144.02	0	0.059	0	0.859
2	2357.33	163.87	163.87	0.065	0.059	0.894
3	2351.06	60.80	102.67	0.020	0.125	0.915
4	2206.02	159.93	0	0.072	0	
5	2187.19	109.28	281.78	0.038	0.072	
6	2170.39	200.59	0	0.092	0	

5.7.2 *Application of the Cycle Function Method*

Example 5.2 Figure 5.22 shows the regenerative cycle of 8.82 MPa/535 ° C, 100 MW single-cylinder condensing steam turbine, with the steam-water enthalpy values of each part marked in the figure (unit kJ/kg), try to find the cycle efficiency η_t of this regenerative cycle.

Solution This regenerative system includes six stages of heaters, which can be divided into three heating units. The first unit consists of 1#, 2# drainage flow heaters and 3# deaerator, the second unit consists of 4# drainage flow heater and 5# heater, and the third unit only has 6# heater.

According to the steam-water enthalpy values marked in the figure, the performance parameters of each heater can be calculated, as shown in Table 5.15.

The exhaust coefficient $\alpha_k = g_{f,1} g_{f,2} g_{f,3} = 0.703$

Extraction Steam Regenerative Cycle Efficiency

$$\eta_t = 1 - \frac{\alpha_k q_c}{h_1 - h_{fw}} = 1 - \frac{0.703 \times (2269.81 - 136.36)}{3479.65 - 974.85} = 0.401$$

The application of the steam turbine cycle function method is mainly reflected in the following three aspects:

1. Used for quantitative calculation of the thermal economy of the steam turbine regenerative cycle or regenerative system;
2. Used to derive the single thermal economic indicators or thermodynamic characteristic coefficients of the steam turbine and power plant auxiliary steam-water cycle, to establish the thermodynamic characteristic equations of the steam turbine and power plant;
3. Used to find the partial derivatives of the cycle function, establish the optimal heating distribution equation of the steam turbine cycle, and solve the numerical value of the optimal heating distribution, so that the thermal efficiency of the steam turbine cycle reaches the highest value or its energy consumption is the lowest.

The cycle function method uses the unit inlet water coefficient to calculate the condensation coefficient, without the need to solve the $(Z + 1)$ element equation simultaneously. As long as the cycle parameters and the thermal system are given, the function equation can be directly listed according to the corresponding general formula, and the relevant steam-water parameter values can be substituted to find out, simplifying the calculation. When choosing different thermal system schemes, if different schemes only involve changes in a heating unit, only the inlet water coefficient of that unit needs to be recalculated, and the inlet water coefficients of other heating units that have not changed in the scheme do not need to be recalculated, greatly reducing the calculation workload. Compared with the conventional thermal calculation method, this point is particularly prominent.

By substituting the relationship formula of the condensation coefficient as the product of the inlet water coefficients of each heating unit into the absolute internal efficiency formula of the thermal cycle, not only can its numerical solution be obtained, but also the optimal regenerative distribution of the regenerative cycle or reheat cycle can be derived and solved by mathematical methods.

5.8 Condensation Method

The condensation method is a new method for the overall calculation and partial calculation of the power plant thermal system. The calculation of the thermal system using the condensation method does not involve the work and enthalpy drop of the thermal process. It starts from the anti-equilibrium angle of the thermal cycle, fully considers the structure and parameter characteristics of the thermal system, establishes the coefficient with clear physical concepts and practical

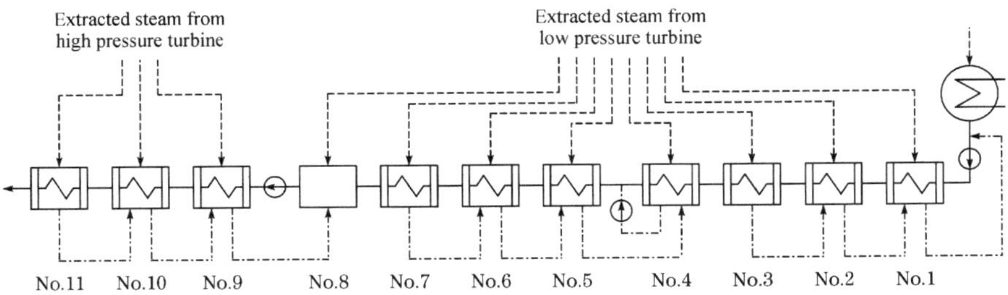

Fig. 5.23 Principle process of thermal system for calculating condensation coefficient

significance - the condensation coefficient, according to the principle of thermal cycle extrusion and the principle of working substance balance. In the cycle, the thermal equipment extrudes 1 kg of steam, and a part of it does work to the steam turbine exhaust port and enters the condenser. The share of the steam entering the condenser is called the condensation coefficient, which is determined with the thermal system and its parameters. It is an important parameter to simplify the overall calculation and partial quantitative calculation of the thermal system. Using it, the thermal equipment and system can be directly analyzed and calculated.

5.8.1 Extraction Condensation Coefficient

When the jth stage reheater of the thermal system obtains heat from the outside and extrudes 1 kg of extraction steam, the share of the 1 kg of extraction steam that reaches the condenser after doing work in the steam engine is called the extraction condensation coefficient, represented by L_j. Obviously, when the jth stage heater reduces the extraction steam and the condensation coefficient represents the increase in the share entering the condenser by obtaining heat q_j from the outside, L_j is a positive value; on the contrary, when the jth stage extraction heat is used by the outside to increase the extraction, the share entering the condenser decreases, and L_j is a negative value.

Now, take the unit thermal system shown in Fig. 5.23 as an example for analysis and calculation.

Assume that No. 1 heater obtains heat q_1 from the outside, just reducing the extraction steam by 1 kg. At this time, the extruded extraction steam will all enter the condenser after doing work in the steam turbine, and its condensation coefficient is equal to the amount of extruded extraction steam, that is, $L_1 = 1$;

Assume that No. 2 heater obtains heat q_2 from the outside, just reducing its extraction by 1 kg, the drain water entering the No.1 heater will also decrease by 1 kg. To compensate for this heating deficiency, the steam extraction of the No. 1 heater will increase by (γ_1/q_1)kg, and the remaining $(1 - \gamma_1/q_1)$kg of extracted steam will continue to flow downstream to expand and do work. The condensation coefficient of the 1 kg of extracted steam at this point is $L_1 = 1$, and the condensation share of the corresponding $(1-\gamma_1/q_1)$kg of extracted steam is $(1-\gamma_1/q_1)L_1$, so the condensation coefficient of the steam extraction of the No.2 heater is $L_2=(1-\gamma_1/q_1)L_1$.

Similarly, the No. 3 heater is a drain type, and its steam extraction condensation coefficient is $L_3=(1-\gamma_2/q_2)L_2$.

The No. 4 heater is a collection type heater. If it obtains heat q_4 from the outside, causing its steam extraction to decrease by exactly 1 kg, this 1 kg of extracted steam will eventually become condensate through different paths and collect in the condenser, causing the main condensate to increase by 1 kg, which will cause the steam extraction of the Nos. 3, 2, and 1 heaters to increase by τ_3/q_3, τ_2/q_2, τ_1/q_1 kg respectively. The condensation coefficients of the steam extraction of these three heaters are L_3, L_2, andL_1, so the condensation coefficient of the steam extraction of the No. 4 heater is

$$L_4=1-\frac{\tau_3}{q_3}L_3-\frac{\tau_2}{q_2}L_2-\frac{\tau_1}{q_1}L_1$$

If the No. 5 heater obtains heat q_5 from the outside and displaces 1 kg of extracted steam, because the No.5 heater is a drain type, only $(1-\gamma_4/q_4)$kg of displaced steam will continue to flow downstream to do work at the steam extraction port of the No. 4 heater. The condensation coefficient of the 1 kg of extracted steam at this point is L_4, and the condensation share of the corresponding $(1-\gamma_4/q_4)$kg of extracted steam is $(1-\gamma_4/q_4)L_4$, so the condensation coefficient of the steam extraction of the No. 5 heater is $L_5=(1-\gamma_4/q_4)L_4$.

Similarly, the condensation coefficients of the steam extraction of the Nos. 6, 7, 8, 9, 10, and 11 heaters are

$$L_6=\left(1-\frac{\gamma_5}{q_5}\right)L_5$$

$$L_7=\left(1-\frac{\gamma_6}{q_6}\right)L_6$$

$$L_8=L_4-\frac{\tau_7}{q_7}L_7-\frac{\tau_6}{q_6}L_6-\frac{\tau_5}{q_5}L_5-\frac{\tau_4}{q_4}L_4$$

$$L_9=\left(1-\frac{\gamma_8}{q_8}\right)L_8$$

$$L_{10}=\left(1-\frac{\gamma_9}{q_9}\right)L_9$$

$$L_{11} = \left(1 - \frac{\gamma_{10}}{q_{10}}\right) L_{10}$$

The above analysis shows that the condensation coefficient of steam extraction has a clear calculation rule. For the drain type heater No.*j*, regardless of whether the adjacent No. ($j - 1$) is a collection type or a drain type heater, the general formula for calculating its steam extraction condensation coefficient is

$$L_j = \left(1 - \frac{\gamma_{j-1}}{q_{j-1}}\right) L_{j-1} \tag{5.112}$$

Its physical meaning is: when 1 kg of steam is displaced in the *j*th section, only $(1 - \gamma_{j-1}/q_{j-1})$kg will continue to flow downstream to do work after the ($j - 1$)th section. The condensation coefficient of the 1 kg of displaced steam at this point is L_{j-1}, so the condensation share of the $(1 - \gamma_{j-1}/q_{j-1})$kg of displaced steam is $(1 - \gamma_{j-1}/q_{j-1})L_{j-1}$, so the condensation coefficient of the steam extraction in the *j*th section is $L_j = (1 - \gamma_{j-1}/q_{j-1})L_{j-1}$.

For the collection type heater, such as No. *j* and No. *m* are two collection type heaters, the general formula for calculating their steam extraction condensation coefficient is

$$L_j = L_m - \sum_{r=m}^{j-1} \frac{\tau_r}{q_r} L_r \tag{5.113}$$

Its physical meaning is: when 1 kg of displaced steam from the collection type heater No. *j* returns to the steam engine to do work to the lower collection type heater No. *m*, the condensation coefficient subtracts the condensation share reduced due to the increase in steam extraction between the two levels $\sum_{r=m}^{j-1} \frac{\tau_r}{q_r} L_r$, that is, $L_m - \sum_{r=m}^{j-1} \frac{\tau_r}{q_r} L_r$ is the condensation coefficient of the*j*th level heater. It is particularly pointed out that if there are other collection type heaters between No. *j* and No. *m*, this formula also applies.

5.8.2 *Fresh Steam Condensation Coefficient*

1 kg the portion of the fresh steam entering the turbine to do work and reaching the condenser is called the fresh steam condensation coefficient. When only considering the main cycle system, it is called the gross fresh steam condensation coefficient, represented by L_{gr}; if the effects of auxiliary components such as gate rods,

shaft seal steam leakage, pump work, and heater heat dissipation are considered, it is called the net fresh steam condensation coefficient, represented by L.

If the evaporator is regarded as a collection-type heater, the gross fresh steam condensation coefficient can be deduced in the same way as the extraction condensation coefficient.

$$L_{gr} = L_8 - \frac{\tau_{11}}{q_{11}} L_{11} - \frac{\tau_{10}}{q_{10}} L_{10} - \frac{\tau_9}{q_9} L_9 - \frac{\tau_8}{q_8} L_8$$

Accordingly, the general formula for calculating the gross fresh steam condensation coefficient is

$$L_{gr} = L_m - \sum_{r=m}^{Z} \frac{\tau_r}{q_r} L_r \tag{5.114}$$

The general formula for calculating the net fresh steam condensation coefficient L is

$$L = L_m - \sum_{r=m}^{Z} \frac{\tau_r}{q_r} L_r + \sum \Delta L \tag{5.115}$$

where m is any collection-type heater, the deaerator is best when actually calculated; $\sum \Delta L$ is the change in the condensation ratio caused by local changes in thermal equipment and systems.

Once the fresh steam condensation coefficient L is known, the absolute internal efficiency of the turbine can be calculated using the following formula.

$$\eta_i = 1 - \frac{L q_c}{q}$$

where q is the cycle heat absorption, kJ/kg; q_c is the heat released by 1 kg of turbine exhaust steam in the condenser, kJ/kg.

5.8.3 Condensation Method Calculation Example

Example 5.3 As shown in Fig. 5.24, the extraction regenerative system, the fresh steam parameters $p_1 = 12.75$ MPa, $T_1 = 535°$ C, $h_1 = 3433.5$ kJ/kg. If the effects of gate rods, shaft seal steam leakage, pump work, heater heat dissipation and other auxiliary components are ignored, try to use the condensation method and the original data of the thermal system listed in Table 5.16 to calculate the absolute internal efficiency of the turbine.

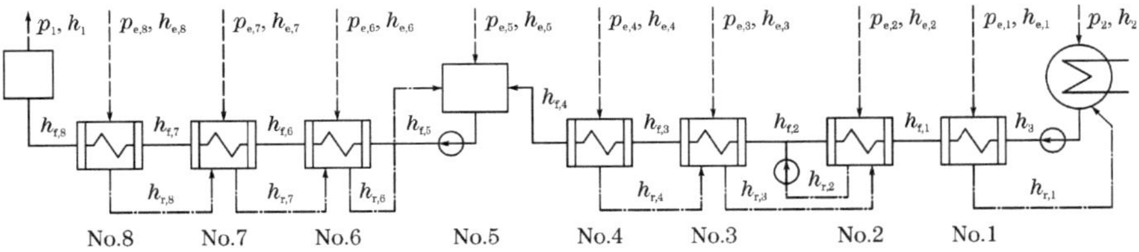

Fig. 5.24 Unit reheat system process

Table 5.16 Nuclear power unit raw data

No.	$p_{e,j}$ (MPa)	$h_{e,j}$ (kJ/kg)	$h_{f,j}$ (kJ/kg)	$h_{r,j}$ (kJ/kg)	Other (kJ/kg)
1	0.045	2691.4	305.8	321.8	$h_2 = 2438.0$
2	0.146	2888.4	439.8	453.8	$h_3 = 140.7$
3	0.245	2979.5	506.0	521.5	
4	0.543	3164.5	636.1	639.8	
5	0.829	3275.3	667.0		
6	1.210	3382.1	793.8	783.8	
7	2.460	3040.2	932.1	890.6	
8	3.750	3137.9	1037.6	1,046.0	

Table 5.17 The calculated values of the correlation coefficient

No.	Enthalpy rise of unit mass feedwater τ_j (kJ/kg)	Heat release of unit mass extraction steam q_j (kJ/kg)	Heat release of unit mass drainage γ_j (kJ/kg)
1	305.8 – 140.7 = 165.1	2691.4 – 140.7 = 2550.7	
2	439.8 – 305.8 = 134.0	2888.4 – 305.8 = 2582.6	521.5 – 305.8 = 215.7
3	506.0 – 439.8 = 66.2	2979.5 – 521.5 = 2458.0	639.8 – 521.5 = 118.3
4	636.1 – 506.0 = 130.1	3164.3 – 639.8 = 2524.5	
5	667.0 – 636.1 = 30.9	3275.3 – 636.1 = 2639.2	783.8 – 636.1 = 147.7
6	793.8 – 667.0 = 126.8	3382.1 – 783.8 = 2598.3	890.6 – 783.8 = 106.8
7	932.1 – 793.8 = 138.3	3040.2 – 890.6 = 2149.6	1,046.0 – 890.6 = 155.4
8	1037.6 – 932.1 = 105.5	3137.9 – 1046.0 = 2091.9	

1. Organization of thermal system and original data of the unit
According to the composition form of the thermal system shown in Fig. 5.24 and the thermal parameters listed in Table 5.16, the following coefficient values are obtained: Table 5.17

$$q_c = h_2 - h_3 = 2438.0 - 140.7 = 2297.3\,\text{kJ / kg}.$$

2. Calculation of extraction condensation coefficient and fresh steam condensation coefficient

$$L_1 = 1$$

$$L_2 = \left(1 - \frac{\tau_1}{q_1}\right) L_1 = 0.935$$

$$L_3 = \left(1 - \frac{\gamma_2}{q_2}\right) L_2 = 0.857$$

$$L_4 = \left(1 - \frac{\gamma_3}{q_3}\right) L_3 = 0.816$$

$$L_5 = L_2 - \frac{\tau_4}{q_4} L_4 - \frac{\tau_3}{q_3} L_3 - \frac{\tau_2}{q_2} L_2 = 0.822$$

$$L_6 = \left(1 - \frac{\gamma_5}{q_5}\right) L_5 = 0.776$$

$$L_7 = \left(1 - \frac{\gamma_6}{q_6}\right) L_6 = 0.744$$

$$L_8 = \left(1 - \frac{\gamma_7}{q_7}\right) L_7 = 0.690$$

$$L_{gr} = L_5 - \frac{\tau_8}{q_8} L_8 - \frac{\tau_7}{q_7} L_7 - \frac{\tau_6}{q_6} L_6 - \frac{\tau_5}{q_5} L_5 = 0.691$$

3. Calculation of absolute internal efficiency of turbine

Cycle heat absorption

$$q = h_1 - h_{\mathrm{f},8} = 2395.9\,\mathrm{kJ/kg}$$

Cycle heat release

$$\sum \Delta q_{\mathrm{c}} = L_{gr} q_c = 1588.6\,\mathrm{kJ/kg}$$

Absolute internal efficiency of turbine

$$\eta_{\mathrm{i}} = 1 - \frac{\sum \Delta q_{\mathrm{c}}}{q} = 0.337$$

5.9 Matrix Analysis Method

The matrix analysis method of the thermal system of the power plant is to establish the heat balance equations of each stage of the regenerative heater, and then complete the calculation of various thermal economic indicators of the thermal system by solving a set of linear equations containing each stage of extraction steam. The matrix analysis method is also a traditional analysis method, which can calculate several or dozens of unknown parameters at one time, and simultaneously obtain the extraction steam of each stage. This method is easy to computerize, has good universality, and is fast and accurate.

5.9.1 Power Equation of Turbine Device

When the structure of the system, the cycle parameters, and the extraction parameters are determined, if the influence of auxiliary components is ignored, the work done by 1 kg of fresh steam in the extraction regenerative unit can be expressed as

$$N = (h_1 - h_2) - \sum_{i=1}^{m} \alpha_i (h_{e,i} - h_2) \tag{5.116}$$

where h_1, h_2 are the enthalpy of fresh steam and condensation steam, respectively, kJ/kg; m is the total number of extractions; α_i is the ratio of the ith level of extraction.

Define h_i^d as the work deficiency caused by 1 kg of extraction, that is $h_i^d = h_{e,i} - h_2$, then the vector form of the equation is

$$N = (h_1 - h_2) - \boldsymbol{\alpha}^T \mathbf{h}^T \tag{5.117}$$

where $\boldsymbol{\alpha} = \begin{bmatrix} \alpha_1 \\ \vdots \\ \alpha_i \\ \vdots \\ \alpha_m \end{bmatrix}$, $\mathbf{h} = \begin{bmatrix} h_{e,1} - h_2 \\ \vdots \\ h_{e,i} - h_2 \\ \vdots \\ h_{e,m} - h_2 \end{bmatrix}$

5.9.2 System Thermal Balance Equation

Figure 5.25 shows a simplified flow diagram of an extraction regenerative unit.

The thermal balance equations for each heater are

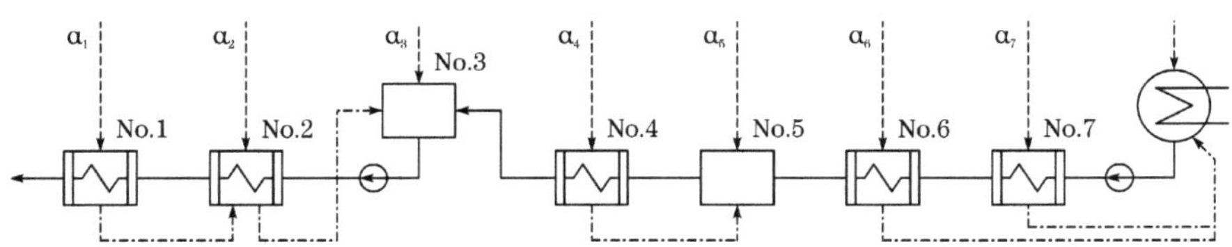

Fig. 5.25 Process diagram of unit extraction and regenerative system

No.1	$\alpha_1 q_1 = \tau_1$
No.2	$\alpha_1 \gamma_2 + \alpha_2 q_2 = \tau_2$
No.3	$\alpha_1 \gamma_3 + \alpha_2 \gamma_3 + \alpha_3 q_3 = \tau_3$
No.4	$\alpha_4 q_4 = (1 - \alpha_3 - \alpha_2 - \alpha_1)\tau_4$
No.5	$\alpha_4 \gamma_5 + \alpha_5 q_5 = (1 - \alpha_3 - \alpha_2 - \alpha_1)\tau_5$
No.6	$\alpha_6 q_6 = (1 - \alpha_5 - \alpha_4 - \alpha_3 - \alpha_2 - \alpha_1)\tau_6$
No.7	$\alpha_7 q_7 = (1 - \alpha_5 - \alpha_4 - \alpha_3 - \alpha_2 - \alpha_1)\tau_7$

By combining the above equations and writing them in standard matrix form, we get

$$\mathbf{A}\alpha_{\mathrm{i}} = \mathbf{b}_{\mathrm{i}} \tag{5.118}$$

where, **A** is the structural matrix of the thermal system, $\boldsymbol{\alpha}_{\mathrm{i}}$ is the extraction coefficient vector, $\mathbf{b}_{\mathrm{i}}$ is the characteristic vector, that is.

$$\mathbf{A} = \begin{pmatrix} q_1 & 0 & 0 & 0 & 0 & 0 & 0 \\ \gamma_2 & q_2 & 0 & 0 & 0 & 0 & 0 \\ \gamma_3 & \gamma_3 & q_3 & 0 & 0 & 0 & 0 \\ \tau_4 & \tau_4 & \tau_4 & q_4 & 0 & 0 & 0 \\ \tau_5 & \tau_5 & \tau_5 & \gamma_5 & q_5 & 0 & 0 \\ \tau_6 & \tau_6 & \tau_6 & \tau_6 & \tau_6 & q_6 & 0 \\ \tau_7 & \tau_7 & \tau_7 & \tau_7 & \tau_7 & 0 & q_7 \end{pmatrix} \tag{5.119}$$

$$\boldsymbol{\alpha}_{\mathrm{i}} = \begin{bmatrix} \alpha_1 & \alpha_2 & \alpha_3 & \alpha_4 & \alpha_5 & \alpha_6 & \alpha_7 \end{bmatrix}^{\mathrm{T}} \tag{5.120}$$

$$\mathbf{b}_{\mathrm{i}} = \begin{bmatrix} \tau_1 & \tau_2 & \tau_3 & \tau_4 & \tau_5 & \tau_6 & \tau_7 \end{bmatrix}^{\mathrm{T}} \tag{5.121}$$

The equation is also applicable to m level extraction, so it can be used as a general representation of the thermal balance equation, where the left side represents heat release and the right side represents heat absorption.

Matrix **A** is a lower triangular matrix, the order of which is the number of regenerative stages. Matrix **A** reflects the basic structure of the power plant thermal system, and the arrangement of matrix elements implies the type of heater and the connection method of the system. Assume that any element of **A** is α_{ij}, where i represents the row and j represents the column. The main diagonal is the enthalpy drop of 1 kg of regenerative extraction in this stage of the heater, that is, $A_{i,\,i} = q_i$; when $i > j$, if the ith stage heater has water from the jth stage heater, then $A_{i,\,i} = \gamma_i$; if the drain of the jth stage heater flows into the collection heater, then $A_{i,\,i} = \tau_i$; if the drain of the jth stage heater flows directly into the condenser, then $A_{i,\,i} = 0$; when $i < j$, because the upper heater is not affected by the lower heater, the values of the elements are all zero

Each row of matrix **A** represents a thermal balance equation, which represents a heater. In the first stage heater, only $\boldsymbol{\alpha}_1$ working substance releases heat, in the second stage heater, $\boldsymbol{\alpha}_1$ and $\boldsymbol{\alpha}_2$ two streams of steam release heat, and so on, in the last stage heater, $\boldsymbol{\alpha}_1$~$\boldsymbol{\alpha}_7$ working substances participate in heat release. Each column of matrix **A** represents the heat release process experienced by each extraction. The first column shows that $\boldsymbol{\alpha}_1$ extraction releases heat q_1 in the first stage heater, releases heat γ_2 and γ_3 in the second and third stage heaters, and releases heat τ_4, τ_5, τ_6, τ_7 in the fourth, fifth, sixth stage heaters, and finally this flow enters the condenser; while the 7th column shows that $\boldsymbol{\alpha}_7$ only releases heat q_7 in the 7th stage heater and then enters the condenser.

Vector $\mathbf{b}_\mathbf{i}$ represents the heating process of condensate water from the condenser through $\tau_1 - \tau_7$'s 7-stage heater.

Questions and Exercises

1. What are the basic steps of thermal balance analysis of NPP thermal system using thermal efficiency method?
2. Briefly describe the basic idea of steam generator thermal balance calculation.
3. The thermal load of the steam generator is Q_s, and it continuously discharges waste during operation, with the discharge amount being 1.05% of the fresh steam output. The discharged water enters the low-pressure blowdown water evaporator, where part of the discharged water is superheated and flashes to produce steam, and the concentrated wastewater is discharged overboard. The feedwater heater uses the steam produced by the blowdown water evaporator to heat the feedwater, and any deficiency is supplemented with fresh steam. The enthalpy values of the working substances at the inlet and outlet of each device are known, and heat dissipation losses are ignored. Try to determine:

   (a) The fresh steam output of the steam generator;
   (b) The amount of fresh steam that needs to be supplemented by the feedwater heater.

4. Briefly describe the basic concept of equivalent enthalpy drop and the characteristics of the equivalent enthalpy drop method.
5. Briefly describe the characteristics of the cycle function method.

Chapter 6
Exergy Analysis of NPP Thermal Systems

An energy balance analysis of the thermal system of a PWR NPP can determine the heat flow distribution and heat dissipation loss in the thermal cycle, but it cannot explain the changes in quality during the energy conversion, transmission, and distribution process, and it is difficult to reveal the possibility, direction, and conditions of thermal power conversion. Therefore, the analysis results have a certain one-sidedness, limiting the guidance for improving the thermal system of the nuclear power device. Only by starting from the first and second laws of thermodynamics and applying exergy analysis can we comprehensively reflect the differences and changes in energy in quantity and quality during the energy use process of the thermal system, objectively evaluate the work loss caused by irreversible factors in various parts of the nuclear power device and the efficiency of the entire thermal cycle, reveal the real reasons for the exergy loss, and point out the correct direction for improving the effective degree of energy use in the thermal system [40].

6.1 Main Equipment Exergy Analysis Model

The structure of the thermal system of a PWR NPP is complex and there are many devices. Before performing exergy analysis, an appropriate analysis model must be selected. Although the material flow changes everywhere in the device after passing through a device, exergy is a state function. Regardless of how the intermediate process changes, only the difference in exergy at the inlet and outlet of the device is calculated to obtain the exergy efficiency and exergy loss of the device. This treatment method is the "black box" model. Because it is simple and intuitive, it has become the most basic and widely used analysis model in practical applications. The exergy analysis of the thermal system of a PWR NPP uses the "black box" model [41].

M. Peng et al., *Thermal Analysis of Nuclear Power Plants*, Nuclear Science and Technology, https://doi.org/10.1007/978-981-96-8983-5_6

According to the working characteristics of the main systems and equipment in the NPP, they can be roughly divided into the following categories:

(1) Reactor;
(2) Pumps (main pump, condensate pump, feedwater pump, etc.);
(3) Heat exchange equipment (steam generator, steam regenerator, feedwater heater, condenser, etc.);
(4) Main steam turbine generator set;
(5) Other equipment (steam-water separator, main steam pipeline, etc.).

The exergy analysis models of these types of equipment are as follows.

6.1.1 Reactor

During the operation of the reactor, nuclear fission occurs in the core of the reactor, releasing a large amount of heat energy; at the same time, the coolant flows through the core, taking the heat released by the core out of the reactor. Therefore, there are two thermal processes in the reactor: the conversion of nuclear fission energy into thermal energy and the transfer of heat released by nuclear fuel to the coolant. The following introduces the exergy loss and its calculation method of these two processes.

6.1.1.1 The Process of Converting Nuclear Energy Into Thermal Energy in the Reactor Core

From the perspective of thermodynamics, nuclear energy is exergy. If we ignore the energy loss in the nuclear fission process, the exergy of nuclear energy should numerically equal the thermal power of the reactor, that is,

$$E_{x,\mathrm{N}} = Q_{\mathrm{R}} \tag{6.1}$$

During the nuclear fission process, the exergy of the heat released by the nuclear fuel in the reactor core depends on the temperature of the nuclear fuel when it releases heat. For any infinitesimal volume of the reactor core, its average temperature is T_u. If the heat release is dQ_R, at an ambient temperature of T_0, its heat exergy is

$$\mathrm{d}E_{x,\mathrm{R}} = \left(1 - \frac{T_0}{T_\mathrm{u}}\right)\mathrm{d}Q_\mathrm{R} \tag{6.2}$$

For the entire reactor, the heat exergy of the heat released by nuclear fission is

$$E_{x,\mathrm{R}} = \int_{V_\mathrm{R}} \left(1 - \frac{T_0}{T_\mathrm{u}}\right)\mathrm{d}Q_\mathrm{R} \tag{6.3}$$

where V_R is the effective volume of the reactor core.

Therefore, in the process of converting nuclear energy into thermal energy in the reactor, the exergy loss is

$$E_{xl,R_1} = E_{x,N} - E_{x,R} = Q_R - \int_{V_R} \left(1 - \frac{T_0}{T_u}\right) dQ_R = T_0 \int_{V_R} \frac{1}{T_u} dQ_R \tag{6.4}$$

For ease of analysis, assume that the reactor core is a cylinder with no reflector layer and uniformly loaded, and the heat release along the axial distribution of the reactor core height follows a sinusoidal rule, as shown in Fig. 6.1, that is,

$$q(z) = q_{max} \sin \frac{\pi z}{H} \tag{6.5}$$

where q_{max} is the maximum linear heat release rate along the axial direction of the reactor core, kW/m; H is the equivalent height of the active area of the reactor core, m.

Take an infinitesimal body with a height of dz at the axial position z of the reactor core, its heat release is

$$dQ_R = q(z)dz = q_{max} \sin \frac{\pi z}{H} dz \tag{6.6}$$

Assume that the heat transfer resistance between this infinitesimal body and the reactor coolant flowing through its surface is $R_t(z)$, then the heat transfer equation is

$$q(z) = \frac{1}{R_t(z)} \left[T_u(z) - T_c(z)\right] \tag{6.7}$$

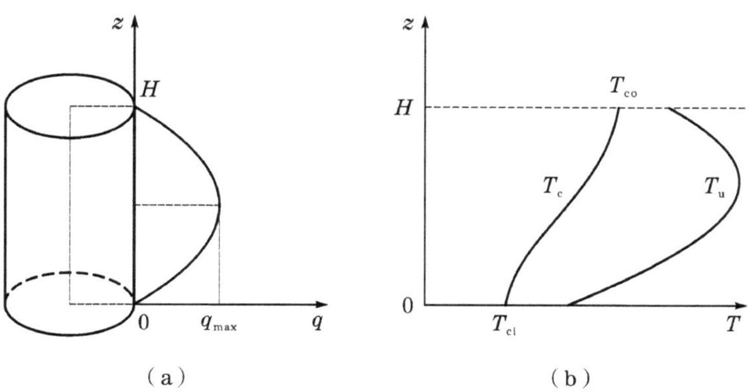

Fig. 6.1 Power distribution and temperature distribution along the axial direction of the core: (**a**) power distribution; (**b**) temperature distribution

Transforming the above equation, we can get the temperature distribution of the nuclear fuel in the reactor core along the axial direction

$$T_u(z) = T_c(z) + R_t(z) q_{max} \sin\frac{\pi z}{H} \tag{6.8}$$

Ignoring the heat dissipation loss of the reactor, the heat absorbed by the coolant flowing through the reactor core is equal to the heat released by the nuclear fuel. The energy balance equation is

$$G_c c_{p,c}\left[T_c(z) - T_{ci}\right] = \int_0^z q_{max} \sin\frac{\pi z}{H} dz = q_{max}\frac{H}{\pi}\left(1 - \cos\frac{\pi z}{H}\right) \tag{6.9}$$

From this, we can get the temperature distribution rule of the coolant along the axial direction of the reactor core

$$T_c(z) = T_{ci} + \frac{q_{max} H}{G_c c_{p,c}\pi}\left(1 - \cos\frac{\pi z}{H}\right) \tag{6.10}$$

At the inlet and outlet of the reactor core, the coolant temperature is T_{ci}, T_{co}, then

$$\begin{cases} T_c(0) = T_{ci} & (z = 0) \\ T_c(H) = T_{ci} + \dfrac{2q_{max} H}{G_c c_{p,c}\pi} = T_{co} & (z = H) \end{cases} \tag{6.11}$$

Therefore, the temperature rise of the coolant flowing through the reactor core can be expressed as

$$\Delta T_c = T_{co} - T_{ci} = \frac{2q_{max} H}{G_c c_{p,c}\pi} \tag{6.12}$$

Substitute Eqs. (6.10) and (6.12) into Eq. (6.8) to get

$$T_u(z) = T_{ci} + \frac{\Delta T_c}{2}\left(1 - \cos\frac{\pi z}{H}\right) + R_t(z) q_{max} \sin\frac{\pi z}{H} \tag{6.13}$$

Substitute Eqs. (6.6) and (6.13) into Eq. (6.4) to get the exergy loss calculation formula for the process of converting nuclear energy into thermal energy in the reactor:

$$E_{xl,R_1} = T_0 \int_0^H \frac{q_{max} \sin\frac{\pi z}{H}}{T_{ci} + \frac{\Delta T_c}{2}\left(1 - \cos\frac{\pi z}{H}\right) + R_t(z) q_{max} \sin\frac{\pi z}{H}} dz \tag{6.14}$$

The exergy loss coefficient of this process is

$$\xi_{R_1} = \frac{E_{xl,R_1}}{E_{x,N}} = \frac{T_0}{Q_R}\int_0^H \frac{q_{max}\sin\frac{\pi z}{H}}{T_{ci} + \frac{\Delta T_c}{2}\left(1-\cos\frac{\pi z}{H}\right) + R_t(z)q_{max}\sin\frac{\pi z}{H}} dz \quad (6.15)$$

In order to find out the main factors affecting the exergy loss in the process of converting nuclear energy into thermal energy, further assume that the temperature of the nuclear fuel at each point in the reactor core is the same, that is, $T_u = \overline{T}_u$, then Eq. (6.4) can be transformed into

$$E_{xl,R_1} = T_0 \int_{V_R} \frac{1}{\overline{T}_u} dQ_R = \frac{T_0}{\overline{T}_u} Q_R \quad (6.16)$$

The corresponding exergy loss coefficient is

$$\xi_{R_1} = \frac{T_0}{\overline{T}_u} \quad (6.17)$$

From this, it can be seen that the exergy loss in the energy conversion process in the reactor depends on the temperature of the nuclear fuel in the reactor core. The higher the temperature, the smaller the exergy loss caused by the energy form conversion.

6.1.1.2 The Process of Transferring the Heat Released by the Nuclear Fuel in the Reactor to the Coolant

The exergy value obtained by the coolant flowing through the reactor core and absorbing heat is

$$E_{x,c} = \int_0^H \left[1 - \frac{T_0}{T_c(z)}\right] dQ_R(z) \quad (6.18)$$

The exergy loss in the heat exchange process is

$$E_{xl,R_2} = E_{x,R} - E_{x,c} = T_0 \int_0^H \left[\frac{1}{T_c(z)} - \frac{1}{T_u(z)}\right] dQ_R(z) \quad (6.19)$$

Substitute the temperature distribution expression (6.10) of the nuclear fuel along the axial direction of the reactor core and the temperature distribution expression (6.13) of the coolant along the axial direction into Eq. (6.19) to get

$$E_{\mathrm{xl,R_2}} = T_0 \int_0^H \frac{R_{\mathrm{t}}(z)\left(q_{\max} \sin \frac{\pi z}{H}\right)^2}{\left[T_{\mathrm{ci}} + \frac{\Delta T_{\mathrm{c}}}{2}\left(1 - \cos \frac{\pi z}{H}\right)\right]\left[T_{\mathrm{ci}} + \frac{\Delta T_{\mathrm{c}}}{2}\left(1 - \cos \frac{\pi z}{H}\right) + R_{\mathrm{t}}(z) q_{\max} \sin \frac{\pi z}{H}\right]} \mathrm{d}z \tag{6.20}$$

Then the exergy loss coefficient of this process is

$$\xi_{\mathrm{R_2}} = \frac{T_0}{Q_{\mathrm{R}}} \int_0^H \frac{R_{\mathrm{t}}(z)\left(q_{\max} \sin \frac{\pi z}{H}\right)^2}{\left[T_{\mathrm{ci}} + \frac{\Delta T_{\mathrm{c}}}{2}\left(1 - \cos \frac{\pi z}{H}\right)\right]\left[T_{\mathrm{ci}} + \frac{\Delta T_{\mathrm{c}}}{2}\left(1 - \cos \frac{\pi z}{H}\right) + R_{\mathrm{t}}(z) q_{\max} \sin \frac{\pi z}{H}\right]} \mathrm{d}z \tag{6.21}$$

In order to identify the main factors affecting the loss of heat transfer process in the pile, it is further assumed that the temperature distribution of the core fuel and coolant in the pile is uniform, that is, $T_{\mathrm{u}} = \overline{T}_{\mathrm{u}}$, $T_{\mathrm{c}} = \overline{T}_{\mathrm{c}}$, then formula (6.19) can be transformed into

$$E_{\mathrm{xl,R_2}} = T_0 \left(\frac{1}{\overline{\overline{T}}_{\mathrm{c}}} - \frac{1}{\overline{\overline{T}}_{\mathrm{u}}} \right) Q_{\mathrm{R}} \tag{6.22}$$

The corresponding exergy loss coefficient is

$$\xi_{\mathrm{R_2}} = T_0 \left(\frac{1}{\overline{\overline{T}}_{\mathrm{c}}} - \frac{1}{\overline{\overline{T}}_{\mathrm{u}}} \right) \tag{6.23}$$

From the above formula, it can be seen that the loss of heat exchange process in the pile mainly depends on the heat transfer temperature difference between the nuclear fuel and the coolant. The larger the temperature difference, the greater the loss of the heat transfer process.

The calculation of definite integrals (6.14), (6.20) or (6.15), (6.21) needs to be carried out with the help of numerical integration methods. Since the integration limits of the two definite integrals are the same, the Simpson integration method can be used to calculate them simultaneously, obtaining satisfactory results.

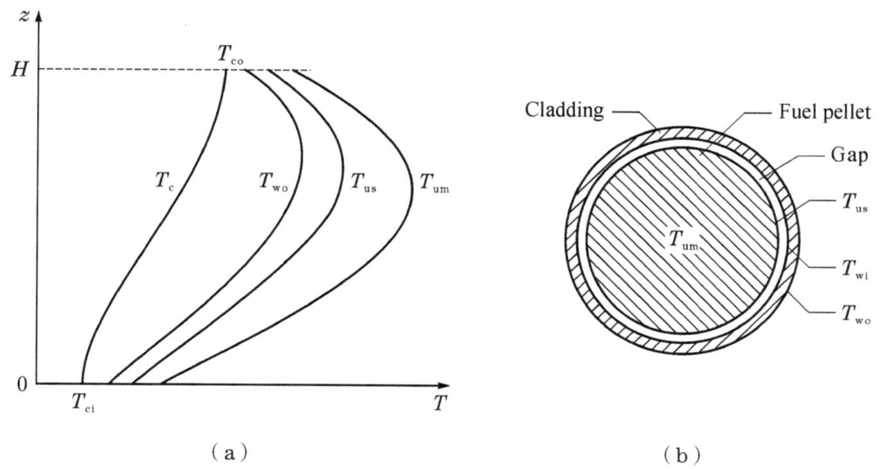

Fig. 6.2 Temperature distribution along the axial direction of fuel elements: (**a**) temperature distribution; (**b**) cross section of fuel element

6.1.1.3 Calculation of Heat Resistance of Core Heat Transfer

The heat released by the nuclear fuel in the core is transferred to the coolant flowing through the core, mainly including the following processes [42]:

1. Conduction of internal heat source in fuel pellet;
2. Conduction of gas gap between fuel pellet and cladding;
3. Conduction of cladding wall;
4. Convective heat exchange between the outer surface of the cladding and the coolant.

Assuming that the axial power distribution of the core follows a sinusoidal rule, the temperature distribution along the axial direction of the fuel element is shown in Fig. 6.2.

The heat transfer resistance between the nuclear fuel and the coolant at the axial position z of the core is

$$R_t(z) = 10^3\left[R_u(z) + R_g(z) + R_w(z) + R_c(z)\right] \quad \text{m}\cdot\text{K}/\text{kW} \tag{6.24}$$

where $R_u(z)$ is the thermal resistance of heat conduction inside the fuel pellet, m·K/W, $R_g(z)$ is the thermal resistance of gas gap conduction, m · K/W, $R_w(z)$ is the thermal resistance of cladding wall conduction, m · K/W, $R_c(z)$ is the heat exchange resistance between the outer surface of the cladding and the coolant, m · K/W.

The calculation methods of these four thermal resistances are discussed separately below.

(1) Convective heat exchange resistance $R_c(z)$ between the outer surface of the cladding and the coolant

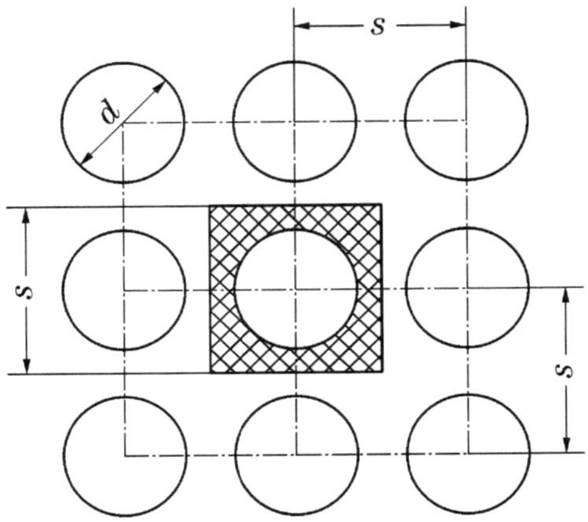

Fig. 6.3 Equivalent cooling channels for square lattice

The convective heat transfer coefficient α_c of the coolant flowing axially over the parallel fuel rod bundle in the core is

$$\alpha_c = C\frac{\lambda}{D_e}\mathrm{Re}^{0.8}\,\mathrm{Pr}^{1/3} \tag{6.25}$$

where C is a constant related to the lattice form of the fuel element, λ is the thermal conductivity of the coolant, W/(m · K), D_e is the equivalent diameter of the flow channel in the lattice, m, *Re* is the Reynolds number of the coolant, *Pr* is the Prandtl number of the coolant.

The fuel elements of the PWR core are usually arranged in a square lattice form, as shown in Fig. 6.3, then the equivalent diameter of the flow channel in the lattice is

$$D_e = \frac{4s^2 - \pi d^2}{\pi d} \tag{6.26}$$

where s is the lattice pitch, m; d is the diameter of the fuel element, m.

The Reynolds number of the coolant is

$$\mathrm{Re} = \frac{D_e \omega \rho}{\mu} \tag{6.27}$$

where ω is the flow speed of the coolant in the channel, m/s, ρ is the density of the coolant, kg/m^3, μ is the dynamic viscosity of the coolant, kg/(m · s).

The constant C related to the lattice form of the fuel element is

$$C=\begin{cases}0.042\dfrac{s}{d}-0.024 & \left(1.1\le\dfrac{s}{d}\le1.3\right)\\ 0.0333\left(1-\dfrac{\pi d^2}{4s^2}\right)+0.0127 & \left(\dfrac{s}{d}>1.3\right)\end{cases}\tag{6.28}$$

The heat transfer equation between the coolant and the outer surface of the fuel element at the axial position z of the core is

$$q(z)=\alpha_{\mathrm{c}}\pi d n_{\mathrm{u}}\left[t_{\mathrm{wo}}(z)-t_{\mathrm{c}}(z)\right]\tag{6.29}$$

where n_{u} is the number of fuel elements in the core, t_{wo} is the temperature of the outer surface of the fuel element, °C, t_{c} is the temperature of the core coolant, °C.

Therefore, the thermal resistance R_{c} of heat transfer between the core coolant and the outer surface of the fuel element is

$$R_{\mathrm{c}}=\frac{1}{\alpha_{\mathrm{c}}\pi d n_{\mathrm{u}}}\tag{6.30}$$

The temperature distribution on the outer surface of the fuel element is

$$t_{\mathrm{wo}}(z)=t_{\mathrm{c}}(z)+R_{\mathrm{w}}\cdot q(z)=t_{\mathrm{ci}}+\frac{\Delta T_{\mathrm{c}}}{2}\left(1-\cos\frac{\pi z}{H}\right)+R_{\mathrm{c}}q_{\max}\sin\frac{\pi z}{H}\tag{6.31}$$

(2) Thermal resistance R_{w} of the fuel element cladding wall

At the axial position z of the core, the heat conduction equation inside the cladding wall of the fuel element is

$$q(z)=\frac{2\pi\lambda_{\mathrm{w}}}{\ln\dfrac{d}{d_{\mathrm{wi}}}}\left[t_{\mathrm{wi}}(z)-t_{\mathrm{wo}}(z)\right]\tag{6.32}$$

The thermal resistance of the cladding wall is

$$R_{\mathrm{w}}(z)=\frac{\ln\dfrac{d}{d_{\mathrm{wi}}}}{2\pi\lambda_{\mathrm{w}}}\tag{6.33}$$

where d_{wi} is the inner diameter of the fuel element cladding, m, λ_w is the thermal conductivity of the cladding material, W/(m · K).

If the cladding material of the fuel element is zirconium-2 alloy, its thermal conductivity is

$$\lambda_{\mathrm{w}}=16.15+9.88\times10^{-3}t_{\mathrm{w}}+1.026\times10^{-5}t_{\mathrm{w}}^2\tag{6.34}$$

where t_w is the average temperature of the cladding, °C.

If it is zirconium-4 alloy, its thermal conductivity is

$$\lambda_{w} = 7.73 + 3.15\times10^{-2} t_{w} - 2.87\times10^{-5} t_{w}^{2} + 1.552\times10^{-8} t_{w}^{3} \tag{6.35}$$

The inner surface temperature of the cladding can be obtained from Eq. (6.32)

$$t_{wi}(z) = t_{wo}(z) + R_{w}\cdot q(z) = t_{ci} + \frac{\Delta T_{c}}{2}\left(1-\cos\frac{\pi z}{H}\right) + (R_{c}+R_{w})\cdot q_{max}\sin\frac{\pi z}{H} \tag{6.36}$$

The distribution of the temperature of the fuel element cladding along the radial direction is

$$t(r,z) = t_{wi}(z) + \frac{t_{wo}(z)-t_{wi}(z)}{\ln\frac{d}{d_{wi}}}\cdot\ln\frac{2r}{d_{wi}} \qquad (d_{wi}\le 2r\le d) \tag{6.37}$$

Let $\delta = \frac{d-d_{wi}}{2}$, the average temperature of the cladding is

$$t_{w}(z) = \frac{1}{\delta}\int_{\frac{d_{wi}}{2}}^{\frac{d}{2}} t(r,z)\,dr = t_{wi}(z) + \frac{t_{wo}(z)-t_{wi}(z)}{\ln(d/d_{wi})} \left[\ln\frac{d_{wi}}{2} - \frac{d\ln(d/2) - d_{wi}\ln(d_{wi}/2)}{d-d_{wi}} + 1\right] \tag{6.38}$$

Assume a value of $t_{wi}(z)$, calculate a corresponding $t_w(z)$ value from the above formula, substitute it into Eq. (6.34) or Eq. (6.35) to calculate λ_w, and then obtain a new $t'_{wi}(z)$ value from Eq. (6.36), if it satisfies

$$\left|t_{wi}(z) - t'_{wi}(z)\right| < 0.01$$

Then $t_{wi}(z)$ can be determined, otherwise, use $t'_{wi}(z)$ to substitute back for iterative calculation until the above requirement is met.

(3) Gap thermal resistance R_g

The gap layer between the fuel core and the cladding is very thin (about 0.08–0.09 mm), and its heat conduction problem can be approximated as the heat conduction of a homogeneous flat plate. The outer surface temperature is the inner surface temperature of the cladding, and the inner surface temperature is the temperature of the fuel surface. According to its heat conduction equation

$$q(z) = \frac{\lambda_{g}\pi(d_{wi}+d_{u})}{d_{wi}-d_{u}}\left[t_{us}(z) - t_{wi}(z)\right] \tag{6.39}$$

where λ_g is the thermal conductivity of the gap, W/(m · K); d_u is the diameter of the fuel core, m; t_{us} is the average temperature of the fuel core surface, °C.

Table 6.1 The value of experimental constant in the formula for calculating helium thermal conductivity

References and dates of publication	GE-TM-66.7-91,966	WAPD-TM-618,1970
$C_1 \times 10^6$	0.396	0.25
C_2	0.645	0.72

The thermal conductivity of inert gas is

$$\lambda_g = C_1 \left(t_g + 273.15\right)^{C_2} \tag{6.40}$$

where C_1, C_2—experimental constants.

If the inert gas filled in the gap is helium, the values of C_1 and C_2 are shown in Table 6.1.

From this, the expression for the gap thermal resistance can be obtained

$$R_g = \frac{d_{wi} - d_u}{\lambda_g \pi \left(d_{wi} + d_u\right)} \tag{6.41}$$

Then the outer surface temperature of the fuel core is

$$t_{us}(z) = t_{wi}(z) + R_g \cdot q(z) = t_{ci} + \frac{\Delta T_c}{2}\left(1 - \cos\frac{\pi z}{H}\right) + \left(R_c + R_w + R_g\right) \cdot q_{max} \sin\frac{\pi z}{H} \tag{6.42}$$

The average temperature of the gap layer is

$$t_g(z) = \frac{t_{wo}(z) + t_{wi}(z)}{2} \tag{6.43}$$

The expression for the gap thermal conductivity $\lambda_g = f(t_g)$ is given in [40]. Use a similar iterative method as described above to determine the values of t_g and R_g.

(4) Thermal resistance $R_u(z)$ within the fuel pellet

The heat conduction within the fuel pellet can be approximated as a one-dimensional steady-state heat conduction problem with a uniform internal heat source, and its heat conduction equation is

$$q(z) = 4\pi\lambda_u \left[t_{um}(z) - t_{us}(z)\right] \tag{6.44}$$

where λ_u is the thermal conductivity of the fuel pellet, W/(m · K); t_{um} is the average temperature at the center of the fuel pellet, °C.

The nuclear fuel commonly used in PWR NPPs is UO_2 ceramic fuel, whose thermal conductivity varies significantly with temperature. To reduce the error caused by calculating its thermal conductivity with the average temperature, the concept of integral thermal conductivity is introduced. Knowing the outer surface temperature of the fuel pellet t_{us}, its corresponding integral thermal conductivity is $\int_0^{t_{us}} \lambda_u(t)\mathrm{d}t$. Assuming that the integral thermal conductivity corresponding to t_{um} is y, then Eq. (6.44) should be written as

$$y - \int_0^{t_{us}} \lambda_u(t)\mathrm{d}t = \frac{q(z)}{4\pi} \tag{6.45}$$

The integral thermal conductivity of UO_2 fuel can be obtained from Table 6.2 according to the fuel temperature.

It can also be calculated according to the following fitting relationship:

$$\int_0^{t} \lambda_u(t)\mathrm{d}t = 1042.336 + 4.106033t - 4.821326 \times 10^{-4} t^2 \quad \mathrm{W/m} \tag{6.46}$$

where t is the temperature of UO_2 fuel, °C.

From the above two equations, the value of y can be obtained. To find the corresponding temperature value t_{um}, let

$$y = at^2 + bt + c$$

This equation is equivalent to Eq. (6.46), where y, a, b, and c are known values. Transform this equation into

$$f(t) = at^2 + bt + (c - y) \tag{6.47}$$

Use Newton's iteration method to find the real root t that satisfies the above equation, which is t_{um}. The first derivative of the above equation is

$$f'(t) = 2at + b \tag{6.48}$$

Newton's iteration formula is

$$t_{k+1} = t_k - \frac{f(t_k)}{f'(t_k)} \tag{6.49}$$

After a finite number of iterative calculations, if

$$|t_{k+1} - t_k| < 0.01$$

Then $t_{um} = t_{k+1}$, and the expression for thermal resistance R_u is obtained from Eq. (6.44)

Table 6.2 Integrated thermal conductivity of UO_2 fuel at different temperatures

Fuel temperature (°C)	200	400	600	800	1000	1200	1400	1600	1800	2000	2200	2400	2800
$\int_0^t \lambda_u(t)\mathrm{d}t$/(W/m)	1544	2642	3497	4204	4806	5341	5840	6285	6860	7180	7570	8017	9000

$$R_u = \frac{1}{4\pi\lambda_u} = \frac{t_{um}(z) - t_{us}(z)}{q(z)} \tag{6.50}$$

6.1.2 Pump

There are various types of pumps used in the thermal system of a NPP, such as the main coolant pump, feedwater pump, circulating water pump, condensate pump, etc. According to the form of the prime mover, they can be divided into electric pumps and steam-driven pumps.

6.1.2.1 Electric Pump

For an electric pump, the driving motor dissipates part of the electrical and mechanical energy as heat during operation due to copper loss, iron loss, and mechanical friction in the armature winding, which is taken away by the equipment cooling water or dissipated into the surrounding environment. The resulting exergy loss is

$$E_{xl,m} = N_m - N_p \tag{6.51}$$

During the operation of the pump, due to hydraulic, flow, and mechanical losses, part of the mechanical energy is dissipated as heat, but this dissipated energy is basically absorbed by the fluid. This part of the exergy loss is

$$E_{xl,p} = \left(N_p - N_e\right)\frac{T_0}{T_f} \tag{6.52}$$

Therefore, the exergy loss of the electric pump is

$$E_{xl,mp} = E_{xl,m} + E_{xl,p} = \left[\frac{1}{\eta_m \eta_p} - \frac{1}{\eta_p} + \left(\frac{1}{\eta_p} - 1\right)\frac{T_0}{T_f}\right]\frac{G_f H}{102} \tag{6.53}$$

6.1.2.2 Steam-Driven Pump

For a steam-driven pump, because the steam turbine works with internal steam expansion and power transmission processes, the exergy loss inside the steam turbine includes two parts:

(1) Exergy loss during the steam expansion process;

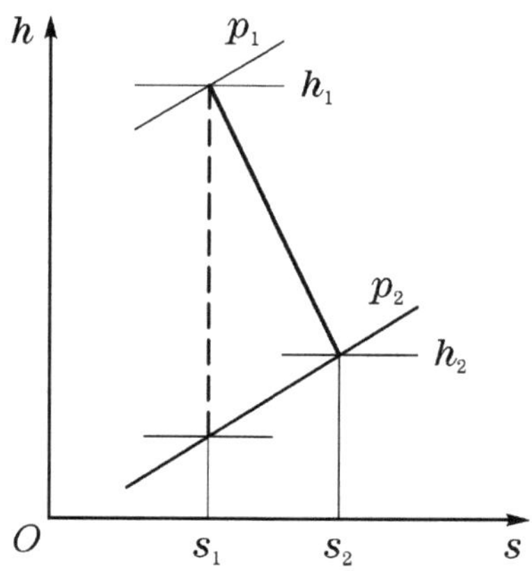

Fig. 6.4 Internal thermal process of steam turbine

Figure 6.4 shows the steam expansion process inside the steam turbine. If the steam flow consumed by the steam turbine is G_s, then the internal power of the steam turbine is

$$N_i = G_s (h_1 - h_2) \tag{6.54}$$

The exergy released by the steam inside the steam turbine is

$$E_{x,s} = G_s [(h_1 - h_2) - T_0 (s_1 - s_2)] \tag{6.55}$$

The exergy loss caused by the irreversibility of the steam expansion process inside the steam turbine is

$$E_{xl,t_1} = E_{x,s} - N_i = G_s T_0 (s_2 - s_1) = G_s T_0 \Delta s \tag{6.56}$$

where Δs is the exergy increase during the steam expansion process inside the steam turbine, $\Delta s = s_2 - s_1$.

(2) Exergy loss in power transmission process.

In the process of power transmission, mechanical energy loss is caused by the mechanical friction of gear reducers and turbine bearings. Assuming the mechanical efficiency of the turbine is η_{mt}, this part of the exergy loss is

$$E_{xl,t_2} = N_i - N_p = \left(\frac{1}{\eta_{mt}} - 1 \right) N_p \tag{6.57}$$

Therefore, the exergy loss of the turbine is

$$E_{xl,t} = E_{xl,t_1} + E_{xl,t_2} = G_s T_0 \Delta s + \left(\frac{1}{\eta_{mt}} - 1 \right) N_p \tag{6.58}$$

Considering the loss of the pump, the exergy loss of the steam-driven pump is

$$E_{\mathrm{xl,tp}} = E_{\mathrm{xl,t}} + E_{\mathrm{xl,p}} = G_{\mathrm{s}} T_0 \Delta s + \left[\frac{1}{\eta_{\mathrm{mt}} \eta_{\mathrm{p}}} - \frac{1}{\eta_{\mathrm{p}}} + \left(\frac{1}{\eta_{\mathrm{p}}} - 1 \right) \frac{T_0}{T_{\mathrm{f}}} \right] \frac{G_{\mathrm{f}} H}{102} \tag{6.59}$$

6.1.3 Heat Exchange Equipment

The main heat exchange equipment in the thermal system of a NPP includes steam generators, condensers, steam reheaters, feedwater heaters, etc. The working processes of these devices are different, but their essence is the same.

6.1.3.1 Steam Generator

The steam generators used in NPPs are divided into natural circulation type and once through type. The natural circulation steam generator can only produce saturated steam, and the heat absorption process of the secondary side working fluid only includes the preheating section and the evaporation section; while the once through steam generator can produce superheated steam, the heat absorption process of the secondary side working fluid includes the preheating section, evaporation section, and superheating section.

In order to make the established steam generator exergy analysis model more representative, the establishment process of the exergy analysis model is introduced here with the once through steam generator as an example.

Figure 6.5 shows the temperature changes of the primary and secondary side working fluids in the heat transfer process of the steam generator.

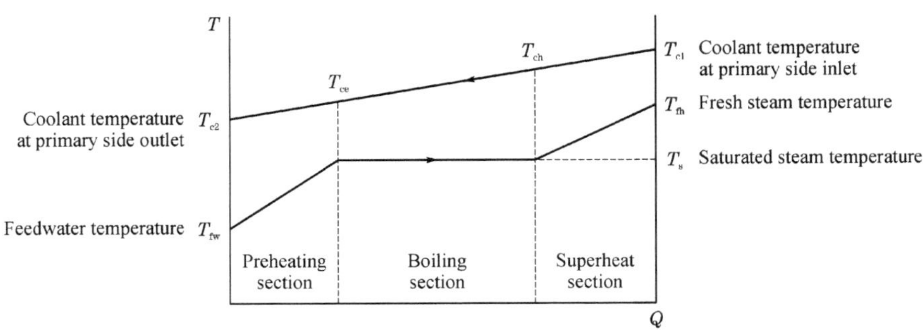

Fig. 6.5 Heat transfer process between the primary and secondary sides of a once through steam generator

In Fig. 6.5, T_{ce} is the coolant temperature in the steam generator corresponding to the boundary point of the secondary side preheating section and evaporation section; T_{ch} is the coolant temperature in the steam generator corresponding to the boundary point of the secondary side evaporation section and superheating section; T_s is the secondary side evaporation temperature.

Preheating Section

The exergy of the primary side coolant heat release is

$$
\begin{aligned}
E'_{x,c} &= \int_{T_{c2}}^{T_{ce}} \left(1-\frac{T_0}{T_c}\right) G_c c_{p,c} \mathrm{d}T_c = G_c c_{p,c}\left(T_{ce}-T_{c2}\right) - G_c c_{p,c} T_0 \int_{T_{c2}}^{T_{ce}} \frac{1}{T_c} \mathrm{d}T_c \\
&= G_c c_{p,c} \Delta T'_c - G_c c_{p,c} T_0 \ln \frac{T_{ce}}{T_{c2}}
\end{aligned}
\tag{6.60}
$$

where $\Delta T'_c$ is the temperature drop of the primary side coolant in the preheating section, $\Delta T'_c = T_{ce} - T_{c2}$.

The exergy of the secondary side working fluid heat absorption is

$$
\begin{aligned}
E'_{x,f} &= \int_{T_{fw}}^{T_s} \left(1-\frac{T_0}{T_f}\right) G_f c_{p,f} \mathrm{d}T_f = G_f c_{p,f}\left(T_s-T_{fw}\right) - G_f c_{p,f} T_0 \int_{T_{fw}}^{T_s} \frac{1}{T_f} \mathrm{d}T_f \\
&= G_f c_{p,f} \Delta T'_f - G_f c_{p,f} T_0 \ln \frac{T_s}{T_{fw}}
\end{aligned}
\tag{6.61}
$$

where $\Delta T'_f$ is the temperature rise of the secondary side working fluid in the preheating section, $\Delta T'_f = T_s - T_{fw}$.

Therefore, the exergy loss of the preheating section heat exchange process is

$$
\begin{aligned}
E'_{xl} &= E'_{x,c} - E'_{x,f} \\
&= \left(G_c c_{p,c} \Delta T'_c - G_f c_{p,f} \Delta T'_f\right) + \left(G_f c_{p,f} T_0 \ln \frac{T_s}{T_{fw}} - G_c c_{p,c} T_0 \ln \frac{T_{ce}}{T_{c2}}\right)
\end{aligned}
\tag{6.62}
$$

If the heat dissipation loss of the preheating section is ignored, the primary and secondary side working fluids have the following heat balance relationship

$$
G_c c_{p,c} \Delta T'_c = G_f c_{p,f} \Delta T'_f \tag{6.63}
$$

From this, it can be obtained

$$
G_f c_{p,f} = G_c c_{p,c} \frac{\Delta T'_c}{\Delta T'_f} \tag{6.64}
$$

Therefore, Eq. (6.62) becomes

$$E_{\mathrm{xl}}^{'}=G_{\mathrm{c}}c_{p,\mathrm{c}}T_{0}\left(\frac{\Delta T_{\mathrm{c}}^{'}}{\Delta T_{\mathrm{f}}^{'}}\ln\frac{T_{\mathrm{s}}}{T_{\mathrm{fw}}}-\ln\frac{T_{\mathrm{ce}}}{T_{\mathrm{c2}}}\right)=G_{\mathrm{c}}c_{p,\mathrm{c}}T_{0}\ln\left[\left(\frac{T_{\mathrm{s}}}{T_{\mathrm{fw}}}\right)^{C_{1}}\frac{T_{\mathrm{c2}}}{T_{\mathrm{ce}}}\right] \tag{6.65}$$

where $C_1=\frac{\Delta T_{\mathrm{c}}^{'}}{\Delta T_{\mathrm{f}}^{'}}$.

The corresponding exergy loss coefficient is

$$\xi_{3}^{'}=\frac{E_{\mathrm{xl},1}}{E_{0}}=\frac{G_{\mathrm{c}}c_{p,\mathrm{c}}T_{0}}{G_{\mathrm{c}}c_{p,\mathrm{c}}\Delta T_{\mathrm{c}}}\ln\left[\frac{T_{\mathrm{c2}}}{T_{\mathrm{ce}}}\left(\frac{T_{\mathrm{s}}}{T_{\mathrm{fw}}}\right)^{C_{1}}\right]=\frac{T_{0}}{\Delta T_{\mathrm{c}}}\ln\left[\frac{T_{\mathrm{c2}}}{T_{\mathrm{ce}}}\left(\frac{T_{\mathrm{s}}}{T_{\mathrm{fw}}}\right)^{C_{1}}\right] \tag{6.66}$$

Evaporation Section or Vaporization Section

The exergy of the primary side coolant heat release is

$$\begin{aligned}E_{x,\mathrm{c}}^{''}&=\int_{T_{\mathrm{ce}}}^{T_{\mathrm{ch}}}\left(1-\frac{T_{0}}{T_{\mathrm{c}}}\right)G_{\mathrm{c}}c_{p,\mathrm{c}}\mathrm{d}T_{\mathrm{c}}=G_{\mathrm{c}}c_{p,\mathrm{c}}\left(T_{\mathrm{ch}}-T_{\mathrm{ce}}\right)-G_{\mathrm{c}}c_{p,\mathrm{c}}T_{0}\int_{T_{\mathrm{ce}}}^{T_{\mathrm{ch}}}\frac{1}{T_{\mathrm{c}}}\mathrm{d}T_{\mathrm{c}}\\&=G_{\mathrm{c}}c_{p,\mathrm{c}}\Delta T_{\mathrm{c}}^{''}-G_{\mathrm{c}}c_{p,\mathrm{c}}T_{0}\ln\frac{T_{\mathrm{ch}}}{T_{\mathrm{ce}}}\end{aligned} \tag{6.67}$$

where $\Delta T_{\mathrm{c}}^{''}$ is the temperature drop of the primary side coolant in the evaporation section, $\Delta T_{\mathrm{c}}^{''}=T_{\mathrm{ch}}-T_{\mathrm{ce}}$.

If the heat dissipation loss of the heat transfer of the primary and secondary side working fluids in the evaporation section is ignored, the heat absorbed by the secondary side working fluid is

$$Q''=G_{\mathrm{c}}c_{p,\mathrm{c}}\Delta T_{\mathrm{c}}^{''} \tag{6.68}$$

The exergy value of this part of the heat is

$$E_{x,\mathrm{f}}^{''}=Q''\left(1-\frac{T_{0}}{T_{\mathrm{s}}}\right)=Q''-G_{\mathrm{c}}c_{p,\mathrm{c}}T_{0}\frac{\Delta T_{\mathrm{c}}^{''}}{T_{\mathrm{s}}} \tag{6.69}$$

Therefore, the exergy loss of the vaporization section is

$$E_{\mathrm{xl}}^{''}=E_{x,\mathrm{c}}^{''}-E_{x,\mathrm{f}}^{''}=G_{\mathrm{c}}c_{p,\mathrm{c}}T_{0}\left(\frac{\Delta T_{\mathrm{c}}^{''}}{T_{\mathrm{s}}}-\ln\frac{T_{\mathrm{ch}}}{T_{\mathrm{ce}}}\right) \tag{6.70}$$

The corresponding exergy loss coefficient is

$$\xi_3'' = \frac{E_{xl}''}{E_0} = \frac{T_0}{\Delta T_c}\left(\frac{\Delta T_c''}{T_s} - \ln\frac{T_{ch}}{T_{ce}}\right) \tag{6.71}$$

Superheating Section

Similar to the preheating section, its exergy loss is

$$E_{xl}''' = G_c c_{p,c} T_0 \ln\left[\left(\frac{T_{fh}}{T_s}\right)^{C_3} \frac{T_{ch}}{T_{c1}}\right] \tag{6.72}$$

where $C_3 = \frac{\Delta T_c'''}{\Delta T_f'''}$. $\Delta T_c'''$ is the temperature drop of the primary side coolant in the superheating section, $\Delta T_c''' = T_{c1} - T_{ch}$; $\Delta T_f'''$ is the temperature rise of the secondary side working fluid in the superheating section, $\Delta T_f''' = T_{fh} - T_s$.

The corresponding exergy loss coefficient is

$$\xi_3''' = \frac{T_0}{\Delta T_c}\ln\left[\frac{T_{ch}}{T_{c1}}\left(\frac{T_{fh}}{T_s}\right)^{C_3}\right] \tag{6.73}$$

Therefore, the exergy loss coefficient of the once through steam generator is

$$\xi_3 = \xi_3' + \xi_3'' + \xi_3''' = \frac{T_0}{\Delta T_c}\left\{\ln\left[\frac{T_{c2}}{T_{c1}}\left(\frac{T_s}{T_{fw}}\right)^{C_1}\left(\frac{T_{fh}}{T_s}\right)^{C_3}\right] + \frac{\Delta T_c''}{T_s}\right\} \tag{6.74}$$

For the U-tube natural circulation steam generator, only the preheating section and the evaporation section exist in the heat transfer process, so its exergy loss and exergy loss coefficient are respectively

$$E_{xl,3} = E_{xl}' + E_{xl}'' = G_c c_{p,c} T_0 \left\{\ln\left[\frac{T_{c2}}{T_{c1}}\left(\frac{T_s}{T_{fw}}\right)^{C_1}\right] + \frac{\Delta T_c''}{T_s}\right\} \tag{6.75}$$

$$\xi_3 = \xi_3' + \xi_3'' = \frac{T_0}{\Delta T_c}\left\{\ln\left[\frac{T_{c2}}{T_{c1}}\left(\frac{T_s}{T_{fw}}\right)^{C_1}\right] + \frac{\Delta T_c''}{T_s}\right\} \tag{6.76}$$

In fact, the natural circulation steam generator not only has heat dissipation to the surrounding environment during operation, but also discharges impurities in the secondary loop water from the steam generator through the sewage system. Therefore, the actual exergy loss of the natural circulation steam generator during operation is greater than the calculated value of formula (6.75). In this case, the method of state exergy can be used to determine the actual exergy loss of the steam generator:

$$E_{\mathrm{xl},3} = \left(E_{x,\mathrm{c1}} - E_{x,\mathrm{c2}}\right) - \left(E_{x,\mathrm{fh}} + E_{x,\mathrm{pw}} - E_{x,\mathrm{fw}}\right) \tag{6.77}$$

where $E_{x,\,\mathrm{c1}}$, $E_{x,\,\mathrm{c2}}$ are the exergy values of the coolant at the inlet and outlet of the primary side of the steam generator respectively; $E_{x,\,\mathrm{fh}}$, $E_{x,\,\mathrm{pw}}$, $E_{x,\,\mathrm{fw}}$ are the exergy values of the new steam, sewage water, and feed water on the secondary side of the steam generator respectively.

6.1.3.2 Surface Heat Exchanger

The exergy of the heat released by the heating medium during the operation of the heat exchanger is

$$E_{x,\mathrm{h}} = G_{\mathrm{h}}\left[\left(h_{\mathrm{h1}} - h_{\mathrm{h2}}\right) - T_0\left(s_{\mathrm{h1}} - s_{\mathrm{h2}}\right)\right] = Q_{\mathrm{h}} - G_{\mathrm{h}} T_0 \Delta s_{\mathrm{h}} \tag{6.78}$$

The exergy obtained from the endothermic medium is

$$E_{x,\mathrm{s}} = G_{\mathrm{s}}\left[\left(h_{\mathrm{s1}} - h_{\mathrm{s2}}\right) - T_0\left(s_{\mathrm{s1}} - s_{\mathrm{s2}}\right)\right] = Q_{\mathrm{s}} - G_{\mathrm{s}} T_0 \Delta s_{\mathrm{s}} \tag{6.79}$$

The exergy loss during the heat exchange process is

$$\begin{aligned} E_{\mathrm{xl,hs}} &= E_{x,\mathrm{h}} - E_{x,\mathrm{s}} = \left(Q_{\mathrm{h}} - Q_{\mathrm{s}}\right) \\ &\quad - T_0\left(G_{\mathrm{h}}\Delta s_{\mathrm{h}} - G_{\mathrm{s}}\Delta s_{\mathrm{s}}\right) = \Delta Q - T_0\left(G_{\mathrm{h}}\Delta s_{\mathrm{h}} - G_{\mathrm{s}}\Delta s_{\mathrm{s}}\right) \end{aligned} \tag{6.80}$$

If the heat dissipation loss is ignored, $\Delta Q = 0$, then the above formula simplifies to

$$E_{\mathrm{xl,hs}} = T_0\left(G_{\mathrm{h}}\Delta s_{\mathrm{h}} - G_{\mathrm{s}}\Delta s_{\mathrm{s}}\right) \tag{6.81}$$

For condensers that use seawater as circulating cooling water, the exergy of the heat released by the waste steam from the turbine and the condensate discharged into the condenser, part of it is lost during the heat transfer process, and most of it is absorbed by the circulating cooling seawater and taken out of the thermal cycle, so it is generally considered that the exergy loss in the condenser is equal to the exergy released by the heating medium.

6.1.3.3 Mixed Heat Exchanger

The mixed heat exchanger in the NPP is mainly the deaerator, where the heating steam and condensate are mixed, and the thermal deoxygenation is achieved by heating the condensate to a saturated state.

The exergy value of the deaerator inlet condensate is

$$E_{x,\mathrm{fwi}} = G_{\mathrm{fwi}}\left[\left(h_{\mathrm{fwi}} - h_0\right) - T_0\left(s_{\mathrm{fwi}} - s_0\right)\right] \tag{6.82}$$

The exergy value of the heating steam is

$$E_{x,\mathrm{hs}} = G_{\mathrm{hs}}\left[\left(h_{\mathrm{hs}} - h_0\right) - T_0\left(s_{\mathrm{hs}} - s_0\right)\right] \tag{6.83}$$

The exergy value of the deaerator exhaust steam is

$$E_{x,\mathrm{ps}} = G_{\mathrm{ps}}\left[\left(h_{\mathrm{ps}} - h_0\right) - T_0\left(s_{\mathrm{ps}} - s_0\right)\right] \tag{6.84}$$

The exergy value of the deaerator outlet feed water is

$$E_{x,\mathrm{fw}} = G_{\mathrm{fw}}\left[\left(h_{\mathrm{fw}} - h_0\right) - T_0\left(s_{\mathrm{fw}} - s_0\right)\right] \tag{6.85}$$

The exergy loss of the deaerator during operation is

$$E_{\mathrm{xl,d}} = E_{x,\mathrm{fwi}} + E_{x,\mathrm{hs}} - E_{x,\mathrm{fw}} \tag{6.86}$$

6.1.4 Steam Turbine Generator

From a thermodynamic point of view, the operation of a steam turbine generator includes the following three energy processes:

(1) The adiabatic expansion process of steam inside the turbine;
(2) The process of transferring mechanical energy output by the turbine to the generator;
(3) The process of converting mechanical energy into electrical energy in the generator.

The exergy loss of the steam expansion process inside the turbine is

$$E_{\mathrm{xl,i}} = E_{\mathrm{xl,ih}} + E_{\mathrm{xl,il}} \tag{6.87}$$

The exergy loss inside the high-pressure turbine is

$$E_{\mathrm{xl,ih}} = E_{x,\mathrm{h1}} - E_{x,\mathrm{eh}} - E_{x,\mathrm{h2}} - N_{\mathrm{ih}} \tag{6.88}$$

where $E_{x,\mathrm{h1}}$ is the exergy of the steam inlet, $E_{x,\mathrm{h1}} = G_{s,\mathrm{h1}}\,[h_{\mathrm{h1}} - h_0 - T_0\,(s_{\mathrm{h1}} - s_0)]$; $E_{x,\mathrm{eh}}$ is the exergy of the steam extraction, $E_{x,\mathrm{eh}} = \sum_{k=1}^{m} G_{\mathrm{seh},k}\left[h_{\mathrm{eh},k} - h_0 - T_0\left(s_{\mathrm{seh},k} - s_0\right)\right]$; $E_{x,}$

$_{h2}$ is the exergy of the steam outlet, $E_{x,\mathrm{h2}} = G_{s,\mathrm{h2}}\,[h_{\mathrm{h2}} - h_0 - T_0\,(s_{\mathrm{h2}} - s_0)]$; N_{ih} is the internal power of the high-pressure steam turbine.

The exergy loss inside the low-pressure steam turbine is

$$E_{\mathrm{xl,il}} = E_{x,\mathrm{L1}} - E_{x,\mathrm{el}} - E_{x,\mathrm{L2}} - N_{\mathrm{il}} \tag{6.89}$$

The meanings of the parameters in the formula are the same as before.

Therefore, formula (6.87) can be written in the following form

$$E_{\mathrm{xl,i}} = \left(E_{x,\mathrm{h1}} + E_{x,\mathrm{L1}} - E_{x,\mathrm{e}} - E_{x,\mathrm{h2}} - E_{x,\mathrm{L2}}\right) - N_{\mathrm{i}} \tag{6.90}$$

where $E_{x,\mathrm{e}} = E_{x,\mathrm{eh}} + E_{x,\mathrm{el}}$; N_{i} is the internal power of the steam turbine.

The exergy loss in the power transmission process is

$$E_{\mathrm{xl,mt}} = N_{\mathrm{i}} - N_{\mathrm{t}} \tag{6.91}$$

where N_{t} is the effective power transferred from the steam turbine to the generator.

The exergy loss in the process of converting mechanical energy into electrical energy inside the generator is

$$E_{\mathrm{xl,me}} = N_{\mathrm{t}} - N_{\mathrm{e}} \tag{6.92}$$

where N_{e} is the effective electrical power of the generator.

In summary, the expression for the exergy loss of the steam turbine generator can be obtained

$$E_{\mathrm{xl,te}} = E_{x,\mathrm{i}} + E_{\mathrm{xl,mt}} + E_{\mathrm{xl,me}} = \left(E_{x,\mathrm{h1}} - E_{x,\mathrm{h2}}\right) + \left(E_{x,\mathrm{L1}} - E_{x,\mathrm{L2}}\right) - E_{x,\mathrm{e}} - N_{\mathrm{e}} \tag{6.93}$$

6.1.5 Exergy Loss in NPPs

After determining the exergy loss of the main equipment of the thermal system of the PWR NPPs, in order to facilitate analysis and comparison, each exergy loss is expressed as an exergy loss coefficient, that is

$$\xi_k = \frac{E_{\mathrm{xl},k}}{Q_{\mathrm{R}}} \tag{6.94}$$

Therefore, the level of thermal system exergy losses in PWR NPPs can be expressed as

$$\xi = \sum_{k=1}^{n} \xi_k \tag{6.95}$$

The exergy efficiency of a NPP can be expressed as

Table 6.3 Exergy analysis results of thermal system of PWR NPPs

No.	Calculation item	Exergy loss coefficient (%)		Main cause of exergy loss
		Qinshan No. 1 NPP	Daya Bay NPP	
1	Energy conversion process in reactor	25.16	21.09	Energy form changes, energy quality decreases
2	Heat transfer process in reactor	25.47	29.61	Heat transfer temperature difference
3	Steam generator	3.51	3.37	Heat transfer temperature difference
4	Steam reheater	0.37	0.37	
5	Feed water heating system	1.24	1.01	
6	Condenser	3.95	4.26	Irreversible heat transfer and energy removal power cycle
7	Steam turbo-generator	7.81	5.52	Irreversible adiabatic expansion process, bearing friction and control system energy consumption, electrical loss
8	Main coolant pump	0.09	0.08	Electrical loss, mechanical friction and fluid loss
9	Feed water pump	0.09	0.23	
10	Motor-driven condensate pump	0.01	0.01	
11	Main air extractor	0.05	0.07	The heat transfer, expansion and mixing processes are irreversible
12	Moisture separator	0.17	0.08	Friction, heat dissipation and leakage
13	Main steam line	0.46	0.26	
14	Exhaust steam line	0.0	0.02	
Total		68.38	65.99	
Plant net efficiency (%)		31.62	34.00	

$$\eta_{ex} = 1 - \xi = 1 - \sum_{k=1}^{n} \xi_k \tag{6.96}$$

6.2 Examples of Exergy Analysis of the Thermal System of a NPP

Based on the above mathematical model, the thermal systems of the Qinshan Phase I and Daya Bay NPPs are analyzed for exergy, and the calculation results are shown in Tables 6.3 and 6.4 [43].

As shown by the values in the above two tables, in the thermal system of the PWR NPP, the reactor is the equipment with the largest energy loss, and its exergy loss accounts for about 50% of the total nuclear energy exergy; the exergy loss of the heat exchange equipment is second only to the reactor, about 9%; the exergy loss

Table 6.4 The ratio of exergy losses for several main types of equipment in thermal systems

No.	Equipment type		Exergy loss coefficient (%)		Exergy loss ratio (%)	
			Qinshan No. 1 NPP	Daya Bay NPP	Qinshan No. 1 NPP	Daya Bay NPP
1	Reactor		50.63	50.70	74.05	76.83
2	Heat exchange equipment	Steam generator	9.07	9.02	13.26	13.67
		Steam reheater				
		Feed water heater				
		Condenser				
3	Steam turbo-generator		7.81	5.52	11.42	8.37
4	Pump	Main coolant pump	0.19	0.32	0.28	0.49
		Feed water pump				
		Condensate pump				
5	Other	Main steam line	0.68	0.43	1.00	0.65
		Moisture separator				
		Main air extractor				
		Exhaust steam line				

of the steam turbine generator is about 5.52%–7.81%, and the sum of the exergy losses of the pump and other equipment is less than 1%.

The exergy loss of the reactor accounts for approximately 74.05%–76.83% of the total exergy loss of the thermal system, which is completely different from the conclusion that the energy loss in the reactor obtained from the energy balance analysis is zero and the heat exchange equipment is the type of equipment with the largest energy loss in the thermal system. Although the amount of energy remains basically unchanged during the energy conversion and transfer process in the reactor, the significant decrease in the ability of energy to do work is caused by the change in the form of energy and the existence of a large heat transfer temperature difference. That is, the energy loss in the reactor is essentially a loss of energy in terms of quality. For heat exchange equipment, its exergy loss only accounts for about 13% of the total system exergy loss. According to the results of the energy balance analysis, the amount of heat dissipation loss in the heat exchange equipment exceeds 97% of the total system heat dissipation loss, and it is mainly caused by the heat discharged into the environment in the condenser. Although the amount of heat dissipation loss in the condenser accounts for about 65% of the reactor thermal power, its exergy loss only accounts for about 4% of the total nuclear exergy.

This is because the temperature and pressure of the spent steam and drainage discharged into the condenser are very low, and the heat released has a very low ability to do work, that is, it has less exergy, so its share of exergy loss is not large. Therefore, from the perspective of exergy analysis, the main reason for the energy loss of the thermal system of the PWR NPP is the existence of a large number of irreversible factors in the energy conversion and transfer process, which leads to a decrease in the ability of energy to do work. The loss of work ability caused by heat dissipation is not the main one. Therefore, the fundamental way to improve the thermal efficiency of the thermal system of the PWR NPP is to reduce the irreversibility of the energy use process, and the main object of improvement is the reactor, not as pointed out by the energy balance analysis, it is to try to reduce the energy loss in the condenser.

According to the results listed in Tables 6.3 and 6.4, a comparison of the exergy loss distribution of the Qinshan Phase I NPP and the Daya Bay NPP shows that, among the five types of equipment, except for the steam turbine generator, the exergy loss coefficients of the other types of equipment in the two NPPs are basically close. In other words, the difference in exergy efficiency between the Qinshan Phase I NPP and the Daya Bay NPP is mainly reflected in the exergy loss of the steam turbine generator.

Through the exergy analysis of the thermal system of the PWR NPP, the following conclusions can be drawn:

(1) In the thermal system of the PWR NPP, the equipment with the largest exergy loss is the reactor, not the condenser as pointed out by the energy balance analysis.
(2) The main reason for the energy loss in the thermal system is the existence of irreversible factors in the energy conversion, transfer, and distribution process, which leads to a decrease in the ability of energy to do work.
(3) The exergy analysis method attributes both the quantity and quality loss of energy in the thermal system of the PWR NPP to the loss of the ability of energy to do work, thus more comprehensively reflecting the imperfection of energy utilization in the thermal system and pointing out the direction for improving the thermal system and increasing exergy efficiency.

6.3 Exergy Optimization of the Thermal System of NPPs

The exergy analysis method is usually used to calculate and determine the distribution of exergy loss in the thermal system, so as to find the part with the largest exergy loss and point out the direction for improving the system and improving thermal economy. The exergy optimization method treats the thermal system as a whole, establishes a mathematical model of the objective function and constraint conditions through the analysis of the relationship between various parameters of the system, and then optimizes the entire thermal system to obtain the optimized values of various parameters under the optimal state or the best structure of the system.

The solution methods for exergy optimization can be roughly divided into the following two categories:

(1) Graphical method. The objective function and constraint equations are drawn into curves according to mathematical operations, and then the optimal points such as maximum or minimum values are intuitively determined based on the shape of the curves in the coordinate system and the restriction range of the constraint equations.
(2) Analytical method. Establish a mathematical model of the objective function and constraints, and obtain the maximum or minimum values and related optimization variables through mathematical operations.

The thermal system of a PWR NPP is large and complex, with various equipment in the system interconnected and influencing each other. According to the viewpoint of system engineering theory, the local optimum of the system is not necessarily the overall optimum, so the minimum exergy loss of a certain device in the thermal system of a NPP does not mean the highest system exergy efficiency.

Since the thermal system of a PWR NPP basically does not change with time when it operates stably under a certain condition, it can be assumed that there exists a set of independent thermal parameters that make the system exergy efficiency the highest. By using engineering optimization methods for analysis, it is possible to determine such a set of independent thermal parameters corresponding to the best system exergy efficiency.

6.3.1 Steps for Exergy Efficiency Optimization

The purpose of optimizing the thermal parameters of the NPP thermal system is to achieve the best system exergy efficiency, so the objective function of optimization can be determined as the lowest system exergy loss coefficient, expressed as

$$\xi = \min\left(\sum_{k=1}^{n}\xi_k\right) \tag{6.97}$$

The determination of optimization variables is the determination of independent variables in the thermal system. Although the thermal system of a PWR NPP is large and complex, the entire system is not "self-contained". For example, the pressure in the condenser is determined by the exhausted steam and drainage pumped into it and the flow rate of circulating seawater used as a cold source. The flow rate of circulating seawater can be adjusted by manually changing the power and speed of the circulating water pump, thereby controlling the pressure in the condenser. Similarly, the feedwater temperature and steam reheating temperature can also be adjusted and controlled within a certain range by changing the heating steam flow rate. According to the characteristics of the research object, the feedwater temperature t_{fw}, condenser pressure p_{cd}, and steam reheating temperature t_{sr2} are taken as optimization variables, expressed as

$$\overline{x} = \left(\overline{x}_1, \overline{x}_2, \overline{x}_3\right) = \left(t_{fw}, p_{cd}, t_{sr2}\right) \tag{6.98}$$

The changes in the above optimization variables are not arbitrary and are constrained by the performance of the relevant equipment in the thermal system. For example, the maximum feedwater temperature t_{fw} cannot exceed the main engine extraction steam temperature t_{se1}, which is the heating medium of the last stage feedwater heater; the saturation water temperature t_{cd} corresponding to the condenser pressure p_{cd} is always higher than the outlet temperature t_{swo} of the circulating seawater; the steam reheating temperature t_{sr2} is not lower than the outlet steam temperature t_{sr1} of the first stage reheater, and it is not higher than the heating steam temperature t_s; in addition, the dryness of the exhaust steam at the end stage of the low-pressure turbine x_{zl} cannot be lower than 0.88. Therefore, based on the characteristics of the research object, the following constraints can be proposed:

$$\begin{cases} g(1) = t_{se,1} - \Delta t_{fw} - t_{fw} \geq 0 \\ g(2) = t_{cd} - \Delta t_{cd} - t_{sw,o} \geq 0 \\ g(3) = t_{sr2} - t_{sr1} \geq 0 \\ g(4) = t_s - \Delta t_{sr2i} - t_{sr2} \geq 0 \\ g(5) = x_{zl} - 0.88 \geq 0 \end{cases} \tag{6.99}$$

There are many calculation methods for nonlinear programming, and they are developing rapidly. For solving optimization design problems with inequality constraints and less than 20 optimization variables, the compound form method that directly handles constraints is quite effective. The basic idea of this method is to take $2n$ vertices in the n-dimensional design space of nonlinear constraints to form a compound shape, then compare the function values of each vertex of the compound shape one by one, constantly discard the worst points, and replace them with new points that can improve the objective function and meet the constraint conditions, gradually approaching the optimal point. Although the compound form method has a large amount of calculation and slow convergence speed, this algorithm is stable in calculation, simple in program, and convenient to use, so it is used more in engineering calculations. The optimization calculation is only based on the thermal system of Qinshan NPP as the research object, and the optimization calculation results are shown in the following Table 6.5.

The optimization calculation results show that compared with the original plan, the exergy efficiency of the PWR NPP has increased by 0.412 3%, and the decrease is only 0.602 2%. Therefore, from the perspective of exergy efficiency, the operating parameters of the Qinshan NPP are close to the optimal values.

Table 6.5 Optimization calculation results of thermal system exergy efficiency in PWR NPPs

No.	Calculation item	Exergy loss coefficient (%)		Increment
		Initial result	Optimization result	
1	Energy form conversion in reactor	25.1648	25.1648	0.0
2	Heat transfer in the reactor	25.4655	25.4655	0.0
3	Main coolant pump	0.0893	0.0928	0.0035
4	Steam generator	3.5062	3.4416	−0.0646
5	Main steam line	0.4607	0.4268	−0.0339
6	Steam turbo-generator	7.8076	7.8053	−0.0023
7	Moisture separator reheater	0.5362	0.4229	−0.1133
8	Condenser	3.9501	3.6105	−0.3396
9	Feedwater regenerative system	1.2385	1.3107	0.0722
10	Exhaust steam line	0.0	0.0071	0.0071
11	Feed water pump	0.0885	0.0866	−0.0019
12	Condensate pump	0.0140	0.0162	0.0022
13	Main air extractor	0.0548	0.0606	0.0058
14	Circulating water pump	0.0848	0.1373	0.0525
Total		68.461	68.0487	−0.4123
Plant net efficiency (%)		31.539	31.9513	0.4123

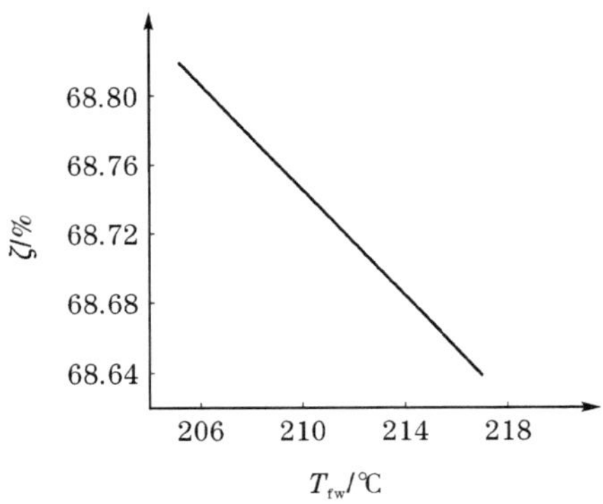

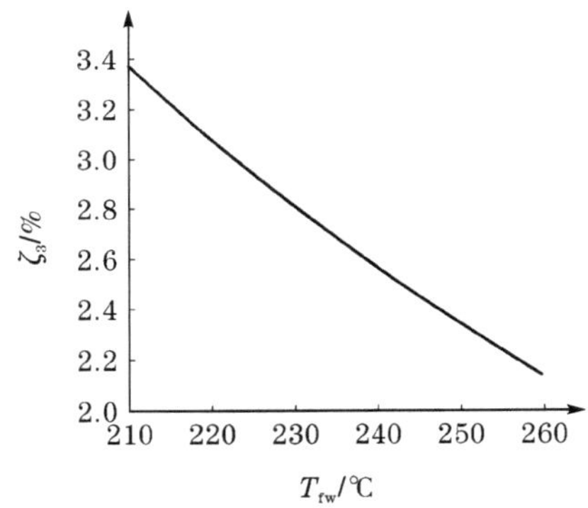

Fig. 6.6 The influence of feedwater temperature on the exergy loss coefficient of NPPs and steam generators: (**a**) variation of NPP loss coefficient with feed water temperature; (**b**) variation of steam generator exergy loss coefficient with feed water temperature

6.3.2 Sensitivity Analysis of Optimization Variables

6.3.2.1 Feedwater Temperature

The effect of feedwater temperature on the exergy loss coefficient of the NPP is shown in Fig. 6.6. As the feedwater temperature increases, the exergy loss coefficient of the NPP shows a decreasing trend. This is because the increase in feedwater temperature raises the average heat absorption temperature of the secondary circuit

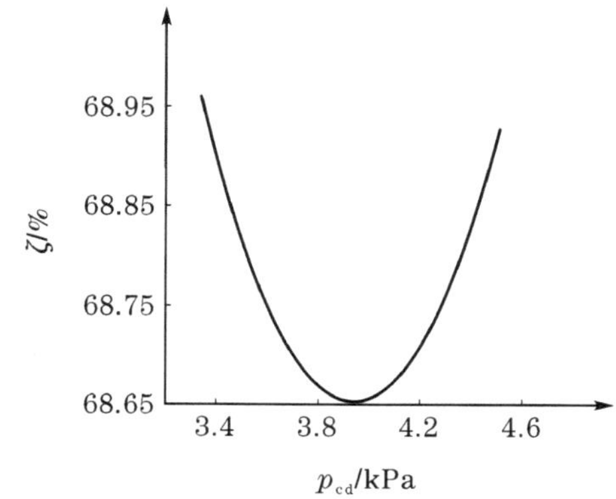

Fig. 6.7 The influence of condenser pressure on the thermal exergy loss coefficient of NPPs

working fluid in the steam generator. With the average temperature of the primary circuit coolant unchanged, the heat exchange temperature difference decreases, reducing the internal exergy loss during the heat exchange process of the steam generator.

6.3.2.2 Condenser Pressure

The exergy loss coefficient of the NPP is quite sensitive to changes in condenser pressure. This is because changes in the final parameters of the steam directly affect the steam consumption rate of the most important steam-consuming equipment in the entire NPP thermal system - the steam turbine generator set. Figure 6.7 shows the influence of condenser pressure on the exergy loss of the NPP thermal system, and there is an optimal value corresponding to the minimum exergy loss coefficient of the NPP.

When the condenser pressure is higher than this value, the exergy loss coefficient of the NPP increases with the increase in pressure, because the increase in the final parameters of the steam will cause the adiabatic enthalpy drop of the steam in the turbine to decrease, the steam consumption to increase, and the exergy loss of the turbine and condenser to increase accordingly; when the condenser pressure is lower than this value, the exergy loss coefficient of the NPP increases with the decrease in pressure, because the condenser pressure is too low, which will cause the power consumption of the circulating cooling water pump to increase significantly, thereby leading to a decrease in the economy of the NPP.

6.3.2.3 Steam Reheat Temperature

The effect of steam reheat temperature on the exergy loss coefficient of the NPP is not significant. When the reheat steam temperature is equal to the steam temperature at the outlet of the first-stage reheater, i.e., there is no steam heating in the

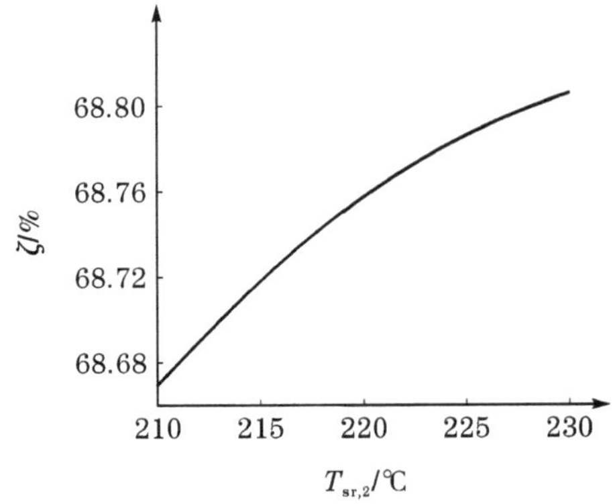

Fig. 6.8 The influence of reheat temperature on the thermal exergy loss coefficient of NPPs

second-stage reheater, the exergy efficiency of the NPP is the lowest under this condition (as shown in Fig. 6.8).

Questions and Exercises

1. What are the two main energy processes in the reactor? What are the main reasons for the exergy loss of these two processes?
2. Try to analyze the reasons for the exergy loss of the condenser during operation.
3. A surface feedwater heater, with a steam mass flow rate of G_{hs}, inlet parameters of h_{hs}, s_{hs}, and outlet parameters of h_{hw}, s_{hw}; the feedwater passes through the heater, and the temperature rises from t_1 to t_2. Assuming the ambient temperature is t_0, and the average specific heat at constant pressure of the feedwater is c_p, try to determine the exergy loss and exergy efficiency of the feedwater heater during operation, and analyze the main reasons for the loss.
4. A shell-and-tube heat exchanger, with a heating fluid mass flow rate of 0.28 kg/s, inlet and outlet temperatures of 200 °C and 50 °C respectively, and the inlet and outlet temperatures of the heat-absorbing fluid are 30 °C and 150 °C respectively. The average specific heat at constant pressure of the heating fluid and the heat-absorbing fluid are 3.349 kJ/(kg · K) and 3.768 kJ/(kg · K) respectively. Assuming the ambient temperature is 25 °C, and neglecting the heat dissipation loss of the heat exchanger, try to calculate the exergy loss of the heat exchange process.

Chapter 7
Energy Saving Analysis of NPP Thermal System

The energy balance method and exergy analysis method are used to conduct thermal analysis on the thermal system of a pressurized water reactor nuclear power plant. Although the net efficiency values obtained by the two methods are basically the same, the essence of the problems revealed is different. The energy balance method believes that the main reason for the energy loss of the thermal system of the pressurized water reactor nuclear power plant is the heat discharge from the thermal cycle during operation, and the energy dissipation caused by mechanical loss and motor loss. The main direction to reduce energy loss is to improve the thermal cycle of the device, reduce the heat discharged in the condenser, and reduce the heat dissipation loss of the thermal equipment. However, the exergy analysis method believes that the main reason for energy loss is the irreversible factors in the energy conversion, transmission and delivery process of the thermal system, which leads to a decrease in the ability of energy to work. The way to improve the efficiency of the thermal system is to try to reduce the irreversibility of the energy use process. Since the analysis results of the two methods on the main parts, quantity and fundamental causes of energy loss in the thermal system are completely different, different methods and ways to improve the thermal economy of the thermal system will inevitably be derived. For example, from the perspective of heat balance, the condenser is the device with the greatest energy loss, but from the perspective of exergy, the heat released by the exhausted steam and drainage into the condenser is of poor energy quality due to low temperature and pressure, and has less exergy, and it is not worth spending a lot of effort to improve it. The main goal should be to reduce the irreversible loss in the energy conversion and transmission process in the thermal system, especially in the reactor.

Because energy has both quantity and quality, the exergy analysis method attributes both the quantity and quality loss of energy to the loss of the ability of energy to work, so it can reveal and evaluate the degree of energy loss in the thermal system more comprehensively than the energy balance method, and point out the correct way to improve the perfection degree of energy utilization in the thermal system.

M. Peng et al., *Thermal Analysis of Nuclear Power Plants*, Nuclear Science and Technology, https://doi.org/10.1007/978-981-96-8983-5_7

7.1 Ways to Improve the Thermal Economy of NPPs [43]

7.1.1 Improve the Exergy Efficiency of the Reactor

The exergy loss in the pressurized water reactor mainly comes from the process of converting nuclear fission energy into thermal energy in the reactor and the process of transferring the heat released by the core nuclear fuel to the main coolant. As known from thermodynamics, when energy exists in the form of heat, its quality depends on the temperature of the heat-carrying medium. The higher the temperature, the higher the quality of energy. All the energy possessed by nuclear fuel is exergy. The quality of the heat released after fission depends on the average temperature of the nuclear fuel. Usually, the temperature of the nuclear fuel will not exceed 3000 °C, so the conversion of nuclear fission energy into thermal energy will inevitably lead to a decrease in energy quality and a reduction in available energy. When the core heat is transferred to the main coolant, the large heat transfer temperature difference between the nuclear fuel and the coolant causes the energy quality to decrease and exergy loss.

For convenience of explanation, the total exergy of nuclear energy, the exergy of nuclear fuel heat release, and the exergy of coolant heat absorption are approximately represented as

$$E_{x,\mathrm{N}} = Q_{\mathrm{R}}$$

$$E_{x,\mathrm{R}} = Q_{\mathrm{R}}\left(1-\frac{T_0}{T_{\mathrm{u,av}}}\right), E_{x,\mathrm{c}} = Q_{\mathrm{R}}\left(1-\frac{T_0}{T_{\mathrm{c,av}}}\right)$$

where *T*u, av, *T*c, av are the average temperatures of the core nuclear fuel and coolant respectively, K.

The corresponding exergy loss is

$$E_{\mathrm{xl,R1}} = E_{x,\mathrm{N}} - E_{x,\mathrm{R}} = Q_{\mathrm{R}}\frac{T_0}{T_{\mathrm{u,av}}}$$

$$E_{\mathrm{xl,R2}} = E_{x,\mathrm{R}} - E_{x,\mathrm{c}} = Q_{\mathrm{R}} T_0 \frac{T_{\mathrm{u,av}} - T_{\mathrm{c,av}}}{T_{\mathrm{u,av}} T_{\mathrm{c,av}}}$$

Then

$$\frac{E_{\mathrm{xl,R1}}}{E_{\mathrm{xl,R1,0}}} = \frac{T_{\mathrm{u,av,0}}}{T_{\mathrm{u,av}}} \tag{7.1}$$

$$\frac{E_{\mathrm{xl,R2}}}{E_{\mathrm{xl,R2,0}}} = \frac{T_{\mathrm{u,av}} - T_{\mathrm{c,av}}}{T_{\mathrm{u,av,0}} - T_{\mathrm{c,av,0}}} \frac{T_{\mathrm{u,av,0}} T_{\mathrm{c,av,0}}}{T_{\mathrm{u,av}} T_{\mathrm{c,av}}} \tag{7.2}$$

Obviously, to reduce the exergy loss in the reactor, the following two aspects should be considered:

(1) Increase the average temperature of the nuclear fuel in the reactor;
(2) Reduce the heat transfer temperature difference between the nuclear fuel and the coolant.

Figure 7.1 reflects the relationship between the exergy loss in the process of converting nuclear fission energy into thermal energy in the reactor and the average temperature of the nuclear fuel. As the average temperature of the nuclear fuel increases, the exergy loss in the energy conversion process in the reactor decreases. If the average temperature of the nuclear fuel doubles, the exergy loss in the energy conversion process decreases by about 50%; when the average temperature of the nuclear fuel approaches infinity, this exergy loss approaches zero.

Figure 7.2 reflects the relationship between the exergy loss in the energy transfer process in the reactor and the heat transfer temperature difference. The smaller the heat transfer temperature difference, the smaller the exergy loss in the energy transfer process in the reactor. For ease of explanation, assume that the average temperature of the nuclear fuel is 1200 °C, and the average temperature of the main coolant is 302 °C. If the heat transfer temperature difference is reduced by 1/2, the exergy loss in the energy transfer process can be reduced by about 2/3; when the heat transfer temperature difference approaches zero, this exergy loss also approaches zero.

Currently, UO_2 ceramic fuel is commonly used in pressurized water reactors, and its main characteristics are shown in Table 7.1.

Considering the impact of radiation in the reactor and the depth of burnup, when the pressurized water reactor operates at full rated power, the highest temperature in the center of the fuel core is about 2000 °C, and the average temperature is only 800–1200 °C. Therefore, for pressurized water reactors using ceramic fuel, further increasing the average temperature of the nuclear fuel is limited by the thermal physical properties of the nuclear fuel and the thermal safety of the reactor. It is necessary to develop new types of nuclear fuel with higher melting points, good thermal stability, radiation stability, chemical stability, and higher thermal conductivity.

Fig. 7.1 Influence of heat transfer temperature difference on core exergy loss

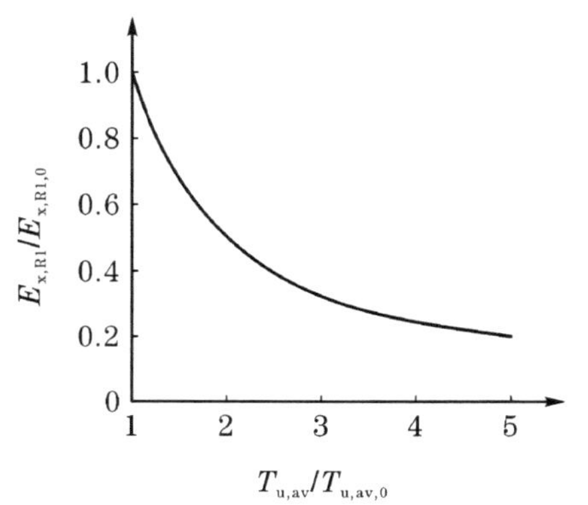

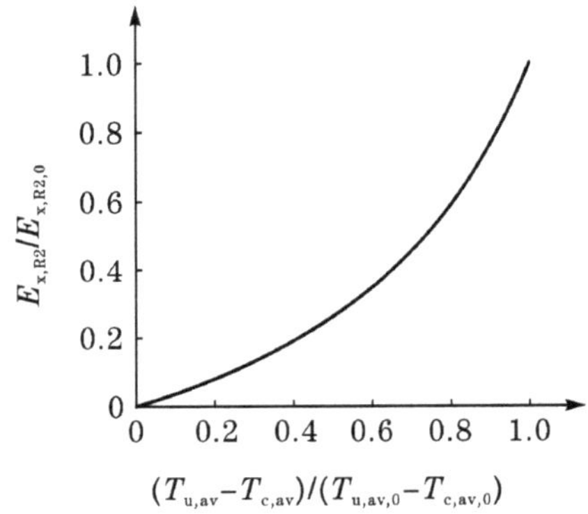

Fig. 7.2 Influence of heat transfer temperature difference on core exergy loss Impact of loss

Table 7.1 Main characteristic parameters of UO_2 fuel

Melting point (°C)	Crystal structure	Crystal lattice parameter (nm)	Theoretical density (g/m^3)
2865 ± 15	Face-centered cube	0.5465	10.96
Thermal conductivity (W/(m·K))	Thermal expansivity (1 °C)	Breaking strength (MPa)	Elasticity modulus (10^5 MPa)
~8.6 (below 20 °C)	~1 × 10^{-5} (0–1500 °C)	~110	2.0 (below 20 °C)

The reactor coolant system pressure of a PWR NPP is generally about 15.2 MPa, the inlet coolant temperature of the reactor is 280–300 °C, the outlet temperature is 310–330 °C, and the highest average temperature of the primary coolant is only about 315 °C. The saturation temperature corresponding to the main system pressure is about 340 °C. If the main system pressure remains unchanged and the average temperature of the main coolant is significantly increased, the outlet coolant temperature of the reactor will inevitably be higher than the saturation temperature, and net steam will be generated in the reactor core heat channel, which may cause deterioration of heat transfer in the reactor core and cause the reactor core to burn out. Therefore, to increase the average temperature of the coolant, the main system pressure must be increased at the same time to ensure the thermal safety of the reactor, but there are several problems:

(1) As the main system pressure increases, the pressure requirements, materials, and manufacturing difficulties of the main equipment increase, resulting in increased investment in power plant construction and affecting the economic operation of the power plant.
(2) The critical temperature of water is 373.99 °C, and the corresponding critical pressure is 22.064 MPa. At any pressure, as long as the temperature of the water is higher than the critical temperature, the water can only exist in the form of steam. The current technical level of pressurized water reactors uses supercooled water as a coolant and moderator. The temperature of the reactor outlet coolant

Table 7.2 The maximum allowable temperature on the surface of several types of fuel cladding

Reactor type	Cladding materials	Permitted maximum surface temperature (°C)
Water cooled reactor	Zircaloy	350
CO_2 cooled reactor	Magnesium alloy/beryllium	405/600
Sodium cooled reactor	Stainless steel	650
Organic liquid cooled reactor	Aluminium alloy	400

can reach up to 360 °C at most, and the average temperature of the coolant will not exceed 350 °C. The room for improvement is extremely limited.

(3) The fuel element cladding material used in pressurized water reactors is mostly zirconium-4 alloy. Zirconium alloy cladding will produce significant creep within the working temperature and pressure range of pressurized water reactors. The creep rate increases significantly with the increase in temperature; in addition, at high temperatures, zirconium will react with water (or steam) as follows:

$$Zr + 2H_2O = ZrO_2 + 2H_2 \uparrow + \text{Heat}$$

This is an exothermic reaction. After 1 kg of zirconium alloy completely reacts with water (or steam), it can release about 6.74×10^3 The heat of kJ. Therefore, the maximum working temperature of the surface of the zirconium alloy casing in contact with water is often limited to below 350 °C. Considering the unevenness of temperature distribution, in pressurized water reactors with zirconium alloy casings, the average temperature of the coolant is not higher than 340 °C.

Table 7.2 gives the maximum temperatures allowed on the surfaces of fuel casings for several types of reactors.

Therefore, for pressurized water reactors at the current technical level, measures need to be taken to increase the average temperature of the nuclear fuel in the reactor, reduce the heat transfer temperature difference between the nuclear fuel and the primary coolant, and significantly reduce the exergy loss in the reactor. However, this is likely to be very limited due to the thermal performance limitations of the fuel, primary coolant, and reactor materials.

The key technologies needed to improve the exergy efficiency of pressurized water reactors are: high melting point, high thermal conductivity nuclear fuel, high temperature corrosion resistant nuclear fuel cladding materials, and the development and selection of appropriate coolant working substances.

Because the higher the temperature of the primary coolant flowing out of the reactor, the higher the quality of the heat carried out, the reduction of losses in the reactor is mainly reflected in the increase in the average temperature of the primary coolant in the first loop. The average temperature of the coolant in a pressurized water reactor nuclear power plant at the current technical level is about 310 °C, with a plant efficiency of 34%; sodium-cooled fast reactors use liquid metal sodium as the coolant, nuclear fuel is PuO_2/UO_2 (a mixture of uranium oxide and plutonium oxide), the average temperature of the coolant reaches 450 °C, with a plant efficiency of about 42.8%; high-temperature gas-cooled reactors use helium as the coolant, graphite as the moderator, highly enriched uranium as the fuel, and the average temperature of the coolant during operation can usually reach 520 ~ 530 °C, and the reactor outlet temperature can reach 740 °C. It can be roughly estimated that, at the same ambient temperature, the relative exergy loss of a pressurized water reactor is about 50%, and the relative exergy loss of a high-temperature gas-cooled reactor is only about 36.7%, with a plant efficiency of up to 41%. Therefore, in the future development of nuclear power, sodium-cooled fast reactors and high-temperature gas-cooled reactors will be widely used as economically superior reactor types.

Table 7.3 gives the main parameters of several built high-temperature gas-cooled reactors, and Table 7.4 gives the temperature levels that can be achieved by the coolants used in different reactor types.

If we break through the concept of existing pressurized water reactors, assume that the reactor loop pressure is increased to 22.12 MPa, and use supercritical temperature water as the coolant, if the average temperature of the first loop is increased to about 600 °C, then the exergy loss of the reactor can be reduced by about 1/2, and when the second loop exergy loss coefficient remains unchanged, the thermal efficiency of the nuclear power plant can reach more than 55%, which will have considerable economic and social benefits. Whether this "supercritical pressurized

Table 7.3 The main parameters of the built HTGR

	Peach bottom	Dragon	AVR	Fort strain	THTR-300
Name of the NPP and the country	United States	United Kingdom	Germany	United States	Germany
Thermal power (MW)	115	20	46	840	750
Electric power (MW)	40	–	15	330	315
Maximum fuel temperature (°C)	1331	1600	1134	1260	1250
Primary loop pressure (MPa)	2.36	2	1.09	4.9	4
Helium temperature at reactor inlet (°C)	344	350	275	406	260
Helium temperature at reactor outlet (°C)	728	750	950	785	750
Secondary loop steam temperature (°C)	538	–	500	540	535
Gross NPP efficiency (%)	34.78	–	32.61	39.28	42

Table 7.4 Coolant temperature of various stack types

Reactor type	HTGR				LMFR	PWR	BWR	Heat supply reactor
	Future type	Process heat supply	Helium turbine	Steam cycle				
Coolant temperature (°C)	>1000	950	950	750	520	320	285	200

water reactor" is technically feasible requires multidisciplinary analysis and research.

7.1.2 Improve the Exergy Efficiency of Heat Exchange Equipment

In the thermal system of a pressurized water reactor nuclear power plant, heat exchangers account for a large proportion, with main equipment including steam generators, condensers, feedwater heaters, and steam reheaters, etc. The exergy loss in heat exchange equipment accounts for about 9.02%–9.07% of the total power of the reactor. It is generally believed that there are two reasons for the exergy loss in heat exchange equipment:

(1) Internal exergy loss caused by the existence of a heat transfer temperature difference, which is the main reason for the exergy loss in heat exchange equipment.
(2) External exergy loss caused by heat dissipation.

Therefore, to improve the energy efficiency of heat exchange equipment, efforts need to be made in two aspects: first, reduce the heat dissipation loss during the heat exchange process to improve heat exchange efficiency; second, enhance heat exchange, while ensuring the smooth progress of heat transfer, reduce the temperature difference in heat transfer.

The principal heat transfer equation for heat exchange equipment is

$$Q = K \cdot F \cdot \Delta T \cdot 10^{-3} \tag{7.3}$$

where Q is the heat exchange capacity, kW; K is the heat transfer coefficient, W/(m^2·K); F is the heat transfer area, m^2; ΔT is the average heat transfer temperature difference, K.

The heat transfer process usually consists of the following parts:

(1) Heat exchange between the heating medium and the heat exchange surface;
(2) Conduction within the heat exchange surface;
(3) Conduction within the dirt layer on the heat exchange surface;
(4) Heat transfer between the heat exchange surface and the heated medium.

The main purposes of applying enhanced heat exchange technology in heat exchange equipment are as follows:

(1) Reduce the heat exchange area while keeping the heat transfer load constant, thereby reducing the size of the heat exchange equipment.
(2) Increase the heat exchange amount of the heat exchanger while keeping the overall size of the heat exchange equipment constant.
(3) Reduce the temperature difference in heat transfer between cold and hot fluids, and increase the outlet temperature of the heat-absorbing medium.
(4) Reduce the pressure loss of the fluid and reduce the power of fluid delivery.

From the perspective of rational energy use, the adoption of enhanced heat exchange measures is to reduce the thermal resistance in each link of the heat transfer process, thereby improving the heat transfer coefficient of the heat exchanger. In this way, when the heat exchange area and heat exchange amount remain constant, the average temperature difference in heat exchange can be reduced, and the internal energy loss in the heat exchange process can be reduced.

The following are the convective heat transfer enhancement methods used in actual engineering:

Passive technology	Active technology
Treated surface	Mechanical agitation
Rough surface	Surface vibration
Development surface	Fluid vibration
Spoiler element	Electromagnetic field
Vortex generator	Spray or suck out
Spiral tube	
Additive	
Jet impact	

The heat exchangers used in the thermal systems of nuclear power plants are mostly traditional shell-and-tube types. The heat exchange tubes used are usually smooth tubes. This is because smooth tubes are cheap, easy to manufacture, install and maintain, and easy to clean, so they are quite common. With the research and application of enhanced heat exchange technology, rough heat exchange tubes, treated surface tubes, or inserts and additives are beginning to be used in heat exchangers to improve heat exchange efficiency.

The rough surface tubes that are currently practically applied include finned tubes and threaded tubes. Finned tubes can enhance heat exchange outside the tube and are mainly used for various air cooling or waste heat utilization equipment where the inside of the tube is a single-phase flow and the outside is a gas (air or flue gas). Their application in nuclear power plant heat exchange equipment is limited. Threaded tubes can enhance heat exchange inside and outside the tube at the same time and are one of the most effective methods of enhancing heat exchange in power device condensers. The U.S. Navy has applied this type of heat exchange tube to the condensers of ship power devices, and this enhanced heat exchange technology has also been successfully used in power plant condensers in Europe and the United States. In the process of researching the miniaturization technology of nuclear power device heat exchange equipment in China, through a large number of experimental studies, it was found that under the same steam load, circulating cooling seawater flow rate, and inlet water temperature conditions, the use of threaded tubes can improve the vacuum degree of the condenser, and the degree of vacuum increases as the steam load increases.

Taking the Qinshan Nuclear Power Plant as an example, its condenser heat transfer tube uses a titanium smooth tube. If a titanium threaded tube is used for condensation enhanced heat exchange, it is assumed:

(1) The evaporation pressure and temperature in the steam generator remain constant, as does the feedwater temperature;
(2) The internal efficiency, mechanical efficiency, and generator efficiency of the steam turbine remain constant;
(3) The pressure and temperature at each extraction point remain constant;
(4) The net electrical output of the power station remains constant;
(5) The power consumption of the circulating cooling water pump remains constant;
(6) The humidity of the exhaust steam at the end of the low-pressure cylinder of the steam turbine does not exceed 12% (this can be achieved by using more effective internal dehumidification and intermediate steam-water separation technologies in the turbine stages).

An increase in the vacuum degree of the condenser (i.e., a decrease in condensation pressure) will improve the exergy efficiency of the power station, as shown in Fig. 7.3.

Under the above assumptions, the reduction in condensation pressure of the condenser affects the thermal system of the nuclear power plant in the following ways:

(1) Due to the reduction in back pressure of the steam turbine, the adiabatic enthalpy drop of steam in the turbine increases, reducing the steam consumption of the turbine and thus reducing the thermal power required from the nuclear steam supply system.
(2) The corresponding reduction in steam turbine exhaust volume can reduce the steam load on the condenser, decrease the absolute value of heat discharged into the environment, reduce the exergy discharged into the environment, and also reduce the size of the condenser, lowering the investment cost of the power station.
(3) The reduction in condensation temperature in the condenser reduces the exergy of the heat released by the working fluid entering the condenser, which can reduce the exergy loss in the condenser.

In addition to adopting some technical measures to enhance heat transfer on one or both sides of the heat transfer surface, methods to enhance heat transfer also include using materials with excellent thermal conductivity to manufacture the heat transfer

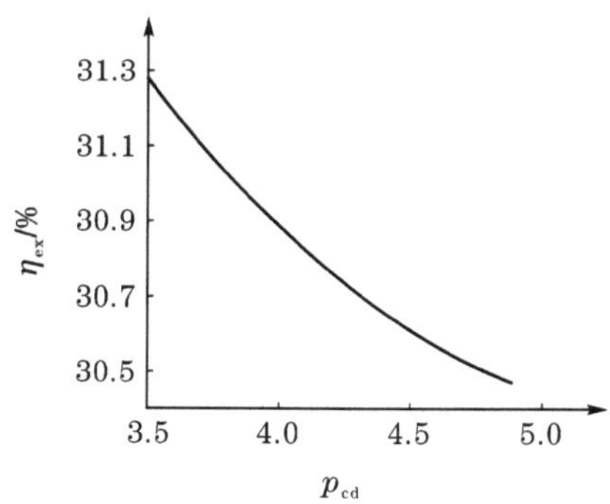

Fig. 7.3 The influence of p_{cd} on η_{ex}

surface and using physical and chemical means to keep the heat transfer surface clean to reduce thermal resistance and fouling resistance during the heat transfer process.

For general heat exchange equipment, assuming the absolute average temperatures of the heating medium and the heat-absorbing medium are *T*1 and *T*2 respectively, and the absolute temperature of the environment is *T*0, ignoring heat dissipation losses, and the heat exchange is *Q*, then the exergy loss in the heat exchange process is

$$E_{\mathrm{xl}} = QT_0 \frac{T_1 - T_2}{T_1 T_2} = QT_0 \frac{\Delta T}{T_1 T_2} \tag{7.4}$$

After adopting measures to enhance heat transfer, if the average heat exchange temperature difference decreases by 30%, i.e., $\Delta' = 70\ \%\ \Delta T$, then the exergy loss of the heat exchange equipment becomes

$$E'_{\mathrm{xl}} = QT_0 \frac{\Delta T'}{T'_1 T'_2} \tag{7.5}$$

If the average temperature of the heating medium remains constant, i.e., $T'_1 = T_1$, then

$$\frac{E'_{\mathrm{xl}}}{E_{\mathrm{xl}}} = \frac{\Delta T'}{\Delta T} \frac{T_2}{T'_2} = 70\% \frac{T_2}{T'_2} \tag{7.6}$$

Because $(T_1 - T_2) > (T'_1 - T'_2)$, and $T'_1 = T_1$, then $T'_2 < T_2$, therefore, under the above assumed conditions, the exergy loss of heat exchange equipment can be reduced by more than 30%.

In summary, the use of enhanced heat transfer technology in the heat exchange equipment of nuclear power plants, especially condensers, can improve the exergy efficiency of nuclear power plants. However, the final use of enhanced heat transfer technology depends on some limiting factors, such as material properties, structural possibilities, safety and reliability, etc.

7.1.3 Improve the Exergy Efficiency of Steam Turbine Generators

The exergy loss coefficient of the steam turbine generator is approximately 7%–8%, it is also one of the devices with larger losses in the thermal system of nuclear power plants. The main reasons for the exergy loss when the steam turbine is working are as follows:

7.1.3.1 Internal Loss of the Steam Turbine

When steam expands and does work in the flow part of the steam turbine, it will generate various internal losses such as blade windage and rotor friction loss, internal steam leakage loss, wet steam loss, and flow loss of steam in the steam inlet, exhaust duct, and high and low-pressure cylinder connection pipes. These increase the irreversibility of the adiabatic expansion of steam, consume the useful energy of steam, and reduce the work done by steam in the steam turbine.

7.1.3.2 Power Transmission Loss

During the power output process inside the steam turbine, the friction resistance of the support bearing and thrust bearing and the energy consumption of the regulation system reduce the effective power output to the generator.

Therefore, the way to improve the exergy efficiency of the steam turbine should be: to optimize the design of the flow part of the steam turbine, choose the best blade type, improve the design and manufacturing level, reduce friction resistance, ensure a higher stage efficiency, reduce the irreversibility of the adiabatic expansion process of steam inside the steam turbine and the mechanical loss of the steam turbine.

In addition, as the capacity of the steam turbine unit increases, the internal efficiency of the steam turbine gradually improves, the proportion of internal loss and mechanical loss of the steam turbine compared to the output power decreases, and the economy of the steam turbine unit improves. From the exergy analysis calculation, it can be seen that the exergy loss coefficient of the steam turbine generator of the Qinshan Nuclear Power Plant is 7.81%, while the energy loss coefficient of the steam turbine generator of the Daya Bay Nuclear Power Plant is only 5.52%. The main reason for this difference, apart from the level of design and manufacturing, is that the latter's capacity is much larger than the former. Since the energy loss of the reactor, various heat exchange equipment, and other thermal equipment is basically unrelated to their capacity. Therefore, increasing the capacity of the nuclear power plant and using large-capacity steam turbine generator units can also improve the thermal economy of the nuclear power plant.

The exergy loss of the generator is mainly due to copper loss, iron loss, and mechanical loss in the armature winding. Therefore, improving the efficiency of the generator should start from optimizing the design of the generator and adopting superconducting technology.

7.1.4 Improve the Exergy Efficiency of Auxiliary Equipment

For some other thermal equipment, since the total relative energy loss is less than 1%, the benefits of modification are not great. For electric pumps, the main thing is to improve motor efficiency and optimize the design of the pump impeller. For steam pipes, the main thing is to wrap insulation materials to reduce heat loss; the pipe layout should be as short and straight as possible to reduce flow friction and local resistance; improve sealing technology to reduce leakage.

7.1.5 Ways and Possibilities to Improve the Thermal Economy of NPPs

Pressurized water reactor nuclear power plants are a complex organic whole, with each subsystem and each thermal equipment interconnected and influencing each other. Even if one device is improved and its energy efficiency is increased, due to the constraints and influences of related devices or systems, the thermal economy of the entire nuclear power plant thermal system may not necessarily be improved. Therefore, it is necessary to start from the perspective of systems engineering and study the methods and ways to improve the thermal economy of nuclear power plants from an overall perspective.

To improve the thermal economic level of pressurized water reactor nuclear power plants, it can be achieved by reducing the energy loss of each device and improving the thermal efficiency of the cycle. The specific ways are:

(1) Increase the average temperature of nuclear fuel during the operation of the reactor, reduce the heat transfer temperature difference between the nuclear fuel and the primary coolant, and increase the average temperature of the primary coolant.
(2) Enhance heat exchange, reduce the heat transfer temperature difference during the heat exchange process in the heat exchange equipment, and improve the heat transfer efficiency.

    This can improve the steam parameters on the secondary side of the steam generator, reduce the condensation pressure in the main condenser, and thus improve the thermal efficiency of the device cycle.
(3) Optimize the design of the main steam turbine and generator, reduce the irreversibility of the steam expansion process in the main steam turbine, and improve the efficiency of energy conversion of the generator.
(4) Develop large-capacity pressurized water reactor nuclear power plants.

By using large-capacity steam turbine generator sets, the internal efficiency and mechanical efficiency of the steam turbine are correspondingly improved, reducing the proportion of the exergy loss of the steam turbine generator set in the total nuclear energy of the power station, and improving the overall economy of the nuclear power plant.

Due to the limitations of the thermophysical properties of nuclear fuel, primary coolant, reactor materials, and the thermal safety rules of the reactor, it is difficult to further increase the average temperature of nuclear fuel and primary coolant in pressurized water reactor nuclear power plants, which limits the improvement of the steam parameters of the secondary circuit and the efficiency of the device, so the thermal economy of pressurized water reactor nuclear power plants will not be significantly improved.

7.1.6 Analysis Conclusion

The following conclusions can be drawn from the above analysis:

(1) The exergy analysis method points out the real reason for the imperfect energy utilization of the energy system, which lies in the irreversibility of the energy conversion and transfer process.
(2) The part with the highest exergy loss rate in the pressurized water reactor nuclear power plant is the reactor, not the condenser, which is usually considered by the energy balance method to be the equipment with the largest energy loss. The main direction to improve the thermal economy of pressurized water reactor nuclear power plants should be to reduce the irreversibility of energy conversion and transfer in the thermal system of nuclear power plants, especially in the reactor.
(3) By using enhanced heat exchange technology to modify the heat exchange equipment of the pressurized water reactor nuclear power plant, the exergy efficiency of the power station can be increased by about 3% under certain conditions.
(4) For pressurized water reactor nuclear power plants at the current technical level, due to the limitations of material thermal performance, it is no longer possible to significantly improve their thermodynamic economy. Only by developing supercritical pressurized water reactors or adopting new reactor types that can more effectively utilize nuclear fission energy can the thermal efficiency of nuclear power plants be raised to a new level.

7.2 Basic Principles of Rational Energy Use [44, 45]

Rational and effective use of energy is the basic premise to ensure that the thermal system has the best thermal economy, and it is also an important condition for the sustainable development of human society.

In practical engineering, the rational and effective use of energy is often simply attributed to energy conservation. The issue of energy conservation involves a wide range, including energy policy, energy management issues, as well as process, equipment, control, materials and other energy-saving technical issues. As for energy-saving technology, each industry has its own characteristics and specific situations, and many energy-saving measures can be proposed. However, the basic principles of energy conservation have common points. The most important of which is that we must follow the basic principles of rational energy use, that is, energy use according to quality. Only by using energy rationally can we achieve a high rate of effective energy use and achieve the purpose of energy conservation.

7.2.1 Principle of Minimum External Loss

Energy losses in the energy system during the process of energy conversion, transmission, and distribution, caused by working fluid leakage, equipment surface heat dissipation, and heat taken away by equipment cooling water, are collectively referred to as external losses. These are tangible losses. Although the exergy losses caused by these external losses are not significant, they are all transformed from the high-level energy input into the system due to the irreversibility of the process. Therefore, in the design and operation of the thermal system, efforts should be made to minimize the heat energy discharged from the system and not utilized, to achieve full utilization of energy.

The main technical measures include the following aspects:

(1) Improve the sealing performance of equipment and pipelines to reduce the running, rising, dripping, and leaking of working fluid.

    The thermal system of a power plant is large and complex, with many devices prone to working fluid leakage. For each specific thermal device, the leaked working fluid may not be much, but if there is leakage in all the thermal devices of the entire power plant thermal system, the energy loss caused by this will be astonishing. According to calculations, if a thermal power plant leaks 15 kg of saturated steam at 130 °C/h, it is equivalent to a loss of 16.4 tons of standard coal per year.

(2) Improve the thermal system, reduce waste heat emissions, and strengthen the recovery and utilization of waste heat.
(3) Improve the level of thermal insulation to reduce heat loss caused by poor insulation.

The thermal system of a nuclear power plant dissipates heat to the environment during operation, including the following processes: convective heat exchange between high-temperature and high-pressure steam inside the turbine and the cylinder wall, convective heat exchange between the working fluid inside the main steam pipeline, reheated steam pipeline and feed water pipeline and the inner wall of the pipeline; conduction of heat within the cylinder wall, pipeline wall, and insulation layer; convective heat exchange and radiation heat exchange between the outer surface of the equipment and pipeline insulation layer and the surrounding environment. If the insulation is poor for a large number of thermal devices, pipelines, and thousands of valves, flanges, etc. in the thermal system, it will lead to huge energy losses.

For example, a valve on a pipe with a nominal diameter of 300 mm, under the condition of a working fluid temperature of 500 °C in the pipe, the heat dissipation to the environment without insulation is 14,537.5 W. After taking insulation measures, the heat dissipation is reduced to 4163.5 W. If calculated based on 7000 h of operation per year, the exposed state loses 2.6×10^{11} J more heat than the insulated state, equivalent to consuming an additional 8.9 tons of standard coal. A pair of flanges on a pipe with a nominal diameter of 300 mm, with a working fluid temperature of 450 °C inside the pipe, dissipates 9154 W without insulation. After taking insulation measures, the heat dissipation is 517.5 W. After 7000 h of operation, the exposed state loses 2.2×10^{11} J more heat than the insulated state, equivalent to consuming an additional 7.4 tons of standard coal.

Table 7.5 shows the heat dissipation losses of exposed pipes and their losses after insulation. As can be seen from the table, the heat dissipation loss of bare pipes is several times to tens of times larger than that of insulated pipes, and taking insulation measures can reduce the heat dissipation loss of bare pipes by more than 90%.

Tables 7.6 and 7.7 compare the heat dissipation losses of valves and flanges before and after insulation.

Table 7.8 shows the relationship between heat dissipation loss and working fluid temperature when the heat dissipation loss of thermal equipment or pipelines and the environmental temperature are constant.

The data in the Table 7.8 shows that under the same ambient temperature, if the heat dissipation is the same, the higher the working fluid temperature, the greater the exergy loss caused by heat dissipation loss.

Table 7.5 Comparison of heat dissipation loss between exposed and insulated pipes

Nominal diameter of the pipe (mm)	Temperature of the medium within the pipe (°C)	Heat loss per meter of pipe length (W/m)		Heat loss ratio
		Exposed pipe	Insulated pipe	
100	100	354.72	52.34	6.78
	200	1133.93	104.67	10.83
	300	2360.89	159.33	14.82
	400	4244.95	211.67	20.06
	450	5466.1	238.42	22.93
200	100	668.73	81.41	8.22
	200	2128.29	153.52	13.86
	300	4512.44	227.95	19.8
	400	8082.85	302.38	26.73
	450	10,467	337.27	31.04
300	100	930.4	98.86	9.41
	200	3023.8	189.57	15.95
	300	6489.54	279.12	23.25
	400	11,630	368.67	31.55
	450	15002.7	412.87	36.34

Table 7.6 Comparison of heat loss between exposed valves and insulated valves

Nominal diameter of the pipe (mm)	Temperature of the medium within the pipe (°C)	Heat loss per valve (W)		Heat loss ratio
		Exposed valve	Insulated valve	
100	100	432.64	162.82	2.66
	200	1226.37	407.05	3.02
	300	2419.04	744.32	3.25
	400	4070.5	1221.15	3.33
	500	6163.9	1791.02	3.44
200	100	732.69	261.68	2.8
	200	2023.62	651.3	3.11
	300	3907.68	1203.71	3.25
	400	6745.4	1977.1	3.41
	500	9886	2826.09	3.5
300	100	1122.3	395.42	2.84
	200	3058.7	965.3	3.17
	300	5873.15	1773.58	3.31
	400	9886	2884.2	3.43
	500	14537.5	4163.5	3.49

Table 7.7 Comparison of heat loss between exposed and insulated flanges

Nominal diameter of the pipe (mm)	Temperature of the medium within the pipe (°C)	Heat loss per pair of flanges (W)		Heat loss ratio
		Exposed flange	Insulated flange	
100	100	188.41	23.26	8.1
	200	601.27	69.78	8.62
	300	1252.55	104.67	11.97
	400	2251.57	162.82	13.83
	450	2900.52	168.64	17.2
200	100	340.76	52.34	6.51
	200	1086.24	139.56	7.78
	300	2302.74	220.97	10.42
	400	4124	314.01	13.13
	450	5340.5	337.21	15.84
300	100	567.54	81.41	6.97
	200	1845.68	215.16	8.58
	300	3541.34	348.9	11.35
	400	7096.63	488.46	14.53
	450	9153.97	517.54	17.69

Table 7.8 Heat dissipation loss corresponding to different temperature under same exergy loss

Medium temperature[a] (K)	473	523	573	623	673	723	773	823	873	923
Heat dissipation loss[b] (W/m^2)	140	121	109	100	94	90	86	83	80	78
Exergy loss (J)	53.3	53.3	53.3	53.3	53.3	53.3	53.3	53.3	53.3	53.3

[a] Approximate to the external surface temperature of the equipment or pipe
[b] The ambient temperature is set to 20 °C

7.2.2 Principle of Optimal Driving Force

Both theory and practice show that all processes where heat phenomena occur, such as the combustion of fuel, chemical reactions, heat exchange under temperature difference, medium throttling and pressure reduction, and flow with friction, are typical irreversible processes, all of which cause a decrease in exergy value and result in exergy loss. From the perspective of energy utilization, all energy-using processes are processes of energy transfer and transformation, all of which are driven by a certain thermodynamic potential difference (temperature difference, pressure difference, electric potential difference, chemical potential difference, etc.). Without a thermodynamic potential difference, there is no driving force, and the thermal process cannot be realized. Therefore, the thermodynamic potential difference is both the driving force of the process and the main reason for the irreversibility of the thermal process.

A basic principle of rational energy use is to take various measures under the premise of technical and economic conditions, seek the optimal driving force for the process, and improve the effective utilization rate of energy.

In the design of heat exchange equipment for nuclear power plants, a larger heat transfer temperature difference is often chosen to ensure the smooth progress of the heat transfer process. Although this can reduce equipment investment costs, it increases the exergy loss of the heat transfer process, thereby increasing the energy consumption cost of long-term operation. This approach is feasible under conditions of abundant energy and relatively low prices, but it is unreasonable from the perspective of energy conservation. If we want to appropriately reduce the heat transfer temperature difference, without increasing the weight and size of the equipment and ensuring sufficient heat transfer rate, we can only try to adopt measures to enhance heat transfer, reduce the thermal resistance of the heat transfer process, and develop new and efficient heat exchange equipment. In addition, preventing fouling and corrosion of the heat transfer surface during operation, maintaining the thermal resistance during operation without increasing, can also avoid the increase in energy consumption caused by the increase in actual driving force.

7.2.3 Principle of Energy Optimization Utilization

Essentially, the rational and effective way to use energy is not to easily devalue exergy in the energy-using process. For example, combustion and chemical reaction processes should be carried out at high temperatures as much as possible; heating, cooling and other heat exchange processes should make the temperature of the heat-releasing and heat-absorbing media as close as possible; strive to avoid medium throttling and pressure reduction and pressure loss caused by friction and eddy currents; in the process flow, try not to repeat the above irreversible processes multiple times. These are some basic principles of rational use of thermal energy.

From the perspective of energy use, an actual process can be seen as a process of energy transformation, transfer, and loss, that is, the transformation, transfer, and loss of exergy. Therefore, energy must be supplied according to the quantity and quality required by the process.

The purpose of energy supply is to meet the needs of users, not only must it meet the user's requirements in quantity, but also in quality. If energy is supplied below the quality standards required by the user, it cannot meet the user's needs, but if energy is supplied above the quality standards required by the user, it will cause waste. Therefore, the supplied energy must exactly meet the quality standards required by the user, that is, energy supply on demand. Another aspect of rational energy use is to use energy according to quality, that is, in the process of using energy, use it according to the quality of energy (i.e., the high and low and size of the exergy value).

In the actual energy usage process, there are always phenomena of energy waste caused by not using energy according to its quality. In thermal power systems, high-parameter steam is often throttled down to low-parameter steam for reuse. Although the quantity of energy does not decrease significantly, the throttling reduces the work capacity of the steam, resulting in significant exergy loss. For example, saturated steam at a pressure of 1.3 MPa has an exergy of 1012.3 kJ/kg. If it is throttled down to 0.3 MPa for reuse, it will result in nearly 200 kJ/kg of exergy loss, accounting for about 19.5% of the original exergy value.

During winter heating, using the heat generated by fuel combustion to directly heat the room is actually a very unreasonable way of using energy. It does not fully utilize the exergy value of the high-temperature heat source, but only considers the convenience of this heating method. Using high-quality thermal energy for heating, which can be fully met by low-quality thermal energy, inevitably wastes a large amount of high-quality thermal energy. Conversely, if the fuel exergy is first converted into mechanical energy through a thermal engine system, and then the mechanical energy is used by the "heat pump" system to provide the low-quality thermal energy needed for heating, theoretically, 1 kJ of fuel exergy can provide 12 times the low-level thermal energy needed for heating.

Therefore, supplying energy on demand and using energy according to its quality are both aimed at avoiding as much as possible the non-functional degradation of energy and achieving energy quality matching. The so-called energy quality matching is to use high-quality energy as much as possible in situations where low-quality energy is powerless, such as converting it into high-level forms of energy like electricity, and using low-quality energy for heating and other low-level uses. In short, the process of converting exergy into anergy should be fully utilized. Energy quality matching is a very important principle of rational energy use. This principle should be adhered to when supplying energy to users and when configuring energy within the system.

The basic principles of energy optimization utilization include the following aspects:

(1) Prevent the degradation of energy use.

For high-quality energy, its quality should not be reduced before it is used, otherwise it will result in the loss of energy quality. For example, using high-temperature and high-pressure steam after throttling and reducing pressure, or using high-temperature working substance mixed with other low-temperature working substances, often results in the degradation of the energy of the high-temperature working substance.

(2) Simplify energy usage methods.

Try to reduce the number of energy conversions or transfers, thereby reducing energy losses during the conversion or transfer process.

(3) Cascade energy use.

The actual process of energy use is the process of energy conversion and transfer, that is, the process of continuous energy degradation. Different processes have different requirements for energy quality. If high-quality energy is used in processes that require high-quality energy, and the degraded energy after use is used in processes that require lower energy quality, and the high-quality energy is used more fully by reusing it multiple times in the order of gradually decreasing energy quality before and after the use of energy, such as cascade use, multi-effect use, etc.

For example, the new steam generated by the steam generator has higher parameters and can be used to drive the steam turbine generator to obtain electrical energy; the medium and low parameter steam is used to heat the steam generator feed water to recover thermal energy and improve the thermal cycle efficiency. If new steam is used to heat the feed water, although it can meet the system's requirements for feed water temperature, it objectively results in the waste of the work capacity of the new steam.

In the thermal system of a nuclear power plant, a multi-stage feedwater reheating system is set up to heat the low-temperature condensate extracted from the condenser to the feedwater temperature required by the steam generator. From the outlet of the condenser to the inlet of the steam generator, the temperature of the condensate feedwater increases step by step. Each stage of the feedwater reheater uses steam extracted from a certain stage of the turbine to heat the feedwater. A feedwater reheating system with good thermal economy matches the extraction parameters with the feedwater temperature at the outlet of each heater. The new steam is not directly used to heat the feedwater, but is first expanded in the turbine to do work, and after reducing to a certain parameter, a part is extracted to heat the feedwater, thus achieving a reasonable use of energy.

In metallurgical enterprises, different process devices require steam of different grades. For example, turbines for power traction require steam with higher parameters, while general process steam requires lower parameters. If a single steam supply system is adopted, it is bound to produce high-parameter steam to meet the requirements, and part of it is used again after cooling and depressurizing, causing high-grade steam to degrade and causing energy loss, which is also unreasonable. If we can start from the perspective of comprehensive utilization of thermal energy, carry out a comprehensive, reasonable, and comprehensive analysis of the energy use process, let the high-parameter steam first work through the turbine to generate power, and then supply the low-parameter steam discharged from the turbine to the production process for heating, this step-by-step utilization of steam can reduce the waste of high-quality thermal energy and save a lot of energy.

This system of comprehensive energy use according to the high and low energy levels is called the total energy system. The tangible loss and irreversible loss of energy will be greatly reduced, and a very high effective utilization rate of energy can be achieved.

(4) Make full use of low-quality energy.

The energy system will produce a large amount of low-quality energy in the process of energy use. For example, the heat released by the turbine exhaust condensation in a condensing steam power plant can account for about 60% of the total heat consumption of the entire power plant. Because the condensation temperature is close to the ambient temperature, the quality is very low, and the heat exergy it has is less than 5% of the total supply of the power plant, so this part of the energy has no value for the power cycle.

However, for some special energy systems, they often only need to supply low-quality energy, so making full use of these low-quality energies is also a very important energy-saving measure.

7.2.4 Principle of Optimal Techno-Economic

It should be pointed out that correctly evaluating the rationality of thermal energy use is a relatively complex task, and many factors need to be considered comprehensively. In general, the first thing to clarify is the theoretical basis for reducing energy loss. Secondly, consider whether it is technically possible to achieve this. If it is possible, further research is needed on what material conditions are needed, how much cost is paid, and then combined with the economic benefits brought by improving exergy efficiency for comparative analysis, so as to get a more reasonable thermal energy use plan. Finally, it is necessary to check whether this plan complies with environmental protection regulations, such as requirements for environmental pollution, noise, etc. Recently, the combination of exergy analysis and economic analysis has formed a new discipline of "thermal economics", providing a new way to design the optimal system.

Questions and Exercises

1. What measures are there to reduce exergy loss in the reactor? What are the main limiting factors?
2. What is the main way to improve the exergy efficiency of heat exchange equipment?
3. What are the basic principles of rational energy use in energy systems?

Chapter 8
The Analysis Method of Thermal Economics and Its Development

Thermal-economics is a burgeoning discipline that has rapidly emerged since the 1970s. It explores the rationality of energy utilization, the economy of operation, the feasibility of schemes, and the optimization of systems from the perspective of thermodynamic analysis combined with systems engineering.

Pure thermodynamic analysis mainly explains the changes in material parameters during the process and the situation of energy transfer with the outside world. Thermodynamic analysis of different schemes is usually based on the first and second laws of thermodynamics, and the economic indicators used are thermal efficiency, exergy efficiency, exergy loss, etc. Economic analysis needs to find a relationship between the above several thermodynamic quantities and the payment of funds, in order to minimize the unit cost of the product and achieve the best economic benefits. Economic activities based on thermodynamic analysis are the research content and methods of thermal-economics [46-48].

8.1 The Significance of Exergy Economic Analysis of NPPs

Thermodynamic analysis of PWR NPPs is to reveal the weak links and real reasons fully and accurately for energy system losses, find effective ways to improve nuclear power efficiency, and the ultimate goal of improving the thermal-economics of NPPs is to reduce power generation costs.

There are relatively fixed patterns of nuclear power costs at home and abroad, and the differences from the costs of coal power and hydropower are:

(1) After the NPP is put into operation, the issue of decommissioning must be considered, and this part of the cost should be included in the power generation cost;
(2) The cost of storage and reprocessing of spent fuel replaced each year during the operation of the NPP should be included in the cost;

M. Peng et al., *Thermal Analysis of Nuclear Power Plants*, Nuclear Science and Technology, https://doi.org/10.1007/978-981-96-8983-5_8

(3) The cost of environmental dose detection and regular health checks for production personnel should also be included in the cost.

The detailed cost composition of nuclear power is:

Total cost of generation	Plant investment cost	Depreciation of fixed assets
		Amortization charge
		Finance expense
		Retirement fund
	Operating maintenance cost	Major repair cost
		premium
		Personnel salaries and welfare expenses
		Management fee
		Low - and medium-level waste disposal fees
		Cost of materials and power
	Nuclear fuel cost	Nuclear fuel cost
		Spent fuel reprocessing costs

Among them, the key factors affecting the cost of electricity are mainly the depreciation cost of fixed assets of NPPs, fuel costs, and operation and maintenance costs, with fuel costs accounting for about 30%–40% of the total cost.

The unit power generation cost of a NPP is

$$\begin{aligned}&\text{unit power generation cost}\\&\quad=\text{total power production cost}\,/\\&\qquad\text{plant power supply}\left(\text{net electricity}\right)\end{aligned}$$

where, plant power supply = power generation × (1-plant power rate); power generation = unit capacity × equipment utilization hours of the year.

The improvement of the thermal-economics of NPPs is mainly manifested in the saving of fuel. However, any measure to improve the thermal-economics of NPPs requires a certain cost in terms of manpower, material resources, and financial resources. For example, in order to reduce the exergy loss of heat exchange equipment, it is necessary to wrap insulation material on the outer surface of heat exchange equipment such as steam generators and feedwater heaters to reduce heat dissipation losses, and reduce internal heat transfer temperature differences by methods such as expanding heat transfer area and adopting heat transfer enhancement measures. While these measures can improve the efficiency of heat exchange equipment, they also increase the weight and size of such equipment, increase the design and manufacturing costs and inspection and maintenance costs, and ultimately manifest as an increase in power station construction investment and operation and maintenance costs.

Obviously, if the saved fuel costs are lower than or just offset the increased costs, not only will the power generation costs not decrease, but they may even increase. From an engineering point of view, these energy-saving measures are ineffective and redundant. Therefore, simply pointing out the ways and measures to improve the thermal efficiency of thermal equipment from the perspective of exergy analysis only has theoretical guiding significance, and it is difficult to judge whether it is cost-effective in engineering.

In order to overcome the shortcomings of exergy analysis methods and narrow the gap between theoretical analysis and engineering application, it is of great practical significance to introduce economic concepts into exergy analysis methods and conduct exergy economic analysis and research on thermal systems.

Exergy economic analysis of NPPs can solve the following problems:

(1) Provide a theoretical basis for determining reasonable electricity generation costs.
(2) Optimization design of the thermal system of the NPP, for example, determining the optimal steam pipe diameter and insulating layer thickness design.
(3) Determine the best operating mode of the thermal system of the NPP.
(4) Feasibility study of the overall scheme of the thermal system of the NPP.

8.2 Exergy Cost Equation

From the perspective of thermodynamics, exergy is a measure of the maximum work potential that energy theoretically possesses, so the exergy possessed by different forms of energy can be compared. However, from the perspective of practical application and economy, in a specific energy system, the unit exergy values of different forms and even the same form of energy are often not equivalent [25].

In a PWR NPP, the unit electrical exergy is not equivalent to the unit nuclear exergy, because there are many irreversible factors in the process of energy conversion, transmission, and distribution, causing various exergy losses. To obtain a unit of electrical exergy, about three units of nuclear exergy must be consumed. For specific thermal equipment, such as steam generators, due to the existence of heat transfer temperature difference, the exergy absorbed by the secondary circuit working fluid is less than the exergy released by the main coolant. Although the form of energy is the same, the unit exergy cost of the former is significantly higher than that of the latter [26].

In order to consider the non-equivalence of exergy in actual processes and economic factors, a feasible method is to assign a value to the physical quantity of exergy. For convenience of description, the exergy input or consumed as the cost of a unit is defined as fuel exergy, and the exergy as the unit's income is defined as product exergy. Thus, for a thermal system, in addition to being able to list mass balance equations, energy conservation equations, and exergy balance equations, an exergy cost equation can also be listed:

$$c_{\mathrm{P}} E_{x,\mathrm{P}} = c_{\mathrm{F}} E_{x,\mathrm{F}} + Z \tag{8.1}$$

where c_P, c_F are the average unit cost of product exergy and fuel exergy respectively, \$/kJ; $E_{x,P}$, $E_{x,F}$ are the product exergy and fuel exergy respectively, kJ; Z is the equipment operating cost, the sum of equipment investment depreciation, maintenance, and operating costs, \$.

The analysis using the value-oriented view of exergy is called thermal economics or exergy economics. This method is applied to the economic evaluation and optimal design of power systems, and has important engineering significance and broad development prospects.

For a thermal system, if its exergy cost equation before improvement is formula (8.1), the system's exergy efficiency is improved after improvement, and the system's operating cost has increased by ΔZ. Assuming that the fuel exergy $E_{x,F}$ and its unit cost c_F remain unchanged before and after the improvement, because the exergy loss is reduced, the product exergy has increased by $\Delta E_{x,P}$, then the system's exergy cost equation after improvement is

$$c_P + \Delta c_P = \frac{c_F E_{x,F} + Z + \Delta Z}{E_{x,P} + \Delta E_{x,P}} \tag{8.2}$$

Considering formula (8.1), the above formula can be simplified as:

$$\Delta c_P = \frac{\Delta Z - c_P \Delta E_{x,P}}{E_{x,P} + \Delta E_{x,P}} \tag{8.3}$$

If $\Delta c_P > 0$, it means that the cost increase due to the improvement of the thermal system is higher than the benefit obtained due to the efficiency improvement, indicating that the improvement measures are not worth the loss; $\Delta c_P = 0$, indicating that the income and expenditure offset each other; $\Delta c_P < 0$, indicating that the improvement measures have achieved energy-saving effects.

Therefore, using formula (8.3) as a standard to evaluate the merits of energy-saving measures for a thermal device or system can make the designed scheme technically feasible, economically reasonable, and in line with the principles of energy level matching and rational energy use.

According to the introduction in literature [24], the calculation method for the annualized operating cost Z of the equipment is as follows:

Assuming the service life of the equipment is n years, the initial investment cost is Z_0, and F_m represents the usage cost spent in the m-th year, the present value of which is

$$P_m = F_m \times PWF(i,m) \quad (m = 1,2,\cdots,n) \tag{8.4}$$

where $PWF(i,m)$ is the cash factor, defined as

$$PWF(i,m) = \frac{1}{(1+i)^m} \tag{8.5}$$

where i is the annual interest rate.

During these n years, the total present value of the annual usage cost is

$$\sum_{m=1}^{n} P_m = \sum_{m=1}^{n} F_m \times PWF(i,m) \tag{8.6}$$

The estimated residual value of the equipment at the end of its service life is SV. Calculated at present value, the total cost required for the operation of the equipment in n years, excluding fuel costs, is

$$P = Z_0 - SV \times PWF(i,n) + \sum_{m=1}^{n} F_m \times PWF(i,m) \tag{8.7}$$

The average annual operating cost of the equipment in n years is

$$Z = P \times CRF(i,n) \tag{8.8}$$

where $CRF(i, n)$ is the capital recovery factor, expressed as

$$CRF(i,n) = \frac{i}{1-(1+i)^{-n}} \tag{8.9}$$

Example 8.1 A cogeneration system using a back-pressure steam turbine, the shaft power of the unit is 10,300 kW; the exergy value of the live steam is 1328.15 kJ/kg, the exergy value of the turbine exhaust steam is 793.17 kJ/kg, the steam flow rate is 81,720 kg/h, the boiler feed water exergy value is 23.26 kJ/kg; the annualized cost of the boiler is known to be $Z_B = 3.75 \times 10^7$ \$/a, the annualized cost of the steam turbine is $Z_T = 2.63 \times 10^6$ \$/a, the exergy unit price of coal $c_f = 1.17$ \$/GJ; the exergy efficiency of the boiler $\eta_{ex,B} = 0.36$; the unit operates for 6132 h a year, try to calculate the exergy cost of the live steam c_{hs} and the exergy cost of the shaft power c_w.

Solution Taking the boiler as the analysis unit, assume that the exergy cost of the new steam is the same as that of the feed water, the exergy cost equation is

$$c_{\text{hs}}\left(E_{x,\text{hs}} - E_{x,\text{fw}}\right) = c_{\text{f}} \frac{E_{x,\text{hs}} - E_{x,\text{fw}}}{\eta_{\text{ex,B}}} + Z_{\text{B}}$$

The exergy cost of the new steam is

$$c_{\text{hs}} = \frac{c_{\text{f}}}{\eta_{\text{ex,B}}} + \frac{Z_{\text{B}}}{E_{x,\text{hs}} - E_{x,\text{fw}}}$$

$$= \frac{1.17}{0.36} + \frac{3.75 \times 10^7 \times 10^6}{81720 \times 6132 \times (1328.15 - 23.26)} = 60.6\$/\text{GJ}$$

Taking the steam turbine as the analysis unit, assuming that the exergy cost of the new steam is the same as that of the turbine exhaust steam, the exergy cost equation is

$$c_{w}W = c_{hs}\left(E_{x,hs} - E_{x,ls}\right) + Z_{T}$$

The exergy cost of the steam turbine shaft power is

$$\begin{aligned} c_{w} &= c_{hs}\frac{E_{x,hs} - E_{x,ls}}{W} + \frac{Z_{T}}{W} \\ &= 60.6\times\frac{81720\times6132\times(1328.15-793.17)}{10300\times3600} + \frac{2.63\times10^{6}\times10^{6}}{10300\times3600\times6132} \\ &= 83.02\$/\text{GJ} \end{aligned}$$

8.3 Exergy Cost Analysis of Main Thermal Equipment in NPPs

In addition to listing the equation of mass conservation, energy balance equation, and exergy balance equation for the thermal system exergy economic analysis, the cost balance equation must also be listed. The following introduces the specific form of the exergy cost equation of the main thermal equipment in a NPP.

8.3.1 Reactor

In the thermal system of a PWR NPP, the reactor is an important subsystem, its fuel exergy $E_{x,N}$ is nuclear fission energy, product exergy $E_{x,c}$ is the exergy obtained by the main coolant flowing through the core, the exergy cost equation is

$$c_{c} = \frac{E_{x,N}}{E_{x,c}}c_{N} + \frac{Z_{R}}{E_{x,c}} \tag{8.10}$$

Because

$$E_{x,N} = Q_{R},\quad E_{x,c} = G_{c}\left[h_{co} - h_{ci} - T_{0}\left(s_{co} - s_{ci}\right)\right] \tag{8.11}$$

where h_{co}, h_{ci} are the enthalpy of coolant inlet and outlet of reactor respectively, kJ/kg; s_{co}, s_{ci} are the entropy of coolant inlet and outlet of reactor respectively, kJ/(kg·K).

The average temperature of the coolant flowing through the reactor is

$$T_{m} = \frac{h_{co} - h_{ci}}{s_{co} - s_{ci}}\ \text{K} \tag{8.12}$$

Ignoring the heat dissipation loss of the reactor, there is $Q_R = G_c(h_{co} - h_{ci})$, then

$$\frac{E_{x,c}}{E_{x,N}} = 1 - \frac{T_0}{T_m} \tag{8.13}$$

Therefore, the unit cost of the reactor product exergy is

$$c_c = \frac{c_N}{1 - \frac{T_0}{T_m}} + \frac{Z_R}{Q_R\left(1 - \frac{T_0}{T_m}\right)} \tag{8.14}$$

For a PWR with a determined power output, if the unit cost of its fuel exergy c_N is known, under a certain environmental temperature T_0, the unit cost of product exergy c_c depends only on the average temperature T_m of the coolant and the operating cost Z_R of the reactor per unit time. Obviously, the higher the average temperature T_m of the coolant, the higher the requirements for the high-temperature corrosion resistance of the fuel element cladding material, the melting temperature of the nuclear fuel, etc. The development and application of new cladding materials and nuclear fuel materials often lead to an increase in the operating cost Z_R of the reactor per unit time, that is, Z_R is to some extent a hidden function of T_m.

It should be noted that the operating cost Z_R of the reactor per unit time is obtained through pure engineering economic analysis. When determining the optimal output cost through optimization calculations, the impact of changes in the average temperature of the coolant on Z_R must be considered.

8.3.2 Heat Exchange Equipment

Various types of heat exchange equipment occupy a large proportion in the thermal system of a PWR NPP, such as steam generators, steam reheaters, feedwater heaters, and condensers. Because they have similar working characteristics, a general discussion is made here.

Assume that the heat Q_1 released by the heating medium in the heat exchanger, the average temperature during the heat release process is T_{m1}, and its unit exergy cost is c_1; the heat Q_2 absorbed by the heat-absorbing medium, the average temperature during the heat absorption process is T_{m2}, and its unit exergy cost is c_2, and the operating cost of the equipment is Z, then the exergy cost equation of the heat exchange equipment is

$$c_2 = \frac{Q_1\left(1 - \frac{T_0}{T_{m1}}\right)}{Q_2\left(1 - \frac{T_0}{T_{m2}}\right)}c_1 + \frac{Z}{Q_2\left(1 - \frac{T_0}{T_{m2}}\right)} = \frac{1 - \frac{T_0}{T_{m1}}}{\eta_t\left(1 - \frac{T_0}{T_{m2}}\right)}c_1 + \frac{Z}{Q_2\left(1 - \frac{T_0}{T_{m2}}\right)} \tag{8.15}$$

where η_t is the thermal efficiency of the heat exchanger, $\eta_t = \frac{Q_2}{Q_1}$.

The input exergy cost c_1 depends on the previous subsystem of the heat exchanger. Assuming that T_{m1} is basically determined, in order to minimize the output exergy cost c_2, it is necessary to:

(1) Reduce the energy loss during the heat exchange process and make the thermal efficiency of the heat exchanger as close to 1 as possible;
(2) Adopt measures to enhance heat transfer, increase the average temperature of the heat-absorbing medium, and reduce the heat transfer temperature difference;
(3) The measures taken to improve the exergy efficiency of the heat exchanger should ensure that the ratio of the operating cost Z of the equipment to the product exergy $Q_2\left(1-\frac{T_0}{T_{m2}}\right)$ is not greater than the level before improvement.

The first two points are consistent with the conclusions drawn from the previous analysis of the heat exchanger; the third point, because considering the problem from an engineering economic perspective, makes the exergy analysis more practical in engineering.

For the condenser, the heat released by the exhausted steam is taken out of the power cycle by the circulating cooling water, and usually the average temperature of the circulating cooling water is the environmental temperature T_0, so there is no need to list the cost equation separately in the condenser, as long as the input cost of the system before the condenser is the lowest, the loss in the condenser can be minimized, of course, the premise is that the power of the circulating water pump does not increase significantly.

8.3.3 *Steam Turbine Generator*

For the steam turbine generator, for the sake of analysis, assume that the turbine under study has no intermediate extraction and is a single cylinder, then

$$c_{pr} = \frac{\left[N_i + G_s T_0 \left(s_z - s_i\right)\right] c_f + Z}{N_i \eta_m \eta_{ge}} \tag{8.16}$$

where N_i turbine internal power, kW; G_s Turbine steam consumption, kg/s; s_z, s_i the entropy values of the initial and final states of the steam expansion process in the turbine, kJ/(kg·K); c_f Unit cost of input for the turbine, \$/s; Z operating cost of the unit, \$/s; η_m, η_{ge} mechanical efficiency of the turbine and effective efficiency of the generator, respectively.

Based on the above formula, in order to achieve lower output exergy cost, we should consider:

(1) Optimizing the design of the turbine, reducing the irreversibility of the adiabatic expansion process within the turbine, and improving the mechanical efficiency during the power transmission process of the turbine;
(2) Optimizing the design of the generator to improve its effective efficiency.

This conclusion is consistent with the previous exergy analysis of the turbine generator unit.

In summary, when conducting an exergy economic analysis of a PWR nuclear power device, on the one hand, we can obtain the same conclusions as those obtained by using the exergy analysis method alone. On the other hand, with a grasp of the calculation of the operating cost Z of each device and system in the nuclear power device, according to the exergy cost equations listed for each system, we can determine how much economic and engineering potential the measures to improve the thermal economy of the device within the allowable range of thermodynamics have.

8.4 Relationship Between Operating Cost and Exergy Efficiency

The exergy cost equation of a certain thermal system is

$$c_{\mathrm{P0}}E_{x,\mathrm{P0}} = c_{\mathrm{F0}}E_{x,\mathrm{F0}} + Z_0 \tag{8.17}$$

Technical modifications are made to this thermal system to improve its exergy efficiency. Assuming that the unit cost of system fuel exergy remains unchanged, the quantity of product exergy remains unchanged, the system operating cost increases, the consumed fuel exergy decreases, and the unit cost of product exergy decreases. Then the exergy cost equation of the improved system is

$$c_{\mathrm{P}}E_{x,\mathrm{P}} = c_{\mathrm{F}}E_{x,\mathrm{F}} + Z \tag{8.18}$$

And it has

$$c_{\mathrm{F}} = c_{\mathrm{F0}}, \quad E_{x,\mathrm{P}} = E_{x,\mathrm{P0}} \tag{8.19}$$

$$E_{x,\mathrm{F}} = E_{x,\mathrm{F0}} - \Delta E_{x,\mathrm{F}}, \quad c_{\mathrm{P}} = c_{\mathrm{P0}} - \Delta c_{\mathrm{P}}, \quad Z = Z_0 + \Delta Z \tag{8.20}$$

Then Eq. (8.18) is further simplified to

$$\Delta Z = c_{\mathrm{F0}}\Delta E_{x,\mathrm{F}} - \Delta c_{\mathrm{P}}E_{x,\mathrm{P0}} \tag{8.21}$$

If the unit cost of product exergy remains unchanged, then $\Delta c_P = 0$, Eq. (8.21) becomes

$$\Delta Z = c_{F0} \Delta E_{x,F} \tag{8.22}$$

Since the system exergy efficiency $\eta_t = \dfrac{E_{x,P}}{E_{x,F}}$, and

$$\Delta E_{x,F} = E_{x,F0} - E_{x,F} = E_{x,F0} - \frac{E_{x,P0}}{\eta_t}$$

So

$$\delta E_{x,F} = \frac{\Delta E_{x,F}}{E_{x,F0}} = 1 - \frac{\eta_{t0}}{\eta_t} = \frac{\eta_t - \eta_{t0}}{\eta_t} = \delta \eta_t \tag{8.23}$$

Combining Eqs. (8.21) and (8.23), we get

$$\delta Z = \frac{\Delta Z}{Z_0} = \frac{c_{F0} \Delta E_{x,F}}{c_{P0} E_{x,P0} - c_{F0} E_{x,F0}} = \frac{\dfrac{\Delta E_{x,F}}{E_{x,F0}}}{\dfrac{c_{P0} E_{x,P0}}{c_{F0} E_{x,F0}} - 1} = \frac{\delta \eta_t}{\dfrac{c_{P0}}{c_{F0}} \eta_{t0} - 1} \tag{8.24}$$

Let $\alpha = \dfrac{1}{\dfrac{c_{P0}}{c_{F0}} \eta_{t0} - 1}$, then

$$\delta Z_{cr} = \alpha \cdot \delta \eta_t \tag{8.25}$$

The above equation can be considered as the critical judgment relationship for whether the improvement measures of the NPP are economical. If the thermal system of the NPP is improved to increase the efficiency of the device, and the increase in total operating costs is lower than the critical value, it means that the improvement measures are economically feasible, otherwise, they are uneconomical.

8.5 Optimization Method of Exergy Economics for Thermal Systems

Exergy analysis points out the fundamental reasons and main parts of energy loss in thermal systems, and indicates the direction for improving equipment and improving system thermal efficiency. Based on this, using the method of exergy economic analysis to optimize the design of the thermal system can avoid the situation of blindly pursuing the improvement of exergy efficiency and ignoring the economy of the improvement plan.

8.5.1 General Steps of Exergy Economic Optimization

The exergy economic optimization design of the thermal system can generally be divided into the following seven steps.

1. Establish the objective function for optimization

   The objective function is a mathematical expression of the goal to be achieved during system design, reflecting the state of the system and the relationship between the main influencing variables. When carrying out the thermal economic optimization design of the energy system, first determine the thermal economic goal pursued by the system design according to the design purpose.

   (a) In order to express the goal of thermal economic optimization with design variables, the system is usually divided into several subsystems based on its structure or process flow;
   (b) Based on the balance of working substance, energy, exergy, and economy of each subsystem, establish their mathematical models and find out the function relationship between each exergy flow and design variables;
   (c) According to the growth of social economy during the system's usage period, find out the function relationship between the construction cost of each subsystem and the design variables;
   (d) According to the growth of energy prices during the system's usage period, determine the unit price of each exergy flow input into the system.

   After completing the above work, the goal of thermal economic optimization can be expressed with design variables, and the objective function of thermal economic optimization can be obtained.

   In practical problems, establishing the objective function is a complex task, and there is no fixed and unified method, only specific analysis of specific problems.

   The objective function includes considerations of current costs and income, as well as long-term costs and income. Therefore, it may need to consider inflation, bank interest rates, equipment depreciation rates, and price forecasts of various raw materials (including fuel), which further complicates the establishment of the objective function.
2. Establish constraint equations

   System design is subject to various physical, technical, legal, and social conditions. During design, it is necessary to correctly analyze the relevant constraints and express them as a series of constraint equations in the form of equalities or inequalities. In general, inequality constraint equations can be transformed into equality form by adding slack variables.

   The objective function and constraint equations of thermal economic optimization constitute the thermal economic optimization model of system design. Determine which variables to use as independent variables and which to use as dependent variables based on the structural characteristics of the system and the convenience of optimization calculation.

3. Select and determine the system through the system flow chart
   Determine the flow chart, clarify the process to be analyzed, annotate the equipment code on the flow chart, correspond each main parameter with the design scheme, and various variables describing the system operation state.
4. Determine the algorithm
   The method of solving simultaneous equations is called an algorithm. The work of solving the constraint equation group depends on what is chosen as the decision variable. It is important to divide the variables into decision variables (independent variables) and state variables (dependent variables). Proper selection can greatly reduce the amount of calculation. Equipment prices should also be listed as equations, or estimated equations.
5. Develop the flow chart and algorithm steps
   At least calculate one operating condition first to check the arranged computer program.
6. Obtain sensitivity.
   Sensitivity refers to the sensitivity of the objective function to changes in decision variables.
7. Use sensitivity to determine what changes should be made to improve the system, and make plans for the second round of analysis, determining what to do next.

After completing the above seven steps, the analysis conclusion will be obtained. Using the sensitivity of multiple variables, a multi-dimensional space vector can be formed in the concept, indicating the degree of improvement in different directions. Therefore, this vector diagram (imaginary vector diagram, because multi-dimensional vectors cannot be drawn) is called an improvement vector. If the improvement vector equals zero, it means that the system is already in an optimal state.

Example 8.2 The system consists of a single device, with one stream of exergy input and one stream of exergy output. It is known that, $E_{x,1} = E_{x,\mathrm{p}}/\eta_{\mathrm{e}}$, $Z = k\eta_{\mathrm{e}}E_{x,\mathrm{p}}$ where k is a constant related to the annual interest rate and usage period. C_0 is the unit price of the input system exergy stream, C_0 and $E_{x,\mathrm{p}}$ are given constants, η_{e} is the exergy efficiency of the system, which serves as the independent design variable of this system. The optimal design variable of this system is sought.

Solution Given $E_{x,\mathrm{p}}$, the objective function of the system's thermal-economic optimization is

$$Y = C_0 E_{x,1} + Z = \frac{C_0 E_{x,\mathrm{p}}}{\eta_{\mathrm{e}}} + k\eta_{\mathrm{e}} E_{x,\mathrm{p}}$$

To find the extremum of the objective function Y, set $\mathrm{d}Y/\mathrm{d}\eta_{\mathrm{e}} = 0$, the optimal design variable η_{e}^* and the minimum annual cost Y^* can be obtained:

$$\eta_{\mathrm{e}}^* = \sqrt{C_0 / k}$$

$$Y^* = 2\sqrt{C_0 k E_{x,\mathrm{p}}}$$

It can be seen that the independent design variable η_e of this system is not the bigger the better, but there is an optimal value. Only when η_e is taken as the optimal value, can the system designed have the minimum annual product cost.

8.5.2 Thermal-Economic Optimization by Lagrange Method

For a thermodynamic system undergoing thermal-economic optimization design, its objective function is

$$Y = \sum_i c_{0,i} E_{x,i}\left(x_1, x_2, \cdots, x_m, y_1, y_2, \cdots, y_n\right) + \sum_j Z_j\left(x_1, x_2, \cdots, x_m, y_1, y_2, \cdots, y_n\right) \quad (8.26)$$

where $E_{x,i}$, $c_{0,i}$ are the annual exergy amount of the i-th input system and its corresponding exergy unit price; Z_j is the annualized cost of the j-th subsystem in the system; x_m is the independent design variable of the system; y_n is dependent variable.

This optimization problem has n constraints, all of which can be represented as equality constraint equations, i.e.,

$$F_k\left(x_1, x_2, \cdots, x_m, y_1, y_2, \cdots, y_n\right) = 0 \quad (k = 1, 2, \cdots, n) \quad (8.27)$$

For the above set of equality constraint equations, if each dependent variable $y_k(x_1, x_2, \cdots, x_m)$ can be solved and substituted into the objective function, then the original optimization model with equality constraints is transformed into an unconstrained optimization model.

Under these conditions, just set $\frac{\partial Y}{\partial x_i} = 0\left(i = 1, 2, \cdots, m\right)$, the extremum point can be determined from these m equations, and the optimal design variables $(x_1, x_2, \cdots, x_m)_{\mathrm{opt}}$ can be obtained; or the unconstrained optimization calculation method can be used for optimization, and the optimal design variables $(x_1, x_2, \cdots, x_m)_{\mathrm{opt}}$ can also be obtained.

In practice, the above optimization method is not always effective, mainly for the following two reasons:

(1) It is sometimes difficult to obtain the explicit expression of each y_k from the equality constraint equations, or even if it can be obtained and substituted into the objective function, the resulting objective function may be too lengthy and complex to find the optimal solution;
(2) The choice of dependent variables is inappropriate, making it difficult to find the optimal solution of the original optimization model.

Therefore, in practice, we generally avoid directly obtaining the explicit expression of each dependent variable and then substituting it into the objective function, and usually use the Lagrange multiplier method.

The Lagrange method is an important means in optimization technology to change the optimization with constraints into unconstrained optimization, and it is a powerful method to study the impact of a decision variable change on the entire system objective function. The basic idea of this method is to introduce a Lagrange multiplier λ, incorporate all the constraint equations into the objective function, obtain a new function, called the Lagrange function, optimize this function and then calculate the Lagrange multiplier λ. Therefore, the Lagrange method is also known as the Lagrange multiplier method.

The Lagrange multiplier is $\lambda_k(x_1, x_2, \cdots, x_m, y_1, y_2, \cdots, y_n)(k = 1, 2, \cdots, n)$, which is multiplied by the constraint equation and then added to the objective function to form the Lagrange function L, i.e.,

$$L\left(x_1, x_2, \cdots, x_m, y_1, y_2, \cdots, y_n\right) = Y + \sum_{k=1}^{n} \lambda_k F_k \tag{8.28}$$

If the original optimization model has an extreme value, it must be in the extreme value obtained by the Lagrange function. To find the extreme point, let

$$\frac{\partial L}{\partial x_i} = \frac{\partial Y}{\partial x_i} + \sum_{k=1}^{n} \lambda_k \frac{\partial F_k}{\partial x_i} = 0 \quad \left(i = 1, 2, \cdots, m\right) \tag{8.29}$$

$$\frac{\partial L}{\partial y_j} = \frac{\partial Y}{\partial y_j} + \sum_{k=1}^{n} \lambda_k \frac{\partial F_k}{\partial y_j} = 0 \quad \left(j = 1, 2, \cdots, n\right) \tag{8.30}$$

And combine with $F_k(x_1, x_2, \cdots, x_m, y_1, y_2, \cdots, y_n) = 0(k = 1, 2, \cdots, n)$, you can find the extreme point from these $m + 2n$ equations with $m + 2n$ variables (x_i, y_j, λ_k), that is, the optimal design variable value.

8.5.3 Thermal-Economic Isolation Principle

The thermodynamic systems encountered in engineering practice are usually very complex and large, and it is often difficult to describe them completely with mathematical methods; even if an optimization model can be established, the optimization calculation is difficult to carry out due to the large number of optimization variables (for example, more than twenty), resulting in a large amount of calculation. If the entire thermodynamic system is decomposed into several subsystems, each subsystem has fewer optimization variables than the entire system, and the optimization calculation of the subsystem is much simpler than the optimization calculation of the entire system. By optimizing each subsystem one by one, the optimization of the entire system can be achieved, which is a clever way to simplify complex problems.

Generally speaking, the optimization of each subsystem does not equal the optimization of the entire system. Only when the optimal variable values obtained from the separate optimization calculations of each subsystem are the optimal variable values obtained from the optimization calculations of the entire system, can the requirements of local optimization and overall optimization be met. According to the thermal-economic isolation theory proposed by Tribus and Evans, as long as it can be proved that the several subsystems into which the entire thermodynamic system is decomposed are mutually independent in thermodynamics and economic relations, that is, they meet the conditions of thermal-economic isolation, the optimization of the entire system can be achieved by optimizing each subsystem.

The principle of thermal-economic isolation is to divide a complex energy system into many subsystems, and isolate each subsystem from the entire system for separate optimization, and the optimization results of all subsystems are the optimization results of the entire system. The principle of thermal-economic isolation points out a direction for the thermal-economic optimization of energy systems from complex to simple.

Isolation means to separate each subsystem from the interaction (i.e., energy interaction) with other subsystems, while also reflecting that it is a part of the entire system and is connected with the system.

As shown in Fig. 8.1, a thermodynamic system is divided into 3 subsystems, the independent design variable of the ith subsystem is x_i, the annualized cost is Z_i, the unit price of the input system exergy flow is C_i, and the annual exergy flow $E_{x,4}$ of the output system is a specified value.

With the annual cost of the system output exergy flow as the optimization goal, the optimization model is expressed as

$$\min Y = C_1 E_{x,1}\left(x_1, E_{x,2}\right) + Z_1\left(x_1, E_{x,2}\right) + Z_2\left(x_2, E_{x,3}\right) + Z_3\left(x_3\right) \tag{8.31}$$

$$E_{x,2} = F_2\left(x_2, E_{x,3}\right) \tag{8.32}$$

$$E_{x,3} = F_3\left(x_3\right) \tag{8.33}$$

Introduce Lagrange multipliers λ_2 and λ_3, transform the constrained optimization model into an unconstrained optimization model, that is,

$$\min L = Y + \lambda_2 \cdot \left(F_2 - E_{x,2}\right) + \lambda_3\left(F_3 - E_{x,3}\right) \tag{8.34}$$

In order to obtain the thermal-economic optimal design variables x_1^*, x_2^*, x_3^* of the system, the partial derivatives of the variables x_1, x_2, x_3, $E_{x,2}$, $E_{x,3}$, λ_2 and λ_3 in the Lagrange function L are taken respectively, and they are set to zero, thus obtaining the following set of equations:

$$\frac{\partial L}{\partial x_1} = 0 \quad \text{i.e.} \quad \frac{\partial\left(C_1 E_{x,1} + Z_1\right)}{\partial x_1} = 0 \tag{8.35}$$

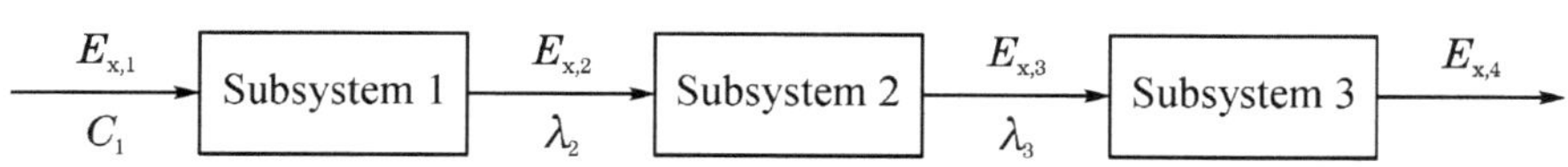

Fig. 8.1 A thermal system composed of three subsystems

$$\frac{\partial L}{\partial x_2} = 0 \quad \text{i.e.} \quad \frac{\partial\left(\lambda_2 E_{x,2} + Z_2\right)}{\partial x_2} = 0 \tag{8.36}$$

$$\frac{\partial L}{\partial x_3} = 0 \quad \text{i.e.} \quad \frac{\partial\left(\lambda_3 E_{x,3} + Z_3\right)}{\partial x_3} = 0 \tag{8.37}$$

$$\frac{\partial L}{\partial E_{x,2}} = 0 \quad \text{i.e.} \quad \lambda_2 = \frac{\partial\left(C_1 E_{x,1} + Z_1\right)}{\partial E_{x,2}} \tag{8.38}$$

$$\frac{\partial L}{\partial E_{x,3}} = 0 \quad \text{i.e.} \quad \lambda_3 = \frac{\partial\left(\lambda_2 E_{x,2} + Z_2\right)}{\partial E_{x,3}} \tag{8.39}$$

$$\frac{\partial L}{\partial \lambda_2} = 0 \quad \text{i.e.} \quad E_{x,2} = F_2\left(x_2, E_{x,3}\right) \tag{8.40}$$

$$\frac{\partial L}{\partial \lambda_3} = 0 \quad \text{i.e.} \quad E_{x,3} = F_3\left(x_3\right) \tag{8.41}$$

From the above set of equations, the thermal-economic optimal design variables x_1^*, x_2^*, x_3^*, $E_{x,2}^*$, $E_{x,3}^*$ of the system can be solved. From this set of equations, it can be seen that:

(1) The Lagrange multiplier λ_i has the dimension of exergy unit price, and it plays the role of exergy unit price of the input exergy flow of each subsystem in the system, because it is derived from differentiation, it is called differential exergy unit price or marginal exergy unit price.

The exergy unit price of the input exergy flow of subsystem 2 is

$$C_2 = \frac{C_1 E_{x,1} + Z_1}{E_{x,2}} \tag{8.42}$$

The marginal exergy unit price of this exergy flow is

$$\lambda_2 = \frac{\partial\left(C_1 E_{x,1} + Z_1\right)}{\partial E_{x,2}}$$

λ_2 and C_2 have the same units. After $E_{x,3}$ is solved, the local optimum of subsystem 2 should be obtained by $\frac{\partial\left(C_2 E_{x,2} + Z_2\right)}{\partial x_2} = 0$, and $\frac{\partial\left(\lambda_2 E_{x,2} + Z_2\right)}{\partial x_2} = 0$ corresponds to it. These two equations indicate that λ_2 and C_2 have the same function. The same applies to λ_3.

(2) If the marginal exergy unit price of each side is constant, each subsystem can be isolated from the entire system, that is, the local thermal economic optimum of each subsystem is the global thermal economic optimum of the entire system.
Because the local thermal economic optimum of subsystem 3 is obtained by Eqs. (8.37) and (8.41). After $E^*_{x,3}$ is obtained by the local thermal economic optimization of subsystem 3, the local thermal economic optimum of subsystem 2 is obtained by Eqs. (8.36) and (8.40). After $E^*_{x,2}$ is obtained by the local optimization of subsystem 2, the local thermal economic optimum of subsystem 1 is obtained by Eq. (8.35). The local thermal economic optimum of each subsystem can be independently obtained, and there is no functional relationship with other subsystems. All local thermal economic optimization models of subsystems just make up the global thermal economic optimization model of the entire system, that is, Eqs. (8.35)–(8.41).
(3) To achieve thermal economic isolation, the input flow of each subsystem needs to be calculated in sequence. This can be seen from Eqs. (8.40) and (8.41). If the system output flow $E_{x,4}$ is given, $E^*_{x,3}$ can be determined by the local optimization of subsystem 3, then the local optimization of subsystem 2 can be carried out in sequence to determine $E^*_{x,2}$, and finally $E^*_{x,1}$ can be determined by the local optimization of subsystem 1.

Summarizing the above three points, the conditions for thermal economic isolation to be established are that the marginal exergy unit price of each subsystem is constant, and the input exergy flow of each subsystem can be determined in sequence.

Example 8.2 As shown in Fig. 8.1, when the unit exergy price of the input system exergy flow C_1 and the output system exergy flow $E_{x,4}$ are given, the independent design variables of the system are x_{11}, x_{21}, x_{22}, x_{31}, x_{32}, to determine whether this system can achieve thermal economic isolation. It is known that a_i, b_i, d_i are constants, F_i, G_i represent the function relationship with the independent design variable x_{ij}, and have

$$E_{x,1} = a_1 + b_1 E_{x,2} + F_1(x_{11})$$

$$Z_1 = d_1 E_{x,2} + G_{11}(x_{11})$$

$$E_{x,2} = a_2 + b_2 E_{x,3} + F_2(x_{21}, x_{22})$$

$$Z_2 = d_2 E_{x,3} + G_2(x_{21}, x_{22})$$

$$E_{x,3} = (a_3 E_{x,4} + b_4 E_{x,4}) \cdot F_3(x_{31}, x_{32})$$

$$Z_3 = (d_3 + d_4 E_{x,4}) G_3(x_{31}, x_{32})$$

Solution With the annual cost of the system output exergy flow as the optimization goal, then the objective function of the system's thermal economic optimization is

$$\min Y = C_1 E_{x,1} + Z_1 + Z_2 + Z_3$$

Introduce the Lagrange multipliers λ_2 and λ_3, then the marginal price of subsystem 2 and subsystem 3 is constant, that is

$$\lambda_1 = \frac{\partial\left(C_1 E_{x,1} + Z_1\right)}{\partial E_{x,2}} = b_1 C_1 + d_1$$

$$\lambda_3 = \frac{\partial\left(\lambda_2 E_{x,2} + Z_2\right)}{\partial E_{x,3}} = b_2 \cdot \left(b_1 C_1 + d_1\right) + d_2$$

When the system's independent design variables x_{ij} and $E_{x,4}$ are known, $E_{x,3}$, $E_{x,2}$ and $E_{x,1}$ can be determined in sequence by the function relationship of each subsystem input flow, so this system can achieve thermal economic isolation. To obtain the system's thermal economic optimal design variables, the local thermal economic optimization of subsystems 3, 2, 1 can be carried out in sequence, and the optimization result is the global thermal economic optimization result of the entire system.

8.5.4 Principle of Quasi-Isolation in Thermal-Economics

The principle of thermal-economic isolation for energy systems points out the direction of simplifying complexity for thermal-economic optimization, but it has many practical difficulties. This is because the conditions for thermal-economic isolation are very strict, requiring the marginal exergy cost of each subsystem to be constant. As can be seen from Example 8.2, for the marginal exergy cost to be constant, the input exergy flow of each subsystem and its annualized cost need to satisfy the linear function of its output exergy flow, and can be divided into two independent parts with the independent design variables of the subsystem. In practice, the marginal exergy cost of each subsystem is not constant, which greatly limits the practicality of thermal-economic isolation. Since the idea of thermal-economic isolation provides an effective direction for system thermal-economic optimization, we should strive in this direction.

The marginal exergy cost of each subsystem in the system is usually not a constant, but a variable. It is a function of the subsystem's independent design variables and output exergy flow, as seen in Eqs. (8.38) and (8.39). When the independent design variables and output exergy flow of each subsystem achieve the thermal-economic optimum of the system, the marginal exergy cost of each subsystem also achieves the thermal-economic optimum. If the marginal exergy cost of each subsystem can be found in advance when optimizing the thermal-economics of the energy system, the system will no longer be restricted by the constant marginal exergy cost in thermal-economic isolation, and thermal-economic isolation optimization can be carried out.

The method of finding the marginal exergy cost λ_i^* of each subsystem when the system reaches the thermal-economic optimum, thus achieving the thermal-economic isolation of the system, is called thermal-economic quasi-isolation. So how can we find λ_i^*? This is the key issue of thermal-economic quasi-isolation.

Now, let's take the system shown in Fig. 8.1 as an example and propose a method of thermal-economic quasi-isolation optimization. The characteristic of this method is that it finds λ_i^* while completing the thermal-economic optimization of the system.

(1) First, complete the basic thermal-economic optimization of the system
 Let $\lambda_1^{(1)} = \lambda_2^{(1)} = \lambda_3^{(1)} = C_1$, under the given C_1 and $E_{x,4}$, complete the thermal-economic optimization of each subsystem in turn, that is, perform the optimization calculation of the following formula

$$\min Y_i^{(2)} = \lambda_i^{(1)} E_{x,i} + Z_i \quad (i = 3,2,1)$$

From this, we can get the basic thermal-economic optimum values of the system $x_i^{(1)}$, $E_{x,i}^{(1)}$, $i = 1,2,3$.

(2) Correction of marginal exergy cost
 Let

$$\lambda_i^{(j)} = \frac{\partial\left(\lambda_{i-1}^{(j-1)} E_{x,i-2} + Z_{i-1}\right)}{\partial E_{x,i}} \quad (i = 2,3)$$

$$\lambda_1^{(j)} = \lambda_1^{(j-1)} = C_1 \quad (j = 1,2,\cdots)$$

Replace the old marginal exergy cost $\lambda_{i(j-1)}$ with the new marginal exergy cost $\lambda_{i(j)}$ for each subsystem, and then complete the thermal-economic optimization of each subsystem in turn, that is,

$$\min Y_i^{(j)} = \lambda_i^{(j)} E_{x,i} + Z_i \quad (i = 3,2,1)$$

(3) Convergence judgment
 If it satisfies

$$\left|\frac{\lambda_i^{(j)} - \lambda_i^{(j-1)}}{\lambda_i^{(j)}}\right| < \varepsilon \quad (i = 2,3)$$

The thermal-economic optimization calculation of the thermodynamic system has reached the required accuracy, and the calculation can be terminated. The results obtained from the jth optimization calculation of each subsystem are the thermal-economic optimum values of the system.

8.6 Discussion on the Thermal-Economic Optimization of NPP Thermodynamic System

For complex and large systems like nuclear power devices, it is necessary to decompose them into many subsystems for easier analysis, based on certain principles. The thermal-economic functional analysis introduced in Yunze et al. [37] is a method for optimizing and improving complex thermodynamic systems. According to this method, a thermodynamic system is considered to be composed of a series of related units, each with a specific function (to achieve a certain purpose or output a certain product). A unit can be a device, or it can be composed of a series of devices. For the optimization of complex thermodynamic systems, decomposing them into smaller systems has many advantages. The analysis of each system is much easier than the analysis of the entire system, and more in-depth research can be conducted.

Yunze et al. [37] points out that complex thermodynamic systems can only be transformed into problems of optimizing each system separately if they meet the conditions of isolation. The complex thermodynamic system is decomposed into multiple subsystems, and the mutual influence between two subsystems with material and energy exchange is represented by marginal cost. Yunze et al. [37] and Qiang et al. [38] both point out that by introducing marginal cost, the entire optimization problem can be transformed into a problem of local system optimization.

To optimize the design of a PWR NPP, it must be based on the principle of isolation. The so-called principle of isolation means that the optimization design of the entire system can be transformed into the problem of optimizing each subsystem separately under certain conditions. This principle provides a method for solving the optimization design of the entire system by starting with the optimization of individual subsystems. Since the factors affecting the optimization design of the subsystem are obviously simpler than the entire system, this provides a convenient and feasible way to solve the optimization design of complex systems. According to Qiang et al. [38], Evans and others have theoretically demonstrated the following conclusion: when analyzing steady flow systems, the requirements of isolation can be met as long as the concept of exergy is used. This is also an important reason why thermal-economics needs to be based on the concept of exergy. When the PWR NPP is operating stably under certain conditions, it can be approximately considered as a steady flow system, thus meeting the conditions of isolation.

Since exergy economics studies the system or device in two environments (physical environment and economic environment), the objective function often includes the following two types:

(1) The criterion of minimum exergy loss.
Its typical function relationship is

$$\Phi_{\text{o.e}} = \sum E_{x,\text{i}} - \sum E_{x,\text{o}} \tag{8.43}$$

where $\Phi_{\text{o.e}}$ is system exergy loss; $\sum E_{x,\text{i}}$, $\sum E_{x,\text{o}}$ are the total input and output exergy of the system, respectively.

(2) The criterion of minimum total system cost.
Its typical function relationship is

$$\Phi_{\text{o.e}} = \sum_{i=1}^{n} c_{\text{f},i} E_{x,i} + \sum_{j=1}^{m} c_j - \sum_{j=1}^{m} c_{\text{p},j} E_{\text{p},j} \tag{8.44}$$

where $\Phi_{o.e}$ total system cost; $c_{f,i}$ the average unit cost of the i-th fuel exergy entering the system; $E_{x,i}$ the i-th fuel exergy entering the system; c_j the operating cost of the equipment in the j-th area of the system; $c_{p,j}$ the average unit cost of the product exergy in the j-th area of the system; $E_{p,j}$ the product exergy in the j-th area of the system.

For the optimization design of the thermal system of a PWR NPP, if the minimum exergy loss is taken as the objective function, it may result in a design scheme with the highest exergy efficiency but economically unprofitable. Therefore, the optimization design of PWR NPPs usually takes the minimum total system cost as the objective function.

The exergy economic analysis of NPP thermal systems allows for practical engineering evaluations of whether measures and approaches to improve the exergy efficiency of thermal systems are economically feasible, thereby bridging the gap between theoretical analysis and engineering application. However, since exergy economics is a new interdisciplinary field combining thermodynamics and economics, with a wide range of content, many basic theories need to be further researched, deepened, and standardized. In addition, due to the numerous and complex structure of equipment in NPP thermal systems, a lot of work is still needed to establish a comprehensive exergy economic analysis method for NPP thermal systems.

Questions and Exercises

1. What is the basic content studied by exergy economics?
2. What is the basic meaning of the exergy cost equation?
3. Briefly describe the exergy cost equation and the impact of exergy efficiency on operating costs for the main equipment of a NPP.
4. What are the basic steps in the optimization design of exergy economics in thermodynamic systems?

Chapter 9
Online Analysis of Thermal Economy of NPPs

Faced with market competition and challenges, NPPs must consider how to improve the economic operation level of the units while ensuring the safe and stable operation of the NPPs, in order to gain more market share in power generation. There are several ways to improve the economy of NPPs. In the design stage of the unit, it is mainly through optimized design, reasonable use of mature and advanced technologies, to improve the thermal economy of the power plant thermal system and the technical economy of the NPP; In the production management stage of the unit, it is mainly through improving the level of operation management to ensure the economic operation of the unit.

In conventional power plants, one of the goals of operation management is to comprehensively monitor the economy of the operation of the power generation units, to calculate the thermal economic indicators of the units in a timely and accurate manner, to analyze the size, location, and reasons for energy loss during operation, so as to guide the operation personnel to operate, reduce controllable energy loss, and improve the economy of the unit operation. The ultimate goal is to guide the operators to choose a reasonable adjustment method, so that the unit is in or close to the optimal operating state, and to minimize controllable losses [49].

Commercially operated NPPs, due to factors such as operating level and equipment status, may experience a decrease in thermal efficiency and inability to output full power. It is necessary to find out the reasons in time and propose reasonable and effective adjustment measures to restore the unit to the optimal operating state. However, due to the complexity of the NPP system and the large number of equipment, and the different operating conditions and environments of each NPP, if there is no quick and effective technical support, it often takes a long time to investigate and analyze the factors affecting the operating state of the unit by manual means alone. This not only leads to excessive consumption of manpower and material resources, but also cannot identify and solve the problem in time, which delays production and causes economic losses. Therefore, the development of an online analysis system for the thermal economy of NPPs, real-time monitoring of the

M. Peng et al., *Thermal Analysis of Nuclear Power Plants*, Nuclear Science and Technology, https://doi.org/10.1007/978-981-96-8983-5_9

operating performance of NPPs, providing decision-making basis and relevant technical data for operation and management personnel in a timely manner, can not only detect potential faults of thermal cycle equipment in advance and handle them in time to avoid shutdown caused by fault expansion; On the other hand, it can calculate and analyze the thermal economy of the full-condition operation of the NPP in real time and online, so that the operators can grasp the various thermal economic indicators of the unit at any time, find out the reasons for the decline of the indicators and improvement measures, and thus take effective measures to fully exploit the efficiency of the unit, which has practical significance for energy saving and consumption reduction and improving the operation management level of NPPs [50].

9.1 Development and Application of Online Analysis Technology for Unit Performance

9.1.1 Development History of Online Analysis Technology for Unit Performance

In the 1950s, thermal power plants began to use computer technology to perform simple functions such as monitoring, recording, and over-limit alarms for operating parameters. In the early 1970s, developed Western countries began to study the performance monitoring technology of power station thermal systems. Among them, the Electric Power Research Institute (EPRI) in the United States concentrated a large number of human and material resources for in-depth research, and cooperated with many power companies to complete a large number of experimental verification work. On this basis, a series of industry norms and standards were formulated, laying a solid foundation for the development of unit operation optimization technology worldwide. The representative thermal system monitoring software of this period mainly includes PEPSE from GE in the United States, PRESTO from Oak Ridge National Laboratory (ORNL), and AHBP from CEGB in the UK.

Due to measurement technology, measurement software reliability, and other technical issues, the early power plant unit operation performance monitoring was only as an adjunct part of the Data Acquisition and Data Processing System (DAS), based on the Power Test Code (PTC) of the American Society of Mechanical Engineers (ASME), to complete a small number of index calculations. Until the early 1980s, with the continuous development of measurement technology, computers, and communication technology, and the continuous improvement of power plant safety and economic management requirements, power station performance monitoring began to develop towards operation optimization that aims to guide on-site operation and can complete the performance and overall performance analysis and diagnosis of various parts of the unit. Expert systems and artificial neural networks began to be introduced into the diagnostic systems of thermal power plants, and countries around the world have conducted in-depth research on this, opening up new ways for power plant performance diagnosis. At this time, the successful development and use include the B&V performance monitoring system from Black&Veatch in the United States, the PERFEXS expert system from ANSALDO in Italy, and so on.

In the past 30 years, the performance monitoring and diagnostic software of thermal power units has been continuously improved, and its functions have gradually become powerful, forming a complete thermal power unit operation optimization software package. So far, most foreign Distributed Control System (DCS) suppliers have developed a varying number of thermal power plant optimization management software, among which the more complete functions mainly include the Sienergy system from Siemens in Germany, the EtaPRO system from General Physics in the United States, the Performer system from Elaghbailey in the United States, the SmartProcess from Westinghouse Electric in the United States, and the Optimax system from ABB in Switzerland. These software have been applied to varying degrees in large thermal power plants that have been put into production abroad in recent years.

In the 1980s, China began to introduce large foreign power units. Due to the expensive equipment and operational level limitations, for a long time, power station personnel were more concerned about the safety of the equipment and less considered the economy of the unit. On the other hand, the operation support system of early DCS products was usually a dedicated operating system developed by each DCS manufacturer, making secondary development difficult and costly, so there was less research on performance monitoring and operation optimization. Until the early 1990s, with the continuous improvement of operational levels, the safety of power station equipment was fully guaranteed, and power station performance monitoring gradually received attention, and power station performance monitoring software appeared, but this software belonged to the category of Power Plant Information Management System (MIS). In the late 1990s, China proposed the "Plant-level Supervisory Information The concept of the "System Information System (SIS)" is centered around unit performance monitoring and operational guidance. In 2000, SIS was officially incorporated into the "Technical Regulations for Thermal Power Plant Design", which clearly states, "When the planned capacity of the power plant is 1200 MW or above, and the single unit capacity is 300 MW or above, a plant-level monitoring information system can be set up". After more than 10 years of laboratory-stage research and development, China has accumulated considerable experience in thermal power unit performance monitoring and operation optimization, and some of the developed thermal power unit operation optimization software has been applied in some domestic units. However, compared with the world's advanced level, China's thermal power unit online analysis software is still relatively backward, such as some modules are just imitations of the existing business functions of the power station and simple computerization, the function realization and user actual demand still have a distance, the system function is still mainly performance monitoring, operation optimization, fault diagnosis and operation guidance and other aspects of development application is relatively insufficient, compared with foreign similar software there is a big gap.

Considering the domestic and foreign development status and trend of the comprehensive thermal power unit performance monitoring and diagnostic system, its hardware is developing from centralized to distributed workstations, and the software function is developing from simple calculation and analysis towards generalization, specialization, and intelligence. The diagnostic content is developing from the overall economic calculation towards quantitative analysis of the reasons for the deviation of unit economy and the optimization scheduling between multiple machines.

9.1.2 Application of SIS in Thermal Power Units

SIS has gradually developed on the basis of thermal performance online monitoring and diagnostic systems. Through real-time monitoring and analysis of the production process data of thermal power units, it realizes the optimization control and economic load distribution of the power plant production process, fully utilizes the capacity of the main and auxiliary equipment within the entire power plant, achieves the purpose of the entire power plant production system operating at the best conditions, and at the same time, SIS provides complete historical/real-time data information of the power plant production process, which can serve as a reliable production information resource for the power company's information network, enabling company management and technical personnel to grasp the production information and auxiliary decision-making information of each power generation enterprise in real time, fully utilize and share information resources, and improve decision-making levels.

9.1.2.1 System Composition

The DCS technology of domestic thermal power units is very mature, it contains all kinds of data, such as real-time data of field system measurement points and other information, and also includes the data needed for thermal system performance monitoring and analysis. Therefore, the SIS of thermal power units is basically established on DCS.

In terms of hardware, the core part of the operation optimization system consists of a database server, application server, and WEB server. Data transmission between switches is carried out through optical cables or network cards; transmission between servers is generally carried out through network cards. Most of the software development platforms use WINDOWS NT, the development language uses C++ programming language supplemented with Visual Basic programming language, and the database uses ORACLE database system or PI (Plant Information System) database, etc.

9.1.2.2 System Functions

The main functions of SIS are focused on the following aspects:

1. Realize the monitoring of the whole plant production process
   By collecting real-time production data from various production process control systems of the power plant, including DCS, DEH, and other auxiliary control systems, unified monitoring and querying of each production process are carried out, and the production data is comprehensively processed to form a whole plant production report. At the same time, the real-time and historical database of the whole plant is used to meet the requirements of the management department for quick and efficient querying and processing of field process data.
2. Load distribution and scheduling
   According to the contracts and agreements between the power plant and the power grid, as well as the medium and long-term maintenance and operation plans formulated by the power plant, the combination and operation mode of the operating units are arranged in the most optimal way by comparing the forecast of daily load. At the same time, based on the actual changes in daily load and the availability and economic characteristics of the main and auxiliary systems of the unit, optimization calculations are carried out to provide a reference for real-time optimization of the economic load distribution of the unit and the operation of the auxiliary system.
3. Real-time processing of production cost information
   Through the DCS of each unit and the information collection of the whole plant auxiliary control system, the relevant production cost information is sent to the real-time database in real time, providing a basis for electricity price calculation.
4. Performance calculation and analysis
   Performance calculations are made through real-time collected data or data within a specified time, including the efficiency and loss of boilers, steam turbine generators, condensers, high-pressure heaters, low-pressure heaters, feedwater pumps and other main equipment and systems, as well as the entire unit. The software also provides a comparison between the expected values and the actual calculated values of performance calculations. More advanced software can analyze the causes of deviations and provide suggestions for improvement.
5. System self-diagnosis and self-learning
   Through related data comparison and analysis, the system can diagnose its own faults, such as sensor faults and software calculation errors. Through theories such as neural networks and memory factors, the system can identify the parameters of unit operation and find the regional optimal point.
6. Equipment fault diagnosis
   Based on real-time data monitoring and performance calculation, when a certain data or calculation parameter deviates from the expected value, it indicates that there is a fault in the equipment or system. The SIS can use fault diagnosis technology or expert systems to analyze the root cause or most likely root cause

of the fault, and propose operation and maintenance suggestions for the equipment. Some systems are designed with remote monitoring functions, which can carry out remote diagnosis and remote technical services, and obtain the diagnosis results and treatment suggestions of experts outside the factory in time.

7. Equipment life management

   Through real-time monitoring, the system can analyze the impact of parameter changes on the metal temperature and its rate of change of main equipment (such as steam turbines and other components), thereby evaluating the life of the equipment. It can track the parameter status during unit start-up and shutdown and significant load changes, and provide the best start-up and shutdown methods.

8. Power plant operation assessment management

   The system has designed small indicators and energy-saving statistical assessment indicators, such as power supply coal consumption, steam turbine heat consumption, indirect economic indicators and various consumption difference statistical results, providing a fair benchmark for the operation assessment of teams and various values, supervising the operation of operators, and promoting the economic operation of units by operators.

9.1.2.3 Application Effect

After the thermal power unit applies the SIS, it can produce quite obvious effects, mainly reflected in the following aspects:

1. Improve unit efficiency

   It can reduce the annual average power supply coal consumption of the thermal power unit by about 15–20 g/(kW·h), greatly improving the power generation efficiency of the unit and reducing production costs.

2. Timely detection of equipment abnormalities

   It can timely detect problems such as improper adjustment of main steam pressure, high-pressure heater end difference, abnormal condenser back pressure, etc. Some software also uses fuzzy theory and neural network theory to develop fault diagnosis subsystems for boiler systems and condenser systems. Although they are still in the initial stage, they can already provide basic fault information.

3. Operation assessment

   It provides a fair basis for the performance assessment of teams and operational values, implementing appropriate rewards and punishments through assessment statistics results, and enhancing the efficiency awareness and economic operation awareness of power plant employees.

4. Performance test

   It can conveniently carry out efficiency tests of the unit as a whole and system equipment, and can give out test result calculation reports in a timely or even real-time manner, completely changing the implementation process of such large-scale tests.

5. Optimized operation

   Using SIS, the optimal load distribution between units can be sought to control production costs. The most common optimized operation is the circulating water system. Through the analysis and calculation of SIS, the operation mode of the circulating water system can be determined, reducing the plant's electricity rate, thereby reducing the coal consumption for power supply. Boiler optimization and the optimization of steam initial pressure during peak shaving periods are also applied in many power plants.

9.2 Online Analysis Framework of Thermal Performance for NPPs

9.2.1 Common Terms

9.2.1.1 Performance Testing and Performance Monitoring

The purpose of performance monitoring is to detect changes in cycle or equipment indicators through regular calculations, and to conduct performance tests under highly controlled conditions, calculating accurate and quantitative performance indicators. Regular monitoring standards and norms are limited to calculating important performance indicators and monitoring their trends compared to reference indicators.

9.2.1.2 Aging and Degradation

Aging is defined as the natural decline of thermal performance over time. Due to various reasons such as mechanical, electrical, chemical, etc., most equipment will age, and aging is inevitable and irreversible.

Degradation is defined as a decline in performance caused by certain factors, which can be removed or corrected to restore the original function of the degraded components.

9.2.1.3 Controllable Loss and Uncontrollable Loss

Controllable loss is defined as the loss caused by degradation, while uncontrollable loss is the loss caused by aging.

Controllable losses come from incorrect settings of operating parameters, including operating states below optimization, undesirable field conditions, and fault channels causing internal or external leaks. For example, a low vacuum in the condenser will increase controllable losses.

For uncontrollable losses caused by equipment aging, it often requires replacement or cleaning of parts to restore it to normal.

9.2.1.4 Achievable Performance and Actual Performance

Achievable performance refers to the performance that a system or device can achieve when it operates without aging or degradation. Actual performance refers to the performance that a system or device can achieve when it operates under different degrees of aging and degradation.

Figure 9.1 shows a schematic diagram of achievable efficiency and actual efficiency under different conditions.

In the figure, the upper curve represents the efficiency that the NPP can achieve without aging or degradation, called achievable efficiency, and the lower curve represents the actual efficiency of the NPP under aging and degradation.

Under reference conditions (or design conditions), the unit's achievable efficiency is the highest, called rated efficiency. Under measurement conditions deviating from the reference conditions, the unit's achievable efficiency is lower than the rated efficiency, called expected efficiency.

Under reference conditions, the actual efficiency of the unit is lower than the achievable efficiency, which is caused by the uncontrollable loss caused by unit aging.

9.2.1.5 Corrected Value and Expected Value

The rated efficiency is the best efficiency that can theoretically be achieved, without any controllable or uncontrollable losses. In practice, the rated efficiency is the guaranteed design efficiency or acceptance test efficiency provided by the supplier.

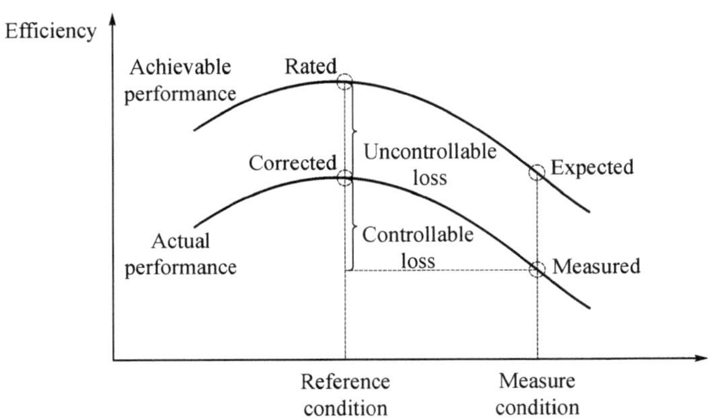

Fig. 9.1 Comparison between achievable efficiency and actual efficiency under different operating conditions [51]

The measured efficiency is the efficiency calculated using the current nuclear power state, including controllable and uncontrollable losses. The deviation between the measured efficiency and the rated efficiency reflects the total loss.

The corrected efficiency is the efficiency obtained based on the rated efficiency, considering uncontrollable losses. By comparing the rated efficiency and the corrected efficiency, the size of uncontrollable losses can be detected. It should be noted that the corrected efficiency is only valid when the deviation between the reference condition and the measured condition is within the specified range.

The expected efficiency only considers controllable losses, and the deviation between the expected efficiency and the measured efficiency reflects the size of uncontrollable losses.

9.2.2 System Framework

The monitoring and diagnosis of the thermal performance analysis of NPPs ultimately aims to fully support condition-based maintenance (also known as predictive maintenance), which can save money by reducing time-consuming and unnecessary maintenance or testing activities, and can also reduce human errors, which is crucial for the safety-critical systems of NPPs.

Figure 9.2 shows the framework of the thermal performance monitoring and diagnosis system of a NPP [51].

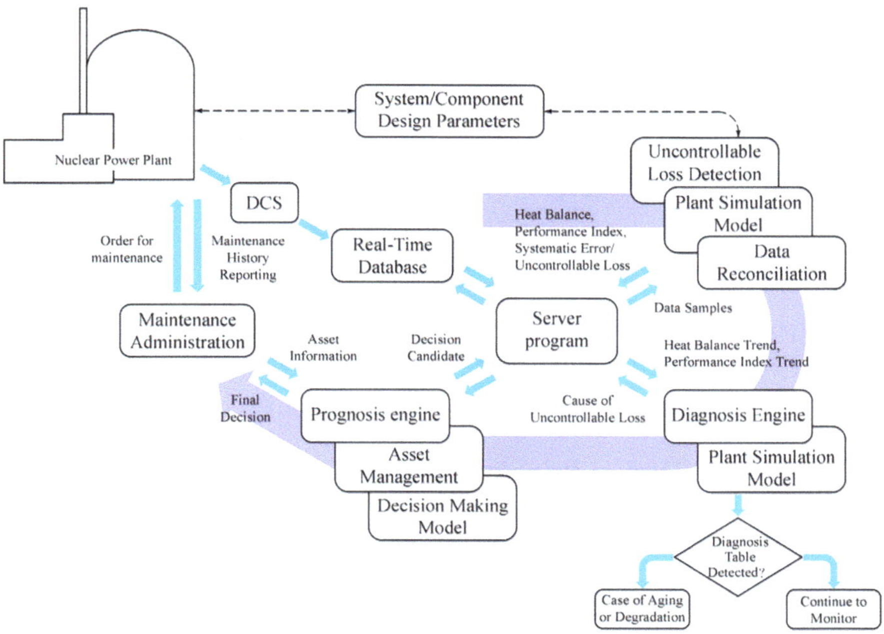

Fig. 9.2 NPP thermal performance monitoring and diagnosis system framework [52]

9.2.2.1 Monitoring Module

Monitoring module used to compare measured performance with expected performance to determine if there are uncontrollable losses. If the deviation between the measured performance and the expected performance is greater than the set value, it is considered that there are uncontrollable losses.

Accurately calculate the expected performance by determining the thermal balance of the clean, brand-new power plant thermal system considering various controllable losses.

Use on-site installed uncalibrated sensors to accurately calculate the measured performance. Signal verification needs to be carried out without operator intervention, that is, data coordination. The purpose of data coordination is to detect signal abnormalities and replace them with appropriate values.

9.2.2.2 Diagnostic Module

When the monitoring module warns, when unexpected residuals are observed in the performance indicators, the diagnostic module starts. Residuals represent changes from the reference conditions due to uncontrollable losses. The purpose of the diagnostic module is to resolve and determine the severity of the root cause of uncontrollable losses.

9.2.2.3 Prediction Module

The prediction module predicts the pattern of uncontrollable losses for future operating conditions, ultimately used to predict the cumulative loss within a certain time interval, to compare the cost-effectiveness of specific maintenance tasks, and to improve asset management.

9.2.2.4 Decision Module

The purpose of the decision module is to provide a cost-benefit analysis for maintenance tasks by integrating information from the diagnostic and prediction modules.

9.2.3 System Design Principles and Functions

9.2.3.1 System Design Principles

Considering the importance of the online monitoring and analysis system for the economic operation of NPPs and future development, the following principles are followed in design.

1. Safety and reliability. Since the real-time data platform is connected to the control system, the design must not destroy the structure and operation of the control system, ensuring the safety of the control system under all circumstances.
2. Easy to use. Adopt B/S or C/S software architecture. If the B/S software architecture is adopted, users can obtain production information through the web without installing special client software.
3. Real-time. The collection of real-time data is synchronized with the online monitoring and analysis system for the economic operation of the unit, providing users with real-time information about production, so that managers can understand the production situation in time and make correct decisions.
4. Advanced technology. The use of process graphic simulation technology allows the company's production managers to see and control the process monitoring screen that is completely consistent with the control system without going to the site, truly realizing the extension of process monitoring from the control room to the desktop, protecting the user's existing investment, and effectively shortening the system construction cycle.
5. Easy to manage. The maintenance and management of the system are centralized on the server, and the management functions are realized through a visual interface.

9.2.3.2 System Functions

The expected functions of the online thermal economic analysis system for NPPs are:

1. Real-time supervision of thermal parameters

   The system will monitor in real time more than 200 parameters related to the thermal cycle of the NPP, and compare them with the expected values preset in the system, so as to timely discover related parameter abnormalities and equipment defects.
2. Online thermal efficiency calculation

   After collecting parameters for a period of time, the system can calculate the thermal efficiency of a single system, a certain equipment, or the entire conventional island, and understand the efficiency status of a single system or the entire conventional island in time.
3. Fault diagnosis

   Through the above two functions, the system or technical personnel can analyze which system or equipment has an abnormality, and can analyze the root cause or possible root cause, and propose timely repair and correction actions. The fault diagnosis before and after the overhaul can also provide maintenance and operation suggestions for the planning of overhaul activities and the follow-up actions after the overhaul, and identify the quality of the overhaul through data comparison.
4. Operation assessment management

   By setting reasonable indicators for statistics, it can supervise the activities of operation and maintenance personnel and strengthen the awareness of economic operation of personnel at all levels of the power plant.

5. Self-update
   The system can automatically recognize the best state of the unit or the best value of the parameters, and update it to the expected value of the system and the thermal model of the system through the system or authorized technical personnel, so as to ensure that the system can follow the best operation state of the unit and the maintenance and technical transformation status of the equipment in time.

9.2.4 System Structure

Figure 9.3 shows the structural principle diagram of the online thermal economic analysis system of the NPP.

As shown in Fig. 9.3, the online thermal economic analysis system of the NPP mainly consists of the following parts:

9.2.4.1 Real-Time Data Acquisition System

One of the main functions is to provide the computer with the original data necessary for thermal economic calculation. Due to the different composition of various unit thermal systems, the original parameters required for performance calculation

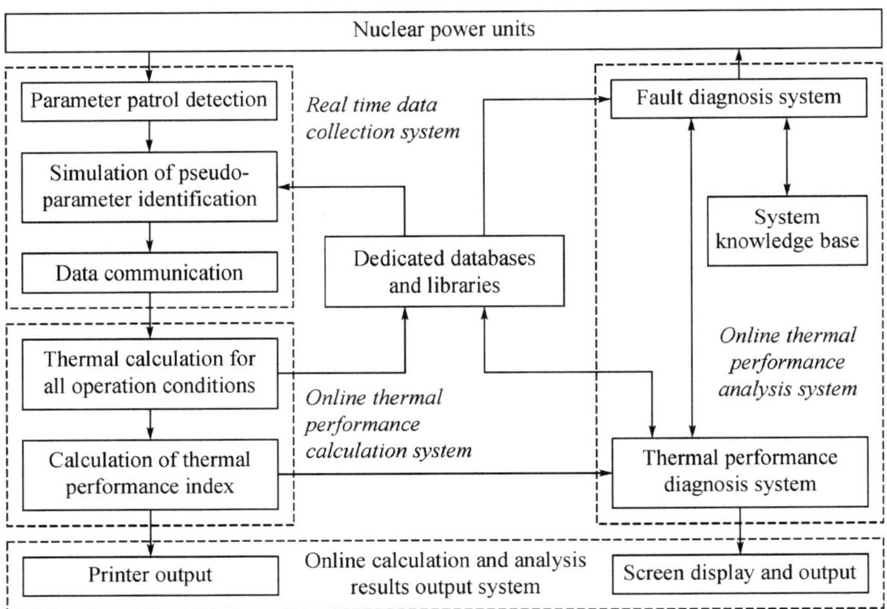

Fig. 9.3 Structure of the online thermal economy analysis system for NPP

will also be slightly different; another function is to identify and analyze the collected data, eliminate bad values, and give simulation values, responsible for transmitting data to the computer.

The original data necessary for unit performance calculation are as follows:

1. Whole plant parameters. Including main steam pressure, temperature and flow, cold/hot reheat steam pressure, temperature and flow, electric power, plant power, feed water pump power, circulating water pump power, condensate pump power, etc.
2. Turbine parameters. Including high-pressure cylinder exhaust pressure and temperature, low-pressure cylinder exhaust pressure and temperature, regulating stage pressure and temperature, turbine regulating valve opening, shaft seal steam pressure and temperature, etc.
3. Condenser parameters. Including circulating water inlet and outlet water temperature, condensate pressure, temperature and flow, etc.
4. Feedwater heater parameters. These include the extraction pressure and temperature at each stage, the inlet and outlet water temperature and drain temperature of each heater (including the deaerator), and the final feedwater pressure, temperature, and flow rate, etc.

Data collection in NPPs is accomplished through KDO (Test Data Collection System), KIT (Central Data Processing System), and other data collection systems. Now, newly built NPPs and technical upgrades of NPPs widely use DCS, making it possible to configure thermal economic online analysis systems for NPPs.

9.2.4.2 Thermal Economic Online Calculation System

It mainly completes the thermal economic calculation and various thermal economic index calculations for the full-condition operation of NPPs, providing analysis data for the thermal economic online analysis system. The thermal economic indicators of the calculated NPPs are as follows:

1. Whole plant performance. This includes reheat steam flow, supply/generation power, steam consumption rate, heat consumption rate, plant power consumption rate, nuclear fuel consumption rate, etc.
2. Turbine performance. This includes low-pressure cylinder exhaust enthalpy/humidity, high/low-pressure cylinder efficiency, turbine cycle efficiency, turbine heat consumption rate, etc.
3. Condenser performance. This includes vacuum, terminal difference, subcooling, circulating water temperature rise, cleanliness factor, etc.
4. Feedwater heater performance. This includes heater terminal difference, feedwater temperature rise, extraction flow, enthalpy drop, pressure loss, drain parameters, etc.

9.2.4.3 Thermal Economic Online Analysis System

Real-time analysis and fault diagnosis analysis are carried out based on the original data and the data obtained from thermal economic calculations, providing decision-making basis and relevant technical data for operation and management personnel.

The economic diagnosis analysis of unit operation includes the economic diagnosis analysis of turbines and other thermal systems. Calculate the fuel loss caused by the deviation of the main monitoring variables of the turbine such as steam initial pressure, steam initial temperature, reheat steam temperature, reheat pressure loss, feedwater temperature, exhaust steam temperature; calculate the fuel loss caused by changes in thermal system parameters such as heater terminal difference, extraction pressure loss.

Figure 9.4 shows the basic process of efficiency deviation calculation for NPPs.

The expected efficiency can be calculated based on the following models and data:

1. Characteristic reference curves provided by equipment suppliers;
2. Historical data, data obtained from acceptance tests, regular tests;
3. Mathematical models based on physical processes;
4. Data-driven models built using verified historical data.

9.2.4.4 Online Calculation and Analysis Result Output System

Provides a graphical, user-friendly result output interface, allowing operation and management personnel to grasp the various thermal economic indicators of the unit at any time, understand the operation status of the equipment, and provide a decision-making basis for operation personnel.

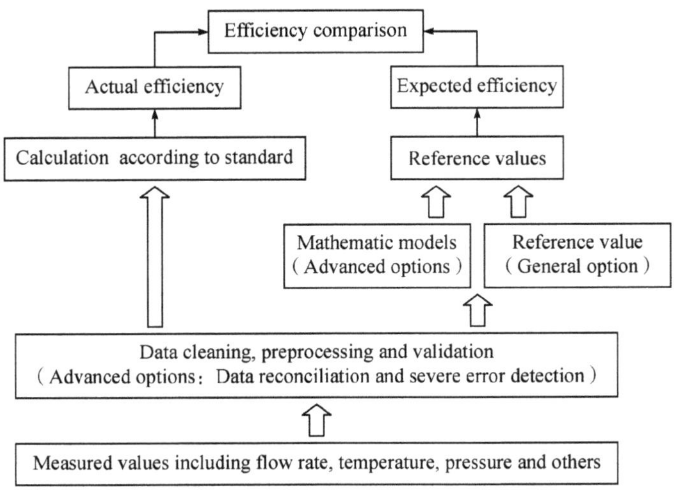

Fig. 9.4 Process for calculating efficiency deviation of NPP

The Web-based publishing method can adopt a B/S structure, the user query interface is fully implemented through the WWW browser, and the query results are displayed on the page in the form of a table. This publishing method is simple and convenient, and can organically integrate the system with the internal LAN.

9.2.4.5 System-specific Data and Function Library System

Includes water and steam thermal parameter function libraries, mathematical calculation function libraries, system-specific databases and historical databases, serving system calculation and analysis. It can provide services in the form of databases or dynamic link libraries, making it easy to store, query, modify, and call, and easy to expand.

9.2.5 Online Calculation Method of Working Substance Property Parameters

9.2.5.1 Online Calculation of Thermal Properties of Water and Steam

In the process of thermal calculation, the thermodynamic properties of water and steam must be expressed in formulas, so as to determine the state parameters of the working substance and the thermal process between the state points by numerical methods; at the same time, sufficient accuracy is required to ensure the accuracy of the final calculation results. There are two common methods:

1. Directly input the parameter table of water and steam into the computer, and then use the interpolation formula to find the parameter relationship at any point. This method has high accuracy, but it requires a large amount of computer memory.
2. Use the IAPWS-IF97 formula formulated by the International Formulation Committee to calculate the thermal characteristic parameters of steam. After giving one or two independent state parameters at a certain state point, other unknown parameter values can be quickly and accurately calculated in the form of function calls.

    (a) In the supercooled water region, among the three parameters of pressure, temperature, and specific enthalpy of supercooled water, two are independent parameters, and any two parameters can be used to find another parameter;
    (b) In the saturated water region, among the three parameters of pressure, temperature, and specific enthalpy of saturated water, only one is an independent parameter, and any one parameter can be used to find the other two parameters;
    (c) In the wet steam region, among the three parameters of pressure (or temperature), dryness, and specific enthalpy of wet steam, two are independent parameters, and any two parameters can be used to find another parameter;

(d) In the saturated steam region, among the three parameters of pressure, temperature, and specific enthalpy of saturated steam, only one is an independent parameter, and any one parameter can be used to find the other two parameters;
(e) In the superheated steam region, among the three parameters of pressure, temperature, and specific enthalpy of superheated steam, two are independent parameters, and any two parameters can be used to find another parameter.

9.2.5.2 Online Determination of Turbine Exhaust Enthalpy

The steam turbine units of PWR NPPs use saturated steam or slightly superheated steam from steam generators. All stages of the high-pressure cylinder of the steam turbine and all stages of the low-pressure cylinder except the first 1–2 stages work in the wet steam region. The enthalpy of wet steam is related to pressure and humidity, and cannot be directly determined based on pressure and temperature. At the current technical level, steam humidity cannot be measured online, so the extraction enthalpy and exhaust enthalpy of the steam turbine cannot be directly determined based on measured values.

The current method of calculating exhaust enthalpy is mostly to first assume or estimate the exhaust enthalpy or other related parameter values, and then approach the actual value of the thermal parameters through iterative calculation of the thermal process of the steam turbine, and finally solve the exhaust enthalpy of the steam turbine. Because the online calculation requires real-time, and the iterative process will extend the calculation time of the program, the online calculation should try to avoid iterative calculation or reduce the number of iterations.

The calculation of the exhaust enthalpy of the steam turbine is carried out in the following steps:

1. Determine the initial value dynamically through curve fitting

   In order to perform system thermal balance calculations, it is necessary to know in advance the enthalpy values of each extraction point (including those in the wet steam area), which requires iterative calculations using a probing method. For units with large operating condition changes, it is difficult to accommodate all operating conditions with this initial value, which may cause the estimated initial value to deviate significantly from the true value, leading to reduced iteration efficiency or even failure. How to determine these initial values is also the focus of the calculation. In this algorithm, based on the known points of the current operating condition, a smooth pressure-enthalpy ($p - H$) curve is fitted using the least squares method, and then extrapolated: that is, the enthalpy values H of the unknown points in the wet steam area are obtained from the pressure p of the unknown points. It is worth noting that when the accuracy of the first calculation of the exhaust steam enthalpy is not enough, a second and third curve fitting is required. This requires that each time, in addition to using the known points, the exhaust steam enthalpy obtained from the previous thermal balance calculation should also be used as a known point in the curve fitting, so as to update the estimated value and achieve a gradual approximation effect. The biggest advantage of curve fitting is that it can track the current operating conditions, fit the curve with the parameters of each point of the current operating conditions, dynamically calculate the initial values of the unknown points closest to the actual values, increase the reliability of the program, and greatly improve the calculation efficiency.

2. Calculate the steam extraction volume at each stage

   When performing thermal balance calculations on the reheating system, it is necessary to calculate the steam extraction volume of each stage of the steam turbine. The calculation of the steam extraction volume is a very complicated task. Using known measured parameters, the steam extraction volume of each stage of the steam turbine can be obtained using matrix analysis.

3. Calculate the exhaust steam enthalpy by thermal power balance

   After obtaining the steam extraction volume of each stage of the unit, the entire reheating system can be subjected to thermal balance calculations to obtain the exhaust steam enthalpy of the steam turbine, and then compare this exhaust steam enthalpy with the exhaust steam enthalpy fitted by the curve to achieve a gradual approximation effect. For a reheating steam turbine with reheating, the thermal balance equation should be

$$N_i = \frac{N_{ge}}{\eta_m \eta_{ge}} \tag{9.1}$$

$$N_i = G_0 h_0 + G_{rh} q_{rh} - \sum_{j=1}^{n} G_{e,j} h_{e,j} - G_c h_c \tag{9.2}$$

where N_i, N_{ge} are the internal power and generator power of the steam turbine, respectively; η_m, η_{ge} are the mechanical efficiency and motor efficiency, respectively; G_0, h_0 are the new steam flow and specific enthalpy, respectively; G_{rh} is the reheating steam flow; q_{rh} is the heat brought to the steam turbine by 1 kg of reheating steam; $G_{e,j}$, $h_{e,j}$ are the extraction flow and specific enthalpy of the *j*th stage, respectively; G_c, h_c are the exhaust steam flow and specific enthalpy, respectively.

Knowing N_i and the extraction volume $G_{e,j}$ of each stage, the exhaust steam enthalpy can be obtained. When the difference between this exhaust steam enthalpy and the exhaust steam enthalpy obtained by curve fitting meets the error accuracy requirements, the calculation loop can be exited.

9.3 Monitoring and Diagnosis Methods of Thermal Performance of Main Equipment in NPPs

The main equipment of the thermal system of a PWR NPP includes a steam turbine generator, a moisture separator reheater, a condenser, a feedwater heater, a condensate pump, and a feed water pump, etc. The operating status changes of these equipment will affect the output electric power of the NPP to varying degrees. Therefore, during the operation of the NPP, it is necessary to monitor the operating status of the main equipment and the output power of the unit to determine the overall performance indicators of the NPP under the current operating conditions [50].

Since most NPPs currently mainly operate at base load and do not participate in grid peak regulation, the performance parameters of the main equipment in the thermal system of the NPP during base load operation remain unchanged, which is convenient for the performance monitoring of the thermal system of the NPP and its key equipment.

The output power of the turbine has significantly decreased compared to the reference power, the main reasons could be:

1. Measurement error of the main variables;
2. Drift in the measurement of the thermal power of the steam generator;
3. Valve leakage or auxiliary steam supply;
4. Severe decline or damage in the performance of key equipment.

9.3.1 Turbine Unit

The power generation turbine unit of a PWR NPP usually consists of a high-pressure cylinder and a low-pressure cylinder, with a moisture separator reheater (MSR) set between the high-pressure cylinder and the low-pressure cylinder. The wet saturated steam discharged from the high-pressure cylinder after doing work enters the steam-water separator reheater, removes the water content in the steam through the separator, and then reheats the steam to a superheated state through two-stage reheaters. The superheated steam enters the low-pressure cylinder to do work, which can reduce the steam humidity of the low-pressure cylinder exhaust, avoid the last stage blades of the low-pressure cylinder from being eroded by water droplets, and thus improve the thermal economy and safety of the turbine unit operation.

The turbine unit is one of the devices that have the greatest impact on the thermal efficiency of the thermodynamic cycle. Every 1% decrease in the efficiency of the high-pressure cylinder will cause the output power of the generator to decrease by about 0.26%–0.41%. Similarly, for the low-pressure cylinder, this effect will cause a power decrease of about 0.59%–0.74%. Since the exhaust of both the high-pressure cylinder and the low-pressure cylinder is in a wet saturated state, it is difficult to determine and monitor the efficiency of each part of the turbine. Therefore, we indirectly use the corrected turbine generator output power, steam expansion ratio (pressure ratio), and steam passing capacity (flow coefficient) as per the ASME PTC 6 [52] standard as the comprehensive performance indicators of the turbine.

9.3.1.1 Main Monitoring Parameters

Corrected Turbine Generator Output Power

In order to accurately evaluate the performance of the turbine generator, it is necessary to consider the change in the turbine extraction steam flow, and adjust the turbine generator output power when the operating conditions and the baseline reference conditions are different. According to the standard given by ASME PTC 6, the correction of the turbine generator output power can be divided into the following two groups of correction factors:

The first group of correction factors mainly includes state quantities involved in the feedwater heating system, with the main parameters as follows:

Feedwater heater terminal difference, feedwater heater drain terminal difference, extraction pipeline pressure drop and heat loss, system water storage change, feedwater enthalpy rise through the condensate pump and feedwater pump, condenser condensate subcooling, makeup water flow, generator power factor.

The more common practice is to use correction curves (or tables) for correction, and on the basis of the measured generator output power, each influencing variable's correction factor is multiplied by the table to obtain the first group of correction factors. At the same time, these correction factors can also be calculated by thermal balance calculation through measuring the design parameters of the turbine, feedwater heating system, and steam-water separator reheater. The correction curve method is more common in the turbine cycle of NPPs.

The second group of correction factors includes variables related to the inlet and outlet steam parameters of the turbine, the purpose of which is to correct the measured generator power, and can reflect the change in the available enthalpy drop of the high-pressure cylinder and low-pressure cylinder due to the steam parameters being different from the baseline conditions, specifically including the following parameters:

High-pressure cylinder inlet steam pressure, high-pressure cylinder inlet steam humidity, low-pressure cylinder exhaust steam pressure, steam-water separator efficiency, reheater terminal difference, steam pressure drop passing through the steam-water separator reheater.

According to ASME PTC 6, the corrected output power of a steam turbine generator can be represented as

$$N_{\mathrm{el},c} = \left(N_{\mathrm{el}} + \Delta N_{\mathrm{f}}\right)\left(\prod_{i=1}^{n} f_{1,i}\right)^{-1}\left(\prod_{i=1}^{n} f_{2,i}\right)^{-1}\frac{1}{f_{\mathrm{Q_R}}}\frac{1}{f_{\Delta p}} \tag{9.3}$$

where N_{el} is the measured output power of the steam turbine generator, kW; ΔN_{f} is the corrected value of the generator power factor, kW; $f_{1,i}$, $f_{2,i}$ are the i-th correction factor of the first group, the i-th correction factor of the second group, respectively; $f_{\mathrm{Q_R}}$ is the correction factor of the reactor thermal power; $f_{\Delta p}$ is the correction factor for the throttling loss of the steam turbine control valve.

Steam Turbine Expansion Linear Efficiency

The expansion linear efficiency of the steam turbine (i.e., the isentropic efficiency of the steam turbine) is a very important performance parameter in the steam turbine cycle, but most of the high-pressure cylinder and low-pressure cylinder are in the wet steam area during the operation of the unit, making it difficult to determine the enthalpy of the steam. Therefore, monitoring this indicator is not realistic.

A more realistic method is to assume the exhaust enthalpy of the high-pressure cylinder using the design efficiency level of the high-pressure cylinder (such as the expansion linear slope), and the exhaust enthalpy of the low-pressure cylinder is determined by the thermal balance calculation of the entire steam turbine cycle, including the shaft power of the steam turbine.

Steam Expansion Ratio (Pressure Ratio)

When diagnosing the performance of a steam turbine, the steam expansion ratio (pressure ratio) of the first stage of the high-pressure cylinder and the last stage of the low-pressure cylinder with the change of steam flow and exhaust pressure of the low-pressure cylinder is a practical monitoring indicator. If the operating conditions of the extraction equipment (such as feedwater heaters, auxiliary steam headers) do not change dramatically, then the steam expansion ratio is always constant. Therefore, if the extraction distribution does not change, but the steam expansion ratio changes, it indicates that there may be a fault in these stages of the steam turbine.

Steam Flow Coefficient

The expression of the steam flow coefficient is

$$K_{\mathrm{stg}} = G_{\mathrm{stg}} \cdot \sqrt{\frac{p_{\mathrm{stg}}}{v_{\mathrm{stg}}}} \tag{9.4}$$

where K_{stg} is the steam flow capacity (flow coefficient) of the steam turbine stage; G_{stg} is the flow of steam entering the stage of the steam turbine, kg/s; p_{stg} is the pressure of steam entering the stage of the steam turbine, MPa; v_{stg} is the specific volume of steam entering the stage of the steam turbine, m^3/kg.

The steam flow coefficient described in Eq. (9.4) at the inlet section of the steam turbine and the subsequent extraction stages is a hardware factor of the steam turbine, which needs to be kept constant regardless of the operating conditions.

This coefficient is used to indicate changes in the geometric shape of the steam flow part of the steam turbine or overestimation of RTP. For example, if the final

Table 9.1 Diagnosis of common turbine problems [50]

Possible cause of turbine problems	Throttle flow	HP turbine pressure	LP turbine pressure	HP turbine efficiency[a]	LP turbine efficiency[a]
Decreased throttle steam pressure	↓	↓	↓	→or↓[b]	↑
Decreased hot reheat steam temperature (MSR outlet)	→	→	↓	→	↓
Increased HP turbine first stage area	→	→	→	↓	→
Decreased HP turbine first stage area	→or↓[b]	→or↓[b]	→or↓[b]	↓	↑
Increased HP turbine second stage area	→	↓	→	↓	→
Decreased HP turbine second stage area	→or↓[b]	↑	↓	↓	↑
Increased LP turbine first stage area	→	→	↓	→	↓
Decreased LP turbine first stage area	→	→	↓	→	↓

Note: The up arrow indicates an increase, the down arrow indicates a decrease, the right arrow indicates basically unchanged

[a] It is not practical to monitor the efficiency of the high-pressure cylinder and low-pressure cylinder in the steam turbine cycle, this only represents the mechanism of steam turbine performance deterioration

[b] Only when the high-pressure control valve cannot open enough to compensate for the change in flow area

feedwater flow is overestimated, it will increase the steam passing capacity of the steam turbine, and vice versa.

9.3.1.2 Diagnosis of Common Problems

Monitoring parameters such as steam turbine high-pressure cylinder pressure (or first stage), low-pressure cylinder pressure, and throttling flow (or final feedwater flow minus steam generator blowdown flow) provides a basis for analyzing and discussing the possible root causes of common problems with steam turbines.

Table 9.1 shows the typical characteristics of some main parameter changes under common problems of steam turbines.

9.3.2 Moisture Separator Reheater

The moisture separator reheater includes a moisture separator and a steam reheater. Currently, commercial PWR NPPs generally use steam-water separator reheaters with two-stage reheaters. When performing thermal performance analysis, the following performance parameters are mainly monitored [53].

9.3.2.1 Determination of Performance Parameters

Efficiency of the Steam-Water Separator

The efficiency of the steam-water separator is defined as the proportion of the total water mass carried in the wet saturated cycle steam flowing through the steam-water separator that is separated out. The expression is

$$\eta_{\mathrm{sp}} = \frac{G_{\mathrm{spw}}}{G_{\mathrm{zh}}\left(1 - x_{\mathrm{zh}}\right)} \tag{9.5}$$

where G_{spw} is the drainage flow rate of the moisture separator, kg/s; G_{zh} is the mass flow rate of the steam exiting the high-pressure cylinder of the steam turbine entering the steam-water separator, kg/s; x_{zh} is the dryness of the steam exiting the high-pressure cylinder of the steam turbine.

Outlet End Difference of the Steam Reheater

The outlet end difference of the steam reheater is defined as the difference between the saturation temperature of the heating steam entering the reheater and the cycle steam temperature of the outlet heating steam from the reheater:

$$\theta_{\mathrm{rh}} = T_{\mathrm{hs}} - T_{\mathrm{rh}} \tag{9.6}$$

where T_{hs} is the saturation temperature of the heating steam entering the reheater, °C; T_{rh} is the temperature of the cycle steam leaving the reheater, °C.

For a single-stage reheater or the second stage of a two-stage reheater, the outlet end difference can be determined by directly measuring the heating steam temperature and the cycle steam temperature. However, since the cycle steam temperature at the outlet of the first stage reheater is generally not measured, the outlet end difference of the first stage reheater cannot be determined. It can be determined based on the energy balance of the second stage reheater.

Relative Pressure Drop of Cycle Steam

The pressure drop of the cycle steam is defined as the percentage of the pressure drop caused by the cycle steam flowing through the steam-water separator reheater, that is

$$\Delta p_{\mathrm{msr,cs}} = \frac{p_{\mathrm{msr,csi}} - p_{\mathrm{msr,cso}}}{p_{\mathrm{msr,csi}}} \tag{9.7}$$

where $p_{\mathrm{msr,\,csi}}$, $p_{\mathrm{msr,\,cso}}$ are the inlet and outlet cycle steam pressures of the moisture separator reheater, respectively, MPa.

9.3.2.2 Diagnosis of Common Problems

By monitoring the key performance parameters and operating parameters of the moisture separator reheater (such as the outlet end difference of the reheater, the water level of the moisture separator reheater, the flow rate of the heating steam, the drainage flow rate and temperature of the reheater, etc.), it can provide a basis for analyzing and discussing the possible root causes of common problems with the moisture separator reheater and proposing corrective measures.

Table 9.2 shows the typical characteristics of these parameter changes when common problems occur in the moisture separator reheater.

9.3.3 Condenser

The efficiency of a NPP largely depends on the integrity and cleanliness of the condenser and cooling system. Any performance defect in the condenser will have a significant impact on the overall performance of the steam turbine cycle. One of the

Table 9.2 The diagnosis of moisture separator reheater [50]

<table>
<tr><th>Possible cause of MSR problems</th><th>Moisture separator effectiveness</th><th>Reheater outlet temp.</th><th>Cycle steam press drop</th><th>Moisture separator level</th><th>Heating steam flow</th><th>Reheater drain temp</th></tr>
<tr><td>MS chevron failure</td><td>↓</td><td>↓</td><td>↑</td><td>→</td><td></td><td></td></tr>
<tr><td>Excess moisture in the incoming steam</td><td rowspan="2">↓</td><td rowspan="2">↓</td><td rowspan="2">↑</td><td rowspan="2">↑</td><td rowspan="2"></td><td rowspan="2"></td></tr>
<tr><td>MS drain system restrictions</td></tr>
<tr><td>Reheater tube bundle uplift</td><td rowspan="2"></td><td rowspan="2">↓</td><td rowspan="2">↓</td><td rowspan="2"></td><td rowspan="2"></td><td rowspan="2"></td></tr>
<tr><td>Reheater bypass due to shroud buckling</td></tr>
<tr><td>Reheater tube leaks</td><td></td><td>↓</td><td>→</td><td></td><td>↑</td><td>↓</td></tr>
<tr><td>Reheater drain system leaks</td><td rowspan="2"></td><td rowspan="2">↓</td><td rowspan="2">→</td><td rowspan="2"></td><td rowspan="2">↑</td><td rowspan="2">↑</td></tr>
<tr><td>Emergency drain valve open or leaking</td></tr>
<tr><td>Excessive leak of heating steam drain valve</td><td rowspan="2"></td><td rowspan="2">↓</td><td rowspan="2">→</td><td rowspan="2"></td><td rowspan="2">↓</td><td rowspan="2">↓</td></tr>
<tr><td>Heating steam flow restrictions on reheater heating steam supply line (supply valves not fully open, dropped valve seat, foreign object damage lodged in valve)</td></tr>
<tr><td>Reheater tube fouling</td><td></td><td>↓</td><td>→</td><td></td><td>↓</td><td>→</td></tr>
</table>

main operating parameters affecting the efficiency of the steam turbine cycle is the pressure of the condenser. An increase in condenser pressure reduces the available enthalpy drop of steam in the low-pressure steam turbine, leading to a decrease in unit output power and an increase in the heat consumption rate of the steam turbine cycle.

The increase in operating pressure of the condenser is the result of one factor or the combination of multiple factors, such as:

1. The decrease in heat transfer performance due to pipe blockage or scaling;
2. The presence of non-condensable gases in the condenser due to excessive air leakage, the decline in performance of the jet pump or vacuum pump;
3. High inlet temperature of circulating cooling water or reduced flow of circulating cooling water.

9.3.3.1 Main Monitoring Parameters

The main parameters for measuring the performance of the condenser include the cleanliness factor (CLF) and the condenser pressure determined based on the heat load of the condenser, the temperature and flow of the circulating cooling water. By performing a condenser performance test to obtain the baseline condenser cleanliness factor, the expected condenser pressure under current operating conditions can be determined and compared with the actual value.

Cleanliness Factor

The following calculation is based on the standard method of heat transfer society, and it is necessary to know in advance the material of the condenser and the inlet flow of cooling water. These information may not be available, in which case it is best to use the Log Mean Temperature Difference (LMTD) method or empirical curves.

To determine the cleanliness factor of the condenser, the following condenser performance factors need to be calculated based on current operating parameters, and the heat load of the condenser under test conditions is:

$$Q_{\text{cd}} = G_{\text{sw}} c_{p,\text{sw}} \left(T_{\text{sw},2} - T_{\text{sw},1}\right) = \sum_{i=1}^{n} Q_{\text{in},i} - \sum_{j=1}^{n} Q_{\text{out},j} \tag{9.8}$$

where G_{sw} is the flow of circulating cooling water passing through the condenser under test conditions, kg/s; $c_{p,\text{sw}}$ is the specific heat at constant pressure of the circulating cooling water, kJ/(kg·K); $T_{\text{sw},1}$, $T_{\text{sw},2}$ are the temperatures of the circulating cooling water at the inlet and outlet of the condenser,respectively, °C; $Q_{\text{in},i}$ is the heat released by the *i*-th working substance entering the condenser, kW; $Q_{\text{out},j}$ is the *j*-th heat leaving the condenser, kW.

The log mean temperature difference of the condenser is

$$\Delta T_m = \frac{T_{sw,2} - T_{sw,1}}{\ln \frac{T_{cd} - T_{sw,1}}{T_{cd} - T_{sw,2}}} \tag{9.9}$$

where T_{cd} is the saturated water temperature under the operating pressure of the condenser, °C.

The total heat transfer coefficient of the condenser is

$$K_{cd} = \frac{Q_{cd}}{F_{cd} \cdot \Delta T_m} \tag{9.10}$$

where Q_{cd} is the heat load of the condenser under test conditions, kW; F_{cd} is the effective heat transfer area of the condenser, m^2.

The log mean temperature difference and the total heat transfer coefficient are basic factors indicating the heat exchange level of the condenser. These factors change with the inlet temperature and flow of the circulating cooling water, so they cannot provide an intuitive judgment of the performance of the condenser.

From this perspective, using the cleanliness factor as the performance parameter of the condenser is more practical and accurate. It corrects for different inlet temperatures and flows (tube flow rates) of circulating cooling water and remains constant throughout the load range. The cleanliness factor is defined as the ratio of the total heat transfer coefficient measured under current operating conditions to the total heat transfer coefficient expected according to the Heat Exchange Institute (HEI) "Surface Steam Condenser Standard", that is

$$C_{cd} = \frac{K_{cd}}{K_{uc,te} F_{w,te} F_m} \tag{9.11}$$

where $K_{uc,\,te}$ is the uncorrected total heat transfer coefficient at the test tube flow rate (HEI Standard Table 9.1); $F_{w,\,te}$ is the correction factor for circulating water temperature at the test cooling water temperature (HEI Standard Table 9.2); F_m is the correction factor for materials and specifications (HEI Standard Table 9.3).

In a NPP, if the condenser cleanliness factor decreases by 3%, the condenser pressure increases by 1 mmHg, and vice versa.

Corrected Condenser Pressure

Based on the predetermined condenser cleanliness factor, the total heat transfer coefficient corrected for the reference (design) cooling water inlet temperature and flow rate can be calculated using the HEI method, namely

$$K_{cd,c} = K_{uc,de} F_{w,de} F_m C_{cd} \tag{9.12}$$

Table 9.3 Diagnosis of common condenser problems [50]

Possible cause of condenser problems	Non-condensable gas (O_2)	LMTD (or TTD)	Cooling water temp. Rise	Cooling water pressure drop
Microfouling	→	↑	→	→
High hotwell level Air binding (air removal problem or air leakage)	↑	↑	→	→
High heat rejection to condenser	→	↑	↑	→
Increased cooling water system resistance (cooling water pump head increase)	→	→	↑	↓
Cooling water pump performance deterioration.	→	→	↑	↓
Motor current				
– increase: damaged casing or impeller				
– fluctuate: pump cavitation				
– decrease: casing or impeller wear/ corrosion				

where $K_{uc,\,de}$ is the uncorrected total heat transfer coefficient at the design tube flow rate (from HEI standard Table 9.1); $F_{w,\,de}$ is the correction factor for circulating water temperature at the design cooling water temperature (from HEI standard Table 9.2).

The number of heat transfer units can be determined as follows:

$$\mathrm{NTU} = \frac{K_{cd,c} F_{cd}}{c_{p,sw} G_{sw}} \tag{9.13}$$

The saturation temperature and corrected condenser pressure can be determined according to formulas (9.12) and (9.13):

$$T_{cd} = \frac{T_{sw,2} - T_{sw,1} e^{-\mathrm{NTU}}}{1 - e^{-\mathrm{NTU}}} \tag{9.14}$$

$$p_{cd} = f\left(T_{cd}\right) \tag{9.15}$$

Expected Condenser Pressure

Under the current operating conditions, the pressure of the condenser can be determined using the HEI method, which can use the benchmark condenser cleanliness factor from previous tests (instead of the measured condenser cleanliness factor), the current cooling water inlet temperature, and flow rate to determine the current Kuc and Fw under the conditions. If there is no available benchmark condenser cleanliness factor, the design cleanliness factor can be adjusted to make the expected condenser pressure reach the measured value at that time, without guessing the heat transfer tube scaling or excessive leakage.

9.3.3.2 Diagnosis of Common Problems

Monitoring of key performance parameters and operating parameters of the condenser (such as cleanliness factor, logarithmic mean temperature difference, circulating cooling water temperature rise, and heat transfer tube pressure drop) can provide a basis for analyzing the possible causes of common problems with the condenser.

Table 9.3 shows the typical characteristics of these parameter changes when common problems occur in the condenser.

9.3.4 Feedwater Heater

The feedwater heater of a PWR NPP mainly uses the steam turbine's extraction steam for heating. The two most commonly used measurements during operation are the top difference and the bottom difference, and their reference values are usually obtained from the design heat balance diagram.

9.3.4.1 Main Monitoring Parameters

Top Difference

The top difference of the feedwater heater is defined as the difference between the saturation temperature of the extraction steam entering the feedwater heater and the feedwater temperature at the outlet of the feedwater heater, namely:

$$\theta_{\mathrm{u}} = T_{\mathrm{e}} - T_{\mathrm{f,o}} \tag{9.16}$$

where T_{e} is the saturation temperature of the extraction steam entering the feedwater heater, °C; $T_{\mathrm{f,o}}$ is the feedwater temperature at the outlet of the feedwater heater, °C.

Bottom Difference

The bottom difference of the feedwater heater is defined as the difference between the drain temperature of the feedwater heater and the feedwater temperature at the inlet of the heater, namely:

$$\theta_{\mathrm{d}} = T_{\mathrm{d}} - T_{\mathrm{f,i}} \tag{9.17}$$

Table 9.4 Diagnosis of common feedwater heater problems [50]

Possible cause of heater problems	TTD	DCA	Feedwater temp. rise	Heater level
Feedwater heater tube leak (rapid level increase)	↑	↓	↓	↑
Heater level controller or normal drain valve malfunction (level high)				
Drain cooling zone flashing (steam induction through snorkelling)	→ or ↓	↑	→ or ↑	↓
Heater level controller or normal drain valve malfunction (level low)				
Emergency drain valve open or excessive leaks	→ or ↓	↑	↓	↓
Tube fouling corrosion	↑	↑	↓	→
Pinhole leaks near steam or drain inlet (rapid)				
Normal vent closed or air leaks into feedwater heater (heaters operating under atmospheric pressure)	↑ Oscillate	↑ or →Oscillate	↓	→
Normal vent closed (heaters operating above atmospheric pressure) Partition plate leaks	↑	→	↓	→

where T_d is the drain temperature of the feedwater heater, °C; $T_{f,i}$ is the feedwater temperature at the inlet of the feedwater heater, °C.

When the NPP is operating normally at stable power, both θ_u and θ_d of the feedwater heater maintain constant values, and the measured values can be directly used for performance monitoring.

9.3.4.2 Diagnosis of Common Problems

For feedwater heaters that import feedwater temperature changes with corresponding condenser pressure and load changes, a more complex method needs to be adopted, which is to calculate the expected θ_u and θ_d using the manufacturer's design data and basic heat transfer relationships, and then compare the calculated results with the measured values to identify performance degradation.

Table 9.4 shows the typical characteristics of the main parameter changes of the feedwater heater under common problems.

9.3.5 Steam Turbine Feedwater Pump System

Most PWR NPPs use steam turbine feedwater pump systems to deliver feedwater to the steam generator. The feedwater pump turbine extracts drive steam from the steam-water separator reheater outlet, and the performance index of the steam turbine feedwater pump system can be represented by the share of the drive steam flow of the feedwater pump turbine in the steam flow at the outlet of the steam generator.

The efficiency of the feedwater pump turbine and the efficiency of the feedwater pump are both indicators of the theoretical performance of the steam turbine feedwater pump system, but these indicators are too sensitive to the measured feedwater temperature and cannot be used for performance monitoring purposes. For example, a measurement error of 0.5 °C in the suction or discharge temperature of the feedwater pump will result in an error of 3.5% in the efficiency of the feedwater pump and an error of 4.3% in the efficiency of the feedwater pump turbine.

Under the reference condition and the measured actual condition, the steam flow consumed by the feedwater pump turbine is respectively

$$G_{\mathrm{fwpt},0} = \frac{N_{\mathrm{fwpt},0}}{H_{\mathrm{i},\mathrm{fwpt},0}} \tag{9.18}$$

$$G_{\mathrm{fwpt}} = \frac{N_{\mathrm{fwpt}}}{H_{\mathrm{i},\mathrm{fwpt}}} \tag{9.19}$$

where $N_{\mathrm{fwpt},0}$, N_{fwpt} are the output internal power of the feedwater pump turbine under the reference condition and the actual condition, respectively, kW; $H_{\mathrm{i,\,fwpt,\,0}}$, $H_{\mathrm{i,\,fwpt}}$ are the internal enthalpy drops of the feedwater pump turbine under the reference condition and the actual condition, respectively, kJ/kg.

Assume that the power output of the feedwater pump turbine under the actual operating condition is equal to the power output of the feedwater pump turbine under the reference condition, i.e., $N_{\mathrm{fwpt},0} = N_{\mathrm{fwpt}}$, then

$$G_{\mathrm{fwpt}} = G_{\mathrm{fwpt},0} \frac{H_{\mathrm{i},\mathrm{fwpt}}}{H_{\mathrm{i},\mathrm{fwpt},0}} \tag{9.20}$$

$$\xi_{\mathrm{fwpt}} = \frac{G_{\mathrm{fwpt}}}{D_{\mathrm{s}}} \times 100\% \tag{9.21}$$

where ξ_{fwpt} is the percentage of the drive steam flow of the feedwater pump turbine in the steam flow at the outlet of the steam generator, %; D_{s} is the sum of the steam flow at the outlet of all steam generators, kg/s.

The measured drive steam flow of the feedwater pump turbine is corrected for the low-pressure turbine exhaust pressure different from the reference (design) thermal balance condition. If this exhaust pressure is higher than the reference value, the drive steam flow of the feedwater pump turbine will increase due to the reduction of the available energy of the feedwater pump turbine, and vice versa.

The relationship between the discharge pressure and speed of the feedwater pump and booster pump (if any) is also an important performance indicator, which can be used to detect pump damage or system operation abnormalities.

Table 9.5 Diagnosis of common steam turbine feedwater pump problems [50]

Possible cause of feedwater pumping system problems	Driving steam flow	Feedwater pump discharge press	Booster pump discharge press
Booster pump impellor worn or damaged Booster pump wear rings worn	→ or ↑	→	↓
Feedwater pump impellor worn or damaged Feedwater pump wear rings worn	↑	↓	→
Flow restrictions on feedwater line Malfunction of FWPT admission valve	↑	→	→
Excessive FWPT blades worn or damage Excessive leaks in driving steam supply line and/or FWPT admission valve seat drain	↑	→	→
Excessive leakage on the feedwater pump minimum flow line	↑	↓	↓

Table 9.5 shows the typical characteristics of the main parameter changes of the steam turbine feedwater pump under common problems.

Questions and Exercises

1. Briefly describe the main functions of the plant-level monitoring information system (SIS).
2. What is the purpose and significance of online thermal economic analysis of NPPs?
3. What are the components of the online thermal economic analysis system of a NPP? What are the functions of each part?
4. How is the exhaust enthalpy of the turbine calculated when conducting online economic analysis of NPP operation?
5. What are the main factors affecting the performance of the condenser? What are the trends in the main parameters of the condenser under common problems?

Appendix

Table A.1 Main variable symbol table

Symbols	Meaning	Unit	Symbols	Meaning	Unit
A	Flow sectional area	m^2	$g_{f,i}$	Intake coefficient of heating unit	kJ/kg
A_n	Anergy	kJ	h'	Specific enthalpy of saturated water	kJ/kg
c_F	Average unit cost of fuel exergy	$/kJ	h''	Specific enthalpy of saturated steam	kJ/kg
c_p	Constant-pressure specific heat	kJ/(kg·K)	h_{fh}	Fresh steam enthalpy	kJ/kg
c_P	Average unit cost of product exergy	$/kJ	h_{fw}	Feed water enthalpy	kJ/kg
d_0	Steam rate	kg/(kW·h)	Δh_a	Adiabat specific enthalpy drop	kJ
d_{si}	Tube inner diameter of steam generator	m	H_a	Theoretical enthalpy drop	kJ/kg
d_{so}	Tube outer diameter of steam generator	m	Δh_i	Actual specific enthalpy drop	kJ
d_u	Fuel pellet diameter	m	H_i	Actual enthalpy drop	
E_x	Exergy	kJ	K	Total heat transfer coefficient	W/(m^2·K)
e_x	Specific exergy	kJ/kg	N_a	Turbine output theoretical power	kW
e_x'	Specific exergy of saturated water	kJ/kg	N_e	Effective power	kW
e_x''	Specific exergy of saturated steam	kJ/kg	N_{el}	Electric generator output power	kW
$E_{x,k}$	Exergy of kinetic energy	kJ	N_{el}'	Generating auxiliary power	kW
$E_{x,p}$	Exergy of potential energy	kJ	N_i	Turbine internal power	kW

M. Peng et al., *Thermal Analysis of Nuclear Power Plants*, Nuclear Science and Technology, https://doi.org/10.1007/978-981-96-8983-5

Symbols	Meaning	Unit	Symbols	Meaning	Unit
$E_{x,\text{ph}}$	Physics exergy	kJ	N_i'	Turbine actual internal power	kW
$E_{x,Q}$	Exergy of heat energy	kJ	p_c	Operating pressure of main coolant system	MPa
$E_{x,Q'}$	Exergy of cooling energy	kJ	p_s	Operating pressure at steam generator secondary side	MPa
$E_{x,\text{eff}}$	Obtained exergy	kJ	Pr	Prandtl number	
$E_{x,\text{sup}}$	Supply exergy	kJ	Q_p	Thermal power added to coolant due to main pump	kW
E_{xl}	Exergy loss	kJ	Q_R	Reactor thermal power	kW
$E_{xl,\text{in}}$	Internal exergy losses	kJ	Q_s	Thermal power of nuclear steam supply system	kW
$E_{xl,\text{out}}$	External exergy losses	kJ	Q_{sg}	Thermal power of steam generator secondary side	kW
F	Surface area	m^2	$Q_{g,h}$	High calorific value of gaseous fuel	
G	Mass flow rate	kg/s	$Q_{l,h}$	High calorific value of liquid fuel	
g	Acceleration of gravity	m/s^2	$Q_{l,1}$	Low calorific value of liquid fuel	
$g_{d,i}$	Drain coefficient of heating unit		$Q_{s,1}$	Low calorific value of solid fuel	
$g_{e,i}$	Extraction coefficient of heating unit		q	Heat flux	W/m^2
q_1	Absorbed heat in thermal cycle	kJ	Y	Underwork factor of extraction steam	
q_2	Heat dissipation in the thermal cycle	kJ	Z	Level of regenerative feedwater heat exchanger	
$q_{h,i}$	Exhaust steam heat release	kJ/kg	Z	Equipment operating cost	$
$q_{h,i}'$	Auxiliary steam heat release	kJ/kg	α	Extraction coefficient	
q_l	Linear power density	W/m	α_1	Heat transfer coefficient of steam generator primary side	W/(m^2·K)
r	Latent heat of vaporization	kJ/kg	α_2	Heat transfer coefficient of steam generator secondary side	W/(m^2·K)
Re	Reynolds number		α_f	Feedwater flow	

Symbols	Meaning	Unit	Symbols	Meaning	Unit
R_s	Fouling thermal resistance	m^2·K/W	Δt_m	Mean heat transfer temperature difference	°C
R_t	Thermal resistance	m·K/W	η_e	Effective efficiency	
R_w	Wall thermal resistance	m^2·K/W	η_{el}	Gross thermal efficiency of NPP	
s'	Specific entropy of saturated water	kJ/(kg·K)	η_{ex}	Exergy efficiency	
s''	Specific entropy of saturated steam	kJ/(kg·K)	η_{ge}	Effective efficiency of electric generator	
T	Thermodynamic temperature	K	η_m	Mechanical efficiency	
t	Centigrade temperature	°C	η_{mp}	Pipeline loss factor	
t_{ci}	Coolant temperature at core inlet	°C	η_{net}	Net thermal efficiency of NPP	
t_{co}	Coolant temperature at core outlet	°C	η_{oi}	Relative internal efficiency of steam turbine	
t_m	Average coolant temperature	°C	η_{sg}	Thermal efficiency of steam generator	
t_{si}	Coolant temperature at steam generator primary side inlet	°C	η_t	Thermal efficiency of cycle	
t_{so}	Coolant temperature at steam generator primary side outlet	°C	γ	Drain heat release	kJ/kg
t_g	Gas gap temperature in fuel element	°C	λ	Energy level	
t_{um}	Fuel pellet center temperature	°C	λ_c	Thermal conductivity of coolant	W/(m·K)
t_{us}	Fuel pellet surface temperature	°C	λ_w	Thermal conductivity of tube material	W/(m·K)
t_{wi}	Fuel cladding inside surface temperature	°C	μ	Dynamic viscosity	N·s/m^2
t_{wo}	Outside surface temperature of fuel cladding	°C	υ	Kinematic viscosity	m^2/s
U	Internal energy	J	τ	Feed water enthalpy rise	kJ/kg
v'	Specific volume of saturated water	m^3/kg	ξ_i	Exergy loss coefficient	
v''	Specific volume of saturated steam	m^3/kg	ξ	Steam generator blowdown rate	
W	Volume work	kJ			
x	Mass vapor content				
x	Steam quality				

Table A.2 Common unit conversion table

1. Energy, work and heat							
J or N·m	kgf·m	kW·h	kcal	atm·L	hp·h	ft·lbf	Btu
1	0.10197	2.778×10^{-7}	2.389×10^{-4}	9.869×10^{-3}	3.777×10^{-7}	0.73757	9.478×10^{-4}
9.80665	1	2.724×10^{-3}	2.342×10^{-3}	9.678×10^{-2}	3.704×10^{-6}	7.2331	9.295×10^{-3}
3.600×10^{6}	3.671×10^{5}	1	859.85	35,529	1.3596	2.655×10^{6}	3414.2
4186.8	426.94	1.163×10^{-3}	1	41.321	1.581×10^{-3}	3088.1	3.9683
101.325	10.332	2.815×10^{-5}	2.420×10^{-2}	1	3.827×10^{-5}	74.734	9.604×10^{-2}
2.648×10^{6}	2.700×10^{5}	0.7355	632.42	26,132	1	1.953×10^{6}	2509.6
1.3558	0.13826	3.766×10^{-7}	3.238×10^{-4}	1.338×10^{-2}	5.121×10^{-7}	1	1.285×10^{-3}
1055.1	107.59	2.931×10^{-4}	0.252	10.413	3.985×10^{-4}	778.17	1
2. Pressure							
Pa	kgf/cm^2	atm	mmHg	mmH_2O	lbf/ft^2	psi	in Hg
1	1.020×10^{-5}	9.869×10^{-6}	7.501×10^{-3}	0.10197	2.089×10^{-2}	1.450×10^{-4}	2.953×10^{-4}
98,067	1	0.96784	735.56	10,000	2048.1	14.224	28.959
101,325	1.0332	1	760	10,332	2116.2	14.696	29.921
133.32	1.360×10^{-3}	1.316×10^{-3}	1	13.595	2.7844	1.934×10^{-2}	3.937×10^{-2}
9.8067	1.000×10^{-4}	9.679×10^{-5}	7.356×10^{-2}	1	20.481	1.422×10^{-3}	2.896×10^{-3}
47.88	4.883×10^{-4}	4.726×10^{-5}	0.35914	4.8826	1	6.944×10^{-3}	1.414×10^{-2}
6894.8	7.031×10^{-2}	6.805×10^{-2}	51.715	703.09	143.99	1	2.036
3386.4	3.453×10^{-2}	3.342×10^{-2}	25.4	345.33	70.723	0.4912	1
3. Specific heat							

W/(m·K)	kcal/(m·h·°C)	Btu/(ft·h·°F)
1	0.8599	0.5778
1.163	1	0.672
1.7307	1.4882	1

4. Thermal conductivity

W/(m·K)	kcal/(m·h·°C)	Btu/(ft·h·°F)
1	0.8599	0.5778
1.163	1	0.672
1.7307	1.4882	1

Table A.3 Thermophysical parameters of water [54]

t	p	ρ	c_p	λ	μ	Pr
°C	MPa	kg/m^2	kJ/(kg·K)	W/(m·K)	N·s/m^2	–
0	0.101325	999.9	4.2127	55.122	1789	13.67
10	0.101325	999.7	4.1917	57.448	1306.1	9.52
20	0.101325	998.2	4.1834	59.89	1004.9	7.02
30	0.101325	995.7	4.175	61.751	801.76	5.42
40	0.101325	992.2	4.175	63.379	653.58	4.31
50	0.101325	988.1	4.175	64.774	549.55	3.54
60	0.101325	983.2	4.1792	65.937	470.06	2.98
70	0.101325	977.8	4.1076	66.751	406.28	2.55
80	0.101325	971.8	4.1959	67.449	355.25	2.21
90	0.101325	965.3	4.2085	68.031	315.01	1.95
100	0.101325	958.4	4.2211	68.263	282.63	1.75
110	0.143266	951	4.2336	68.496	259.07	1.6
120	0.198543	943.1	4.2504	68.612	237.48	1.47
130	0.270132	934.8	4.2671	68.612	217.86	1.36
140	0.36138	926.1	4.2881	68.496	201.17	1.26
150	0.476	917	4.3132	68.38	186.45	1.17
160	0.61806	907.4	4.3467	68.263	173.69	1.1
170	0.79202	897.3	4.3802	67.914	162.9	1.05
180	1.00266	886.9	4.4179	67.449	153.09	1
190	1.25512	876	4.4597	66.984	144.25	0.96
200	1.55488	863	4.5058	66.286	136.4	0.93
210	1.90774	853.8	4.5561	65.472	130.52	0.91
220	2.31983	840.3	4.6147	64.542	124.63	0.89
230	2.7976	827.3	4.6817	63.728	119.72	0.88
240	3.34783	813.6	4.7571	62.797	114.81	0.87
250	3.9776	799	4.845	61.751	109.91	0.86
260	4.69434	784	4.9497	60.472	105.98	0.87
270	5.50581	767.9	5.0882	58.963	102.06	0.88
280	6.42018	750.7	5.2303	57.448	98.135	0.9
290	7.44607	732.3	5.4857	55.82	94.21	0.93
300	8.5971	712.5	5.737	53.959	91.265	0.97
310	9.8697	691.1	6.072	52.331	88.321	1.03
320	11.289	667.1	6.5745	50.587	85.377	1.11
330	12.863	640.2	7.2445	48.377	81.452	1.22
340	14.605	610.1	8.1658	45.702	77.526	1.39
350	16.535	574.4	9.5058	43.028	72.62	1.6
360	18.675	528	13.986	39.539	66.732	2.35
370	21.054	450.5	40.326	33.724	56.918	6.79

Table A.4 Saturated water and saturated steam with parameters (by temperature) [54]

t	p	v'	v''	h'	h''	s'	s''	e_x'	e_x''
°C	MPa	m³/kg		kJ/kg		kJ/(kg·K)		kJ/kg	
10	0.001227	0.0010003	106.43	41.99	2519.9	0.151	8.902	0.75	88.2
20	0.0023366	0.0010017	57.838	83.86	2538.2	0.2963	8.6684	2.92	170.3
30	0.0042415	0.0010043	32.929	125.66	2556.4	0.4365	8.4546	6.43	246.9
40	0.007375	0.0010078	19.546	167.45	2574.4	0.5721	8.2583	11.17	318.5
50	0.012335	0.0010121	12.0457	209.26	2592.2	0.7035	8.0776	17.08	385.7
60	0.01992	0.0010171	7.6785	251.09	2609.7	0.831	7.9108	24.1	448.8
70	0.031162	0.0010228	5.0463	292.97	2626.9	0.9548	7.7565	32.15	508.2
80	0.04736	0.0010292	3.4091	334.92	2643.8	1.0753	7.6132	41.2	564.1
90	0.070109	0.0010361	2.3613	376.94	2660.1	1.1925	7.4799	51.19	616.9
100	0.101325	0.0010437	1.673	419.07	2676	1.3069	7.3554	62.08	666.8
110	0.143266	0.0010519	1.2099	461.32	2691.3	1.4185	7.2388	73.84	714
120	0.198543	0.0010606	0.89152	503.72	2706	1.5276	7.1293	86.44	758.5
130	0.270132	0.00107	0.66814	546.31	2719.9	1.6344	7.0261	99.86	800.7
140	0.36138	0.0010801	0.50849	589.11	2733.1	1.739	6.9284	114.08	840.5
150	0.476	0.0010908	0.39245	632.15	2745.4	1.8416	6.8358	129.09	878.1
160	0.61806	0.0011022	0.30676	675.47	2756.7	1.9425	6.7475	144.87	913.6
170	0.79202	0.0011145	0.24255	719.12	2767.1	2.0416	6.663	161.42	947
180	1.00266	0.0011275	0.1938	763.12	2776.3	2.1393	6.5819	178.75	978.4
190	1.25512	0.0011415	0.15632	807.52	2784.3	2.2356	6.5036	196.84	1007.7
200	1.55488	0.0011565	0.12716	852.37	2790.9	2.3307	6.4278	215.73	1035.1
210	1.90774	0.0011726	0.10424	897.74	2796.2	2.4247	6.3539	235.41	1060.6
220	2.31983	0.00119	0.086038	943.68	2799.9	2.5178	6.2817	255.91	1084
230	2.7976	0.0012087	0.07145	990.27	2802	2.6102	6.2107	277.27	1105.4
240	3.34783	0.0012291	0.059654	1037.6	2802.2	2.702	6.1406	299.5	1124.8
250	3.9776	0.0012513	0.050037	1085.8	2800.4	2.7935	6.0708	322.7	1142.1
260	4.69434	0.0012756	0.042134	1134.9	2796.4	2.8849	6.001	346.9	1157.2
270	5.50581	0.0013025	0.035588	1185.2	2789.9	2.9764	5.9304	372.2	1169.9
280	6.42018	0.0013324	0.030126	1236.8	2780.4	3.0683	5.8586	398.7	1180.1
290	7.44607	0.0013659	0.025535	1290	2767.6	3.1611	5.7848	426.5	1187.5
300	8.5927	0.0014041	0.021649	1345.1	2751	3.2552	5.7081	455.9	1191.8
310	9.87	0.001448	0.018334	1402.4	2730	3.3512	5.6278	487	1192.7
320	11.289	0.0014995	0.01548	1462.6	2703.7	3.45	5.5423	520.2	1189.7
330	12.863	0.0015615	0.012989	1526.5	2670.2	3.5528	5.449	556	1181.7
340	14.605	0.0016387	0.01078	1595.5	2626.2	3.6616	5.3427	595.3	1166.7
350	16.535	0.0017411	0.008799	1672	2567.7	3.78	5.2176	639.4	1142.4
360	18.675	0.0018959	0.00694	1764.2	2485.5	3.921	5.06	693.1	1103.3
370	21.054	0.0022136	0.004973	1890.2	2342.8	4.1108	4.8144	767.3	1027.7
372	21.562	0.0023636	0.004439	1935.6	2287	4.1794	4.724	793.9	996.5
374	22.081	0.0028426	0.003466	2046.7	2156.2	4.3493	4.5185	858.7	921.9
374.15	22.12	0.00317	0.00317	2107.4	2107.4	4.4429	4.4429	893.8	893.8

Table A.5 Saturated water and saturated steam with parameters (by pressure) [54]

p	t	v'	v''	h'	h''	s'	s''	e_x'	e_x''
MPa	°C	m³/kg		kJ/kg		kJ/(kg·K)		kJ/kg	
0.001	6.9828	0.0010001	129.209	29.34	2514.4	0.106	8.9767	0.37	62.3
0.01	45.8328	0.0010102	14.675	191.83	2584.8	0.6493	8.1511	14.48	358.2
0.1	99.632	0.0010434	1.6937	417.51	2675.4	1.3027	7.3598	61.66	665
0.3	133.54	0.0010735	0.60556	561.43	2724.7	1.6716	6.9909	104.8	815
0.5	151.844	0.0010928	0.37468	640.12	2747.5	1.8604	6.8192	131.94	884.8
1	179.884	0.0011274	0.19429	762.61	2776.2	2.1382	6.5828	178.54	978
1.5	198.289	0.0011539	0.13166	844.67	2789.9	2.3145	6.4406	212.44	1030.6
2	212.375	0.0011766	0.099536	908.59	2797.2	2.4469	6.3366	240.2	1066.3
2.5	223.943	0.0011972	0.079905	961.96	2800.9	2.5543	6.2536	264.23	1092.7
3	233.841	0.0012163	0.066626	1008.4	2802.3	2.6455	6.1837	285.7	1113.1
3.5	242.54	0.0012345	0.057025	1049.8	2802	2.7253	6.1228	305.3	1129.4
4	250.333	0.0012521	0.049749	1087.4	2800.3	2.7965	6.0685	323.5	1142.7
4.5	257.411	0.0012691	0.044037	1122.1	2797.7	2.8612	6.0191	340.6	1153.5
5	263.911	0.0012858	0.039429	1154.5	2794.2	2.9206	5.9735	356.7	1162.5
5.5	269.933	0.0013023	0.035628	1184.9	2789.9	2.9757	5.9309	372	1169.8
6	275.55	0.0013187	0.032438	1213.7	2785	3.0273	5.8908	386.8	1175.9
6.5	280.82	0.001335	0.029719	1241.1	2779.5	3.0759	5.8527	400.9	1180.8
7	285.79	0.0013513	0.027373	1267.4	2773.5	3.1219	5.8162	414.6	1184.7
7.5	290.496	0.0013677	0.025327	1292.7	2766.9	3.1657	5.781	427.9	1187.8
8	294.968	0.0013842	0.023525	1317.1	2759.9	3.2076	5.7471	440.9	1190
8.5	299.231	0.0014009	0.021926	1340.7	2752.5	3.2479	5.7141	453.6	1191.6
9	303.306	0.0014179	0.020495	1363.7	2744.6	3.2867	5.682	465.9	1192.5
9.5	307.211	0.0014351	0.019208	1386.1	2736.4	3.3242	5.6506	478.1	1192.8
10	310.961	0.0014526	0.018041	1408	2727.7	3.3606	5.6198	490.1	1192.6
10.6	315.274	0.0014741	0.016778	1433.7	2716.9	3.4029	5.5835	504.2	1191.7
11	318.045	0.0014887	0.016006	1450.6	2709.3	3.4304	5.5595	513.5	1190.7
11.6	322.059	0.0015113	0.01494	1475.4	2697.4	3.4708	5.5239	527.3	1188.5
12	324.646	0.0015268	0.014283	1491.8	2689.2	3.4972	5.5002	536.5	1186.7
12.6	328.401	0.0015507	0.013367	1516	2676.1	3.5361	5.4646	550.1	1183.4
13	330.827	0.0015672	0.012797	1532	2667	3.5616	5.4408	559.1	1180.8
13.6	334.357	0.0015928	0.011996	1555.8	2652.5	3.5993	5.4047	572.6	1176.2
14	336.642	0.0016106	0.011495	1571.6	2642.4	3.6243	5.3803	581.6	1172.7
14.6	339.972	0.0016385	0.010786	1595.3	2626.3	3.6613	5.3431	595.2	1166.8
15	342.131	0.0016579	0.01034	1611	2615	3.6859	5.3178	604.2	1162.4
15.6	345.282	0.0016886	0.009707	1634.7	2597.3	3.7226	5.2793	617.8	1155.2
16	347.328	0.0017103	0.009308	1650.5	2584.9	3.7471	5.2531	627	1149.9
17	352.263	0.0017696	0.008371	1691.7	2551.6	3.8107	5.1855	650.8	1135.2
18	356.957	0.0018399	0.007498	1734.8	2513.9	3.8765	5.1128	675.9	1117.3
19	361.431	0.001926	0.006678	1778.7	2470.6	3.9429	5.0332	701.6	1095.8
20	365.702	0.002037	0.005877	1826.5	2418.4	4.0149	4.9412	729.8	1068.6

Table A.6 Element chemistry and temperature correction factors of Kameyama Yoshida environmental Model

Element symbols	Standard chemistry exergy	Reference materials	Temperature correction factor	Element symbols	Standard chemistry exergy	Reference materials	Temperature correction factor
	10^3kJ/ kmol		kJ/(kmol·K)		10^3kJ/ kmol		kJ/(kmol·K)
H	117.61	H_2O (Liquid)	−84.89	Mo	714.42	$CaMoO_4$	−45.27
He	30.125	Air (5.24×10^{-6} atm)	101.09	Ru	0	Ru	0
Li	371.96	$LiCl{\cdot}H_2O$	−485.13	Rh	0	Rh	0
Be	594.25	$BeO{\cdot}Al_2O_3$	−103.26	Pd	0	Pd	0
B	610.28	H_3BO_3	−185.6	Ag	85.32	AgCl	326.6
C	410.53	CO_2	57.07	Cd	304.18	$CdCl{\cdot}2.5H_2O$	−759.94
N	0.335	Air (0.756 atm)	1.17	In	412.42	InO_2	−169.41
O	1.966	Air (0.203 atm)	6.61	Sn	515.72	SnO_2	217.53
F	308.03	$Ca_{10}(PO_4)_6F_2$	81.21	Sb	409.7	Sb_2O_5	−255.98
Ne	27.07	Air (1.8×10^{-5} atm)	90.83	Te	266.35	TeO_2	−188.49
Na	360.79	$NaNO_3$	−400.83	I	25.61	KIO_3	56.82
Mg	618.23	$CaCO_3{\cdot}MgCO_3$	−360.58	Cs	390.9	CsCl	−364.25
Al	788.22	Al_2O_3	−166.57	Ba	784.17	$Ba(NO_3)_2$	−697.6
Si	852.74	SiO_2	−195.27	La	982.57	$LaCl_2{\cdot}7H_2O$	−1224.45
P	865.96	$Ca_3(PO_4)_2$	86.36	Hf	1023.24	HfO_2	−202.51
S	602.79	$CaSO_4{\cdot}2H_2O$	−116.69	Ta	950.69	Ta_2O_5	−242.8
Cl	23.47	NaCl	268.82	W	818.22	$CaWO_4$	−45.44
Ar	11.673	Air (0.009 atm)	39.16	Os	297.11	OsO_4	−325.22
K	386.85	KNO_3	−354.97	Au	0	Au	0
Ca	712.37	$CaCO_3$	−338.74	Hg	431.71	$HgCl_2$	−690.61
Sc	906.76	Sc2O_3	−159.87	Ti	169.7	Ti_2O_4	
Ti	885.59	TiO_2	−198.57	Pb	337.27	PbClOH	
V	704.88	V_2O_5	−236.27	Bi	296.73	BiOCl	−425.68
Cr	547.43	$K_2Cr_2O_7$	30.67	Th	1164.87	ThO_2	−168.78
Mn	461.24	MnO_2	−197.23	U	1117.88	U_3O_8	−247.19
Fe	368.1	Fe_2O_3	−147.38	La	982.57	$LaCl_2{\cdot}7H_2O$	−1224.45
Co	288.4	$CoFe_2O_4$	−19.84	Ce	1020.73	CeO_2	−227.94
Ni	243.47	$NiCl_2{\cdot}6H_2O$	−865.63	Pr	926.17	$Pr(OH)_3$	
Cu	143.8	$Cu_4(OH)_6Cl_2$		Nd	967.05	$NdCl_3{\cdot}6H_2O$	−1214.78
Zn	337.44	$Zn(NO_3)_2{\cdot}6H_2O$	−852.37	Sm	962.86	$SmCl_3{\cdot}6H_2O$	−1215.7
Ga	496.18	Ga_2O_3	−162.09	Eu	872.49	$EuCl_3{\cdot}6H_2O$	−1231.06
Ge	493.13	GeO_2	−194.1	Gd	958.26	$GdCl_3{\cdot}H_2O$	−1220.06
As	386.27	As_2O_5	−255.27	Tb	947.38	$TbCl_3{\cdot}6H_2O$	−1230.26
Br	34.35	$PtBr_2$	−19.92	Dy	958.26	$DyCl_3{\cdot}6H_2O$	−1234.03
Rb	389.57	$RbNO_3$	−353.8	Ho	966.63	$HoCl_3{\cdot}6H_2O$	−1235.2
Sr	771.15	$SrCl_2{\cdot}6H_2O$	−841.61	Er	960.77	$ErCl_3{\cdot}6H_2O$	−1234.82
Y	932.4	$Y(OH)_3$		Tm	894.29	Tm_2O3	−167.65

Element symbols	Standard chemistry exergy	Reference materials	Temperature correction factor	Element symbols	Standard chemistry exergy	Reference materials	Temperature correction factor
	10^3kJ/kmol		kJ/(kmol·K)		10^3kJ/kmol		kJ/(kmol·K)
Zr	1058.59	$ZrSiO_4$	−215.02	Yb	935.67	$YbCl_3·6H_2O$	−1224.45
Nb	878.1	Nb_2O_5	−240.62	Lu	917.58	$LuCl_3·6H_2O$	−1235.2

Table A.7 Chemistry of common inorganic compounds and their temperature correction factors

Molecular formula	Chemical value	Temperature correction factor	Molecular formula	Chemical value	Temperature correction factor
	kJ/mol	J/(mol·K)		kJ/mol	J/(mol·K)
$AlCl_3$	229.83	892.07	$NaHCO_3$	44.69	−39.3
$Al_2(SO_4)_3$	308.36	539.44	MgO	50.79	−218.78
BaO	261.04	−596.01	$MgCl_2$	73.39	343.21
$BaSO_4$	32.55	−415.3	$MgCO_3$	22.59	62.3
$BaCO_3$	63.01	−356.77	$MgSO_4$	58.24	−67.8
CaO	110.33	−227.74	MnO	100.29	−115.94
$Ca(OH)_2$	53.01	−201.46	Mn_2O_3	47.24	−113.51
$CaCl_2$	11.25	349.74	Mn_2O_4	108.37	−213.13
$CaOSiO_2$	21.34	−228.15	NO	88.91	−4.6
$CaOAl_2O_3$	88.03	−251.08	NH_3 (Gas)	336.69	−154.22
CO	275.35	−25.61	Na_2O	346.98	−585.76
CO_2	20.13	67.4	NaCl	0	0.38
FeO	118.66	−71.76	Na_2SO_4	62.89	−413.5
Fe_3O_4	96.9	−70.29	Na_2CO_3	89.96	−364.22
$Fe(OH)_3$	30.29	43.85	Na_3AlF_6	581.95	138.95
Fe_2SiO_4	220.41	−125.35	SO_2 (Gas)	306.52	−114.64
$FeAl_2O_4$	103.18	−66.36	SO_3 (Gas)	239.7	−14.02
H_2O(Vapour)	8.62	−118.78	H_2S (Gas)	804.46	−329.66
HF	152.42	46.61	ZnO	21.09	−745.63
HCl (Gas)	45.77	173.72	$ZnSO_4$	73.68	−587.52
Na_2S	962.86	−798.31	$ZnCO_3$	22.34	−503.46

Table A.8 Chemistry of common organic compounds and their temperature correction factors

Molecular formula	Chemical value	Temperature correction factor	Molecular formula	Chemical value	Temperature correction factor
	kJ/mol	J/(mol·K)		kJ/mol	J/(mol·K)
CH_4 (Gas)	830.19	−201.96	C_5H_{10} (Liquid)	3265.11	−86.44
C_2H_6 (Gas)	1493.77	−221.63	C_6H_{12} (Liquid)	3901.16	−62.93
C_3H_8 (Gas)	2148.99	−248.36	C_6H_6 (Liquid)	3293.18	85.65
C_4H_{10} (Liquid)	2803.2	−540.11	C_8H_{16} (Liquid)	5243.89	−73.51

Molecular formula	Chemical value kJ/mol	Temperature correction factor J/(mol·K)	Molecular formula	Chemical value kJ/mol	Temperature correction factor J/(mol·K)
C_5H_{12} (Gas)	3455.61	−270.29	C_4H_6 (Gas)	2522.53	−130.08
C_5H_{12} (Liquid)	3454.52	−152.38	C_8H_{10} (Liquid)	4580.1	50.92
C_6H_{14} (Gas)	4109.48	−286.14	C_8H_{18} (Liquid)	5407.78	−211.42
C_6H_{14} (Liquid)	4105.38	−193.84	C_9H_{20} (Liquid)	6058.81	−220.87
C_7H_{16} (Gas)	4763.44	−355.81	$C_{10}H_{22}$ (Liquid)	6710.05	−743.08
C_7H_{16} (Liquid)	4756.45	−209.58	$C_{11}H_{24}$ (Liquid)	7361.33	−238.03
C_2H_4 (Gas)	1359.63	−172.38	$C_{12}H_{26}$ (Liquid)	8013.03	−247.07
C_3H_6 (Gas)	1999.95	−196.27	$CH_3C_6H_5$ (Liquid)	3928.36	61.63
C_2H_2 (Gas)	1265.49	−114.43	CH_3OH (Liquid)	716.72	−33.26
CH_3CCH (Gas)	1896.48	−138.2	C_2H_5OH (Liquid)	1354.57	−43.68

References

1. British Petroleum. Statistical review of world energy 2022. bp.com/en/global/corporate/energy-eco-nomics/statistical-review-of-world-energy.html.
2. International Atomic Energy Agency. Energy, electricity and nuclear power estimates for the period up to 2050. Vienna: IAEA-RDS-1/41, IAEA; 2021.
3. Gang S. The "three steps development strategy" of China nuclear power science and technology. Sci Technol Rev. 2016;34(15):33–41.
4. Beibei L, Yanzhou Y. Principles of nuclear reactor engineering. Beijing: Atomic Energy Press; 1982.
5. Zhonhqi S, Dong F, Guichun Y. Environmental impact of nuclear power plants. Beijing: Water Resources and Electric Power Press; 1984.
6. Yongqian S. Benefits of nuclear power generation and status of world nuclear power development. Energy Eng. 2007;1:1–6.
7. Nuclear Power Assessment Department. National nuclear power operation overview (January–December 2022). China Nuclear Energy Industry Association., http://www.china-nea.cn/site/content/42324.html; 2023.
8. International Atomic Energy Agency. The potential role of nuclear energy in national climate change mitigation strategies: final report of a coordinated research project. Vienna: IAEA-TECDOC-1984, IAEA; 2021.
9. Strategic Research Department. Development and prospect of nuclear energy in China (2022). China Nuclear Energy Industry Association., http://www.china-nea.cn/site/content/41473.html; 2022.
10. International Atomic Energy Agency. Nuclear power reactors in the world (2022 edition). Vienna: IAEA-RDS-2/42, IAEA; 2022.
11. Peisheng L. Atomic energy industry. Beijing: Atomic Energy Press; 2002. p. 5.
12. Tikuan Z. Thermal power plant. Beijing: Water Resources and Electric Power Press; 1986.
13. Weidao S, Peizhi Z, Danan J. Engineering thermodynamics. 2nd ed. Beijing: Higher Education Press; 1983.
14. Zohuri B, McDaniel P. Thermodynamics in nuclear power plant systems. 2nd ed. Springer International Publishing AG, part of Springer Nature; 2015. 2019
15. Yuhan M, Hui D, Changpu S. Can we build a heat engine with both high power and high efficiency?—the development and prospects of finite-time thermodynamics. Physics. 2021;50(1):1–9.
16. Jincan C, Zijun Y. The characteristics of finite time thermodynamic theory and some important marks in development. J Xiamen Univ (Nat Sci). 2001;40(2):232–41.
17. Curzon FL, Ahlbom B. Efficiency of a carnot engine at maximum power output. Am J Phys. 1975;43(22):22–4.

M. Peng et al., *Thermal Analysis of Nuclear Power Plants*, Nuclear Science and Technology, https://doi.org/10.1007/978-981-96-8983-5

18. Kaushik SC, Tyagi SK, Kumar P. Finite time thermodynamics of power and refrigeration cycles. Springer International Publishing; 2017.
19. Wenwei C, Huping L. Thermodynamic analysis and energy saving. Beijing: Science Press; 1999.
20. de Souza GFM. Thermal power plant performance analysis. Springer-Verlag London Limited; 2012.
21. Dunbar WR, Moody SD, Lior N. Exergy analysis of an operating boiling water reactor nuclear power station. Energy Convers Manag. 1995;36(3):149–59.
22. Brozynski BM. Exergy method and its application. Beijing: China Electric Power Press; 1996.
23. Lior N, Zhang N. Energy, exergy, and second law performance criteria. Energy. 2007;32:281–96.
24. Verkhivker GP, Kosoy BV. On the exergy analysis of power plant. Energy Convers Manag. 2001;42:2053–9.
25. Mingshan Z. Exergy analysis of energy system. Beijing: Tsinghua University Press; 1988.
26. Ahern JE. The exergy method of energy systems analysis. A Wiley-Interscience Publication; 1980.
27. Daxie C. Power cycle analysis. Shanghai: Shanghai Science and Technology Press; 1981.
28. Zeshao C. Advanced engineering thermodynamics. 2nd ed. Hefei: University of Science and Technology of China Press; 2014. p. 376–84.
29. Yuxuan J, Kaixiang X, Kefa C, et al. A review on supercritical carbon dioxide Brayton cycle. J Chin Soc Power Eng. 2022;42(1):1–9.
30. Gosset J, Gicquel R, Lecomte M, et al. Optimal design of the structure and settings of nuclear HTR thermodynamic cycles. Int J Therm Sci. 2005;44:1169–79.
31. Khaliq A, Kaushik SC. Second law based thermodynamic analysis of Brayton/Rankine combined power cycle with reheat. Appl Energy. 2004;78:179–97.
32. Zohuri B, McDaniel P. Combined cycle Driven efficiency for next generation nuclear power plants: an innovative design approach. 2nd ed. Springer International Publishing AG; 2015. 2018
33. Xinyuan H. Thermal power plant course design. Beijing: China Electric Power Press; 2004.
34. Fangli M. Analysis principle of energy saving in power plant thermal system: functions and equations of power plant steam cycle. Beijing: Water Resources and Electric Power Press; 1992.
35. Pengju C, Bin G. A study on cyclical function method theory for thermal system of the second circuit of PWR nuclear power plant. Proc CSEE. 2004;24(3):206–9.
36. Yunze L, Junjie Y, Wanchao L, et al. The equivalent heat drop method for the second-circuit analysis of PWR nuclear power plant system. Turb Technol. 2000;42(1):31–5.
37. Yunze L, Junjie Y, Wanchao L. Matrix analysis of secondary loop of PWR nuclear power unit. Therm Power Gen. 2000;4:26–8.
38. Qiang L, Lin Y, Hong-Xiang X, et al. A matrix analytic method for the second loop thermodynamic system of a pressurized water reactor-based nuclear power plant. J Eng Therm Energy Power. 2009;24(3):391–4.
39. Hanjun X, Shenghua D. Nuclear energy power plant. Beijing: Atomic Energy Press; 1990.
40. Ze D. Several issues in thermodynamic analysis of PWR nuclear power plants. Nucl Sci Eng. 1990;10(3)
41. Durmayaz A, Yavuz H. Exergy analysis of a pressurized water reactor nuclear power plant. Appl Energy. 2001;69:39–57.
42. Gongzu R. Thermal hydraulic analysis of power reactor. Beijing: Atomic Energy Press; 1982.
43. Minjun P, Ze D, Zhongning S. Research on exergy analysis of PWR nuclear power plants. Nucl Sci Eng. 1996;16(1):18–25.
44. Minyi L. Analysis and evaluation of adiabatic energy saving in thermal power plants. Beijing: China Electric Power Press; 1996.
45. Ruhui L, Dezhang L, Shiwu L. Efficient utilization of energy. Beijing: China Machine Press; 1992. p. 376.
46. Shilu P. Basis of economic analysis and evaluation of nuclear energy industry. Beijing: Atomic Energy Press; 1995.

47. Qindao L, Zhizhen L. Calculation and analysis of thermal economy of thermal power plant[M]. Beijing: China Electric Power Press; 2008.
48. Junjie Y, Yunze L, Wanchao L, et al. Research on thermo-economics diagnosis theory for thermal system of secondary circuit of PWR nuclear power plant. Nucl Power Eng. 2000;21(1):81–5.
49. Pu H, et al. Computer supervisory control and monitoring of thermal power plants. Beijing: China Water Resources and Electric Power Press; 2005.
50. International Atomic Energy Agency. Thermal performance monitoring and optimization in nuclear power plants: experience and lessons learned. Vienna, Austria: IAEA-TECDOC-1971; 2021.
51. Kim H, Na MG, Heo G. Application of monitoring, diagnosis, and prognosis in thermal performance analysis for nuclear power plants. Nucl Eng Technol. 2014;46(6):737–52.
52. American Society of Mechanical Engineers. Steam turbines. ASME; 2006.
53. American Society of Mechanical Engineers. Moisture separator reheaters. ASME; 1992.
54. Shiming Z, Mengle W, Zhongyuan F. Handbook of water and steam properties parameter with exergy parameter. Beijing: Water Resources and Electric Power Press; 1989.

MIX
Papier aus verantwortungsvollen Quellen
Paper from responsible sources
FSC® C105338

If you have any concerns about our products,
you can contact us on
ProductSafety@springernature.com

In case Publisher is established outside the EU,
the EU authorized representative is:
Springer Nature Customer Service Center GmbH
Europaplatz 3, 69115 Heidelberg, Germany

Printed by Libri Plureos GmbH
in Hamburg, Germany

MIX
Papier aus verantwortungsvollen Quellen
Paper from responsible sources
FSC® C105338

If you have any concerns about our products,
you can contact us on
ProductSafety@springernature.com

In case Publisher is established outside the EU,
the EU authorized representative is:
Springer Nature Customer Service Center GmbH
Europaplatz 3, 69115 Heidelberg, Germany

Printed by Libri Plureos GmbH
in Hamburg, Germany